Military Avionics Systems

Military Avionics Systems

Ian Moir

Allan Seabridge

Displays chapter contributed by Malcolm Jukes

John Wiley & Sons, Ltd

Other Wiley Editorial Offices

John Wiley & Sons Inc., 111 River Street, Hoboken, NJ 07030, USA

Jossey-Bass, 989 Market Street, San Francisco, CA 94103-1741, USA

Wiley-VCH Verlag GmbH, Boschstr. 12, D-69469 Weinheim, Germany

John Wiley & Sons Australia Ltd, 42 McDougall Street, Milton, Queensland 4064, Australia

John Wiley & Sons (Asia) Pte Ltd, 2 Clementi Loop #02-01, Jin Xing Distripark, Singapore 129809

John Wiley & Sons Canada Ltd, 22 Worcester Road, Etobicoke, Ontario, Canada M9W 1L1

Wiley also publishes its books in a variety of electronic formats. Some content that appears in
print may not be available in electronic books.

Library of Congress Cataloging-in-Publication Data
Moir, I. (Ian)
   Military avionics systems/Ian Moir.
    p. cm.
   "Displays chapter contributed by Malcom Jukes."
   Includes bibliographical references and index.
   ISBN   0-470-01632-9 (cloth : alk. paper)
   1. Avionics.   2. Airplanes, Military–Electronic equipment.
   3.   Electronics in military engineering.   I. Jukes, Malcom.   II. Title.
   UG1420.M565 2006
   623.74′6049–dc22   2005031935

British Library Cataloguing in Publication Data

A catalogue record for this book is available from the British Library

ISBN-13   978-0-470-01632-9 (HB)
ISBN-10   0-470-01632-9 (HB)

Typeset in 10/12pt Times by Thomson Press (India) Limited, New Delhi.
Printed and bound in Great Britain by Antony Rowe Ltd, Chippenham, Wiltshire
This book is printed on acid-free paper responsibly manufactured from sustainable forestry
in which at least two trees are planted for each one used for paper production.

Contents

Series Preface

The field of aerospace is wide ranging and covers a variety of products, disciplines and domains, not merely in engineering but in many related supporting activities. These combine to enable the aerospace industry to produce exciting and technologically challenging products. A wealth of knowledge is contained by practitioners and professionals in the aerospace fields that is of benefit to other practitioners in the industry, and to those entering the industry from University.

The Aerospace Series aims to be a practical and topical series of books aimed at engineering professionals, operators, users and allied professions such as commercial and legal executives in the aerospace industry. The range of topics spans design and development, manufacture, operation and support of aircraft as well as infrastructure operations, and developments in research and technology. The intention is to provide a source of relevant information that will be of interest and benefit to all those people working in aerospace.

Ian Moir and Allan Seabridge

Acknowledgements

This book has taken a long time to prepare and we would not have completed it without the help and support of colleagues and organisations who gave their time and provided information with enthusiasm.

We would especially like to thank Malcolm Jukes, Kevin Burke and Keith Atkin who reviewed a number of chapters, and to Leon Skorczewski who bravely reviewed the entire manuscript and provided valuable comments.

The following organisations kindly provided information and images:

BAE SYSTEMS	Lockheed Martin
Brilliant Technology	Lockheed Martin Fire Control
Cambridge Display Technology	Martin Baker
Eurofighter GmbH	Microvision Inc
Federation of American Scientists	Northrop Grumman Corporation
for UAV pictures and Nimrod R	Raytheon
Mk 1	Rockwell Collins
Honeywell	Royal Aeronautical Society
Honeywell Aerospace Yeovil	SAAB Avitronics
InfraRed 1	Smiths Group
Kaiser Electronics	Texas Instruments
Kentron	Thales
Korry Electronics	Thales Optronics
L-3 WESCAM	VSI

Special thanks to Marc Abshire, Randy Anderson, Myrna Buddemeyer, Ron Colman, Francesca De Florio, Joan Ferguson, Alleace Gibbs, Charlotte Haensel-Hohenhausen, Karen Hager, Dexter Henson, Clive Marrison, Katelyn Mileshosky, Ian Milne, Marianne Murphy, Shelley Northcott, Beth Seen, Kevin Skelton and Greg Siegel for their kind help in securing high quality images for inclusion in the book.

Aircraft pictures were obtained from the US Department of Defence Air Force Link website at www.af.mil: B-2 Spirit – US Air Force Photo by Tech Sgt Cecilio Ricardo; B-1B Lancer – US Air Force Photo by Senior Airman Michel B.Keller; A-10 Thunderbolt – US Air Force Photo by Senior Airman Stephen Otero; T-38 Talon US Air Force Photo by Staff Sgt Steve Thurow; C-5 US Air Force Photo by Master Sgt Clancy Pence, C-17 US Air Force

Photo by Airman 1st Class Aldric Bowers; Predator US Air Force Photo by Master Sgt Deb Smith, Global Hawk by George Rohlmaller; E-JSTARS US Air Photo by Staff Sgt Shane Cuomo; F-117A Nighthawk US Air Force Photo by Staff Sgt Aaron D Allmon II; B-2 Spirit Bomber US Air Force Photo by Master Sgt Val Gempis;

VP-45 US Navy Squadron is acknowledged for the photograph of a MX-20 turret on P-3 aircraft.

We would like to thank the staff at John Wiley who took on this project part way through its progress and guided us to a satisfactory conclusion.

About the Authors

After 20 years in the Royal Air Force, **Ian Moir** went on to Smiths Industries in the UK where he was involved in a number of advanced projects. Since retiring from Smiths he is now in demand as a highly respected consultant. Ian has a broad and detailed experience working in aircraft avionics systems in both military and civil aircraft. From the RAF Tornado and Apache helicopter to the Boeing 777, Ian's work has kept him at the forefront of new system developments and integrated systems implementations. He has a special interest in fostering training and education in aerospace engineering.

   Allan Seabridge is the Chief Flight Systems Engineer at BAE SYSTEMS at Warton in Lancashire in the UK. In over 30 years in the aerospace industry his work has included avionics on the Nimrod MRA 4 and Joint Strike Fighter as well as a the development of a range of flight and avionics systems on a wide range of fast jets, training aircraft and ground and maritime surveillance projects. Spending much of his time between Europe and the US, Allan is fully aware of systems developments worldwide. He is also keen to encourage a further understanding of integrated engineering systems.

Introduction

Evolution of Avionics

Avionics is a word coined in the late 1930s to provide a generic name for the increasingly diverse functions being provided by AVIation electrONICS. World War II and subsequent Cold War years provided the stimulus for much scientific research and technology development which, in turn, led to enormous growth in the avionic content of military aircraft. Today, avionics systems account for up to 50% of the fly-away cost of an airborne military platform and are key components of manned aircraft, unmanned aircraft, missiles and weapons. It is the military avionics of an aircraft that allow it to perform defensive, offensive and surveillance missions.

A brief chronology of military avionics development illustrates the advances that have been made from the first airborne radio experiments in 1910 and the first autopilot experiments a few years later. The 1930s saw the introduction of the first electronic aids to assure good operational reliability such as blind flying panels, radio ranging, non-directional beacons, ground-based surveillance radar, and the single-axis autopilot. The 1940s saw developments in VHF communications, identification friend or foe (IFF), gyro compass, attitude and heading reference systems, airborne intercept radar, early electronic warfare systems, military long-range precision radio navigation aids, and the two-axis autopilot. Many of these development were stimulated by events leading up to World War II and during the war years.

The 1950s saw the introduction of tactical air navigation (TACAN), airborne intercept radar with tracking capability and Doppler radar, medium pulse repetition frequency (PRF) airborne intercept radar, digital mission computers and inertial navigation systems. The 1960s saw the introduction of integrated electronic warfare systems, fully automated weapon release, terrain-following radar, automatic terrain following, the head-up display laser target marketing technology and the early digital mission computer.

Over the years, as specialist military operational roles and missions have evolved, they have often driven the development of role-specific platforms and avionics. Looking across the range of today's airborne military platforms, it is possible to identify categories of

Military Avionics Systems   I. Moir and A. Seabridge
© 2006 John Wiley & Sons, Ltd

avionics at system, subsystem and equipment levels that perform functions common to all platforms, or indeed perform unique mission-specific functions.

Technology improvements in domestic markets have driven development in both commercial and military systems, and the modern military aircraft is likely to contain avionic systems that have gained benefit from domestic computing applications, especially in the IT world, and from the commercial aircraft field. This has brought its own challenges in qualifying such development for use in the harsh military environment, and the challenge of meeting the rapid turnround of technology which leads to early obsolescence.

Avionics as a Total System

An avionics system is a collection of subsystems that display the typical characteristics of any system as shown in Figure I.1. The total system may be considered to comprise a number of major subsystems, each of which interacts to provide the overall system function. Major subsystems themselves may be divided into minor subsystems or equipment which in turn need to operate and interact to support the overall system. Each of these minor subsystems is supported by components or modules whose correct functional operation supports the overall system. The overall effect may be likened to a pyramid where the total system depends upon all the lower tiers.

Avionics systems may be represented at a number of different levels as described below:

1. A major military task force may comprise a large number of differing cooperating platforms, each of which contributes to the successful accomplishment of the task force

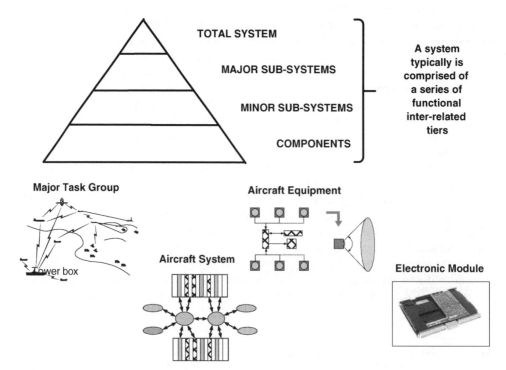

Figure I.1   Avionics as a 'system of systems'.

A military avionics system

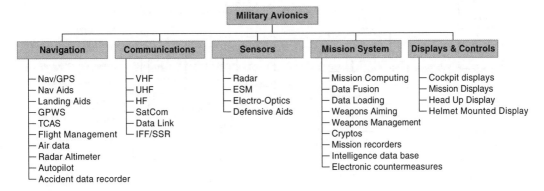

Figure I.2   Product breakdown structure of a military aircraft system.

mission. Within this context an individual strike aircraft or surveillance platform avionics system may represent one component of many within the task force.

2. At the individual platform or aircraft level, a collection of subsystems and components or modules operate to support the successful completion of the primary role of the platform, be it reconnaissance, strike, support or surveillance.

3. The individual equipment that supports the overall system of the platform is a collection of units or modules, control panels and displays, each of which has to operate correctly to support subsystem and overall system operation.

4. Finally, the electronic modules that form the individual components of the aircraft avionics systems may be regarded as systems within their own right, with their own performance requirements and hardware and software elements.

In general within this book, most discussion is centred upon the aircraft-level avionics system and upon the major subsystems and minor subsystems or equipment that support it. Passing reference to the higher-level system is made during brief coverage of network centric operations. In some cases the detailed operation of some components such as data buses is addressed in order that the reader may understand the contribution that these elements have made to advances in the overall integration of platform avionics assets.

The product breakdown structure of a military aircraft system is shown in Figure I.2.

Increasing Complexity of Functional Integration

As avionics systems have evolved, particularly over the past two or three decades, the level of functional integration has increased dramatically. The nature of this increase and the accompanying increase in complexity is portrayed in Figure I.3.

In the early stages, the major avionics subsystems such as radar, communications, navigation and identification (CNI), displays, weapons and the platform vehicle could be

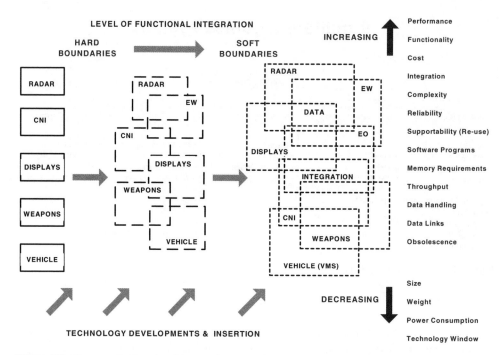

Figure I.3 Increase in functional integration over time.

considered as discrete subsystems, the function of which could be easily understood. The performance requirements could be relatively easily specified and captured, and, although there were information interchanges between them, each could stand alone and the boundaries of each subsystem was 'hard' in the sense that it was unlikely to be affected by the performance of a neighbouring subsystem.

As time progressed, the functionality of each subsystem increased and some boundaries blurred and functions began to overlap. Also, the number of subsystems began to increase owing to the imposition of more complex mission requirements and because of the technology developments that furnished new sensors. Improved data processing and higher bandwidth data buses also contributed to providing much higher data processing capabilities and the means to allow the whole system to become more integrated.

Further technology developments added another spiral to this trend, resulting in greater functionality, further increasing integration and with a blurring of functional boundaries as subsystems became able to share ever greater quantities of data. This evolution has been a continual process, although it is portrayed in three stages in Figure I.3 for reasons of simplicity.

The outcome of this evolution has been to **increase**: performance; sensor types; functionality; cost; integration; complexity; supportability (reuse); software programs in terms of executable code; memory requirements; throughput; reliability; data handling; data links; and obsolescence.

The result has been to **decrease**: size; weight; power consumption; and technology windows.

Organisation of the Book

This book is organised to help the reader to comprehend the overarching avionics systems issues but also to focus on specific functional areas. Figure I.4 shows how the various chapters relate to the various different major functional subsystems.

The book provides a military avionics overview aimed at students and practitioners in the field of military avionics.

Chapter 1 lists and describes the roles that military air forces typically need to perform. It is the understanding of these roles that defines the requirements for a particular suite of avionics, sensors and weapons for different platforms.

Chapter 2 examines the technology that has led to different types of system architecture. This technology has resulted in sophisticated information processing structures to transfer high volumes of data at high rates, and has resulted in greatly increased functional integration.

The subject of radar is covered in Chapter 3 which describes radar basic principles, while Chapter 4 explains some of the advanced features that characterise different types of radar used for specific tasks.

Chapter 5 deals with electrooptical (EO) sensors and their use in passive search, detection and tracking applications. This includes a description of the integration of EO sensor applications in turrets and pods, as well as personal night vision goggles.

Chapter 6 looks at the sensitive, and often highly classified, field of electronic warfare and the gathering of intelligence by aircraft using sophisticated receiving equipment and processing techniques.

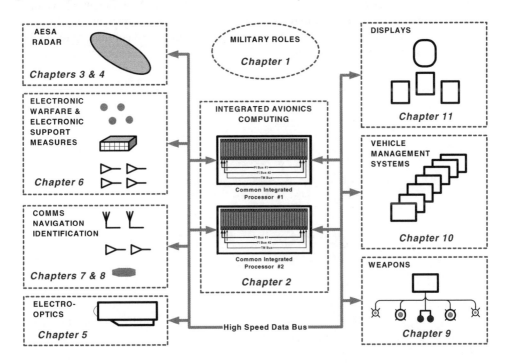

Figure I.4 Military avionics functional subsystems.

Chapter 7 is concerned with communications and identification; this describes the mechanisms by which an aircraft is identified to other stakeholders such as air traffic control and to friendly forces, as well as the different form of communications available for speech and encrypted data.

Chapter 8 covers the subject of navigation and the means by which pilots are able to navigate precisely to their engagement zones, understand their location during and after an engagement and return safely to home base. This makes maximum use of military and civilian navigation aids, using state-of-the-art on-board systems.

Chapter 9 addresses the subject of weapons carriage and guidance to give an understanding of the integrated weapon system. Individual weapons types are described, together with the systems required to ensure that they can be aimed and released to maximum effect.

Chapter 10 deals with the vehicle management system; those systems that provide the platform with power, energy and management of basic platform control functions. Although provided as a separate control system today, it is inevitable that these functions will be absorbed into mission system processing in the future.

Chapter 11 covers part of the human–machine interface – the displays in the cockpit and the mission crew areas that enable the crew to prosecute the operational mission. This chapter deals with the technology of displays and provides numerous examples of display systems in military applications.

The authors believe that this volume will complete the set of companion volumes that describe the aircraft general, avionic and mission systems, as well as the way in which they are developed. This series provides a guide to the interested public, to students and to practitioners in the aerospace field. It should be recognised that this book, like its companion volumes, only scratches the surface of a series of complex topics. Within the book we have provided a comprehensive bibliography as a guide to specialised volumes dealing in detail with the topics outlined here, and it is to be hoped that the reader will continue to read on to understand aviation electronic systems.

1 Military Roles

1.1 Introduction

The military were quick to seize upon the opportunities offered to them by an ability to leave the ground and gain an advantage of height. The initial attempts to make use of this advantage were by using tethered balloons as observation posts, and then as positions from which to direct artillery. The advent of a moving and powered platform allowed guns and, later, bombs to be carried, which led to air war between aircraft, and upon ground troops. Thus, fairly early in the history of the aircraft the main military roles of observation, interception and ground attack had been firmly established. These initial roles increased in sophistication and led to the development of more capable aircraft weapons, aircrew and tactics.

Today the military are called upon to perform a wide variety of aviation roles using fixed-wing and rotary-wing aircraft. The roles largely define the type of aircraft because of the specialist nature of the task; however, there are a number of aircraft types that have been designed as multirole aircraft, or designed to change roles during the prosecution of a mission, the so-called swing-role type.

The military roles that are in place today have emerged over many years of aerial combat experience. The long development timescales of the complex military aircraft have resulted in many types remaining in service long after their original introduction. Consequently, aircraft have adopted new roles as a result of role-fit weapons or mid-life updates. Many of the roles, particularly the intelligence gathering roles, have persisted after combat into the post-war stabilisation period and peacekeeping operations.

The flexibility of weapons and methods of carrying weapons and the adaptability of sensors and avionic systems are what enables this situation to persist. Although many of the 'traditional' roles still exist, there are signs that the changing nature of conflict may lead to new roles or alternative solutions.

To a large extent these new roles and alternative solutions are being driven by advances in the technology of sensors and avionics. Ever more sensitive and effective sensor systems are capable of detecting targets, the use of stealth techniques increases the effectiveness of delivery platforms and the increased capability of on-board computing systems is extending

Military Avionics Systems   I. Moir and A. Seabridge
© 2006 John Wiley & Sons, Ltd

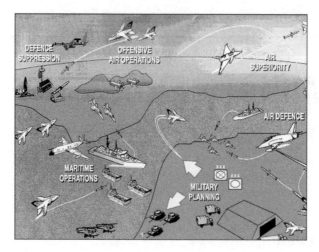

Figure 1.1 Typical battlefield scenario and the major players.

and speeding up the processing of data. The existence of these advances in the hands of enemies spurs on further development.

This chapter will describe the roles that are required in the military defence environment. Some examples of avionic architectures will be described, along with examples of the types of aircraft in service today that perform the various roles. Other chapters in this book will deal with the detail of a number of military avionic systems.

1.2 Air Superiority

1.2.1 Role Description

The primary aim of this role is to deny to an enemy the airspace over the battlefield, thus allowing ground attack aircraft a free rein in destroying ground targets and assisting ground forces, secure in the knowledge that the airborne threat has been suppressed.

The air superiority aircraft is typically designed to enable the pilot to respond rapidly to a deployment call, climb to intercept or loiter on combat air patrol (CAP) and then to engage enemy targets, preferably beyond visual range. The aircraft should also have the capability to engage in close combat, or dogfight, with other aircraft should this prove to be necessary. For this to be successful, an extremely agile machine is necessary with 'carefree handling' capability.

The systems must allow for accurate navigation, accurate identification of targets, prioritisation of targets, accurate weapon aiming capability and the ability to join the tactical communications network.

A typical mission profile is shown in Figure 1.2.

1.2.2 Key Performance Characteristics

The air superiority aircraft is usually a highly manoeuvrable aircraft with a high Mach number capability and rapid climb rate. Many fighters are equipped with afterburning to allow Mach 2 capability, a power to weight ratio greater than 1, allowing acceleration in a climb, and the ability to climb to beyond 60 000 ft. Some types are designed to operate from

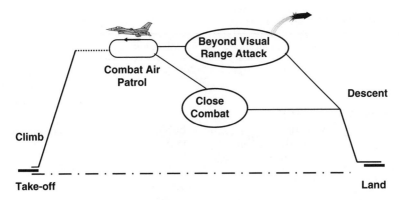

Figure 1.2 Air superiority mission profile.

carriers and will be equipped for catapult launch and for steep approaches and arrestor wire decelerations.

Many modern fighters are unstable and have full authority flight control systems that are designed to allow the pilot to execute manoeuvres to envelope limits without fear of losing control or damaging the aircraft. This is known as 'carefree handling' capability.

1.2.3 Crew Complement

Usually single pilot, but some types employ a pilot and a rear-seat air electronics officer or navigator depending on the role. Trainers or conversion aircraft will have two seats for instructor and student.

1.2.4 Systems Architecture

A typical air superiority platform architecture is shown in Figure 1.3. Typical air superiority systems are listed in Table 1.1.

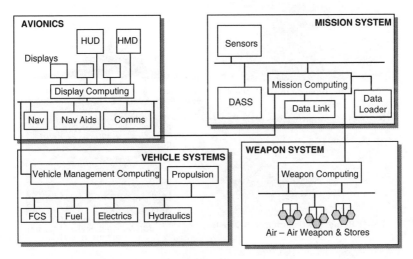

Figure 1.3 Typical air superiority platform architecture.

Table 1.1 Typical air superiority systems

Avionics	Communications	Mission system	Weapons
	VHF	Radar	Air-to-air missile – ASRAAM, AMRAAM
Navigation/GPS	UHF		
FMS	HF	ESM	Internal gun
Autopilot	SHF SatCom	DASS	
ADF	Link 16		
DME			
TACAN		Mission computer	
TCAS		Data loader	
Landing aids			
GPWS			
LPI RadAlt			
Air data			
Digital map			
MDP			
Displays and controls			
Head-Up-display			
Helmet-mounted display			
IFF/SSR			
Avionics data bus		Mission system data bus	Weapons bus

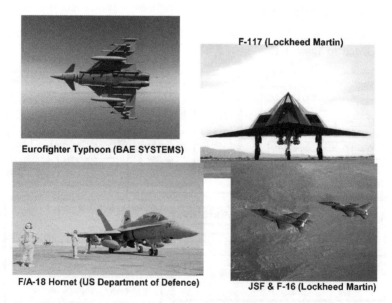

Eurofighter Typhoon (BAE SYSTEMS)

F-117 (Lockheed Martin)

F/A-18 Hornet (US Department of Defence)

JSF & F-16 (Lockheed Martin)

Figure 1.4 Air superiority aircraft types.

1.2.5 Air Superiority – Aircraft Types

The various types of air superiority aircraft are as follows (Figure 1.4):

- McDonnell Douglas F-4 Phantom;
- English Electric Lightning;
- Eurofighter Typhoon;
- Panavia Tornado F-3;
- Dassault Rafale;
- Dassault Mirage 2000;
- SAAB Gripen;
- F-15;
- F-16;
- F-18;
- Mig-21 Fishbed;
- Mig-23 Flogger;
- F-117.

1.3 Ground Attack

1.3.1 Role Description

The ground attack role has been developed to assist the tactical situation on the battlefield. The pilot must be able to identify the right target among the ground clutter and multiplicity of targets and friendly units on the battlefield. The ability to designate targets by laser has enabled precision bombing to be adopted by the use of laser-guided bombs or 'smart' bombs. The role must enable fixed targets such as buildings, radar installations and missile sites, as well as mobile targets such as tanks, guns, convoys, ships and troop formations, to be detected, positively identified and engaged.

This role includes close air support (CAS), where support is given to ground forces, often under their direction, where weapons will be deployed in close proximity to friendly forces.

1.3.2 Key Performance Characteristics

Depending on the target and the on-going military situation, the ground attack role may be performed by either fixed-wing or rotary-wing aircraft. A fixed-wing aircraft usually needs very fast, low-level performance with good ride qualities. It should also be reasonably agile to perform attack manoeuvres and take evasive action. Rotary-wing aircraft benefit from extreme low-level nap of the earth penetration, and the ability to loiter in natural ground cover – popping up when required to deliver a weapon.

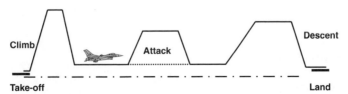

Figure 1.5   Ground attack mission profile.

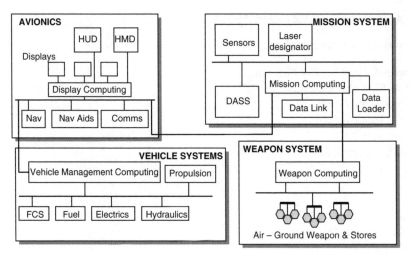

Figure 1.6 Typical ground attack platform architecture.

1.3.3 Crew Complement

This role is usually conducted by two crew members, a pilot and a crew member to operate the sensors and weapons systems. The advent of smart weapons or cooperative target designation means that the mission can be conducted by a single crew, often a role designated to a fighter aircraft as a secondary role.

1.3.4 Systems Architecture

A typical ground attack platform architecture is shown in Figure 1.6. Typical ground attack systems are listed in Table 1.2.

1.3.5 Ground Attack – Aircraft Types

The various types of ground attack aircraft are as follows (Figure 1.7):

- Sepecat Jaguar;
- Panavia Tornado GR4;
- Fairchild A-10 Thunderbolt;
- Apache;
- Sukhoi Su-24 Fencer.

1.4 Strategic Bomber

1.4.1 Role Description

The role of the strategic bomber is to penetrate deep into enemy territory and to carry out strikes that will weaken defences and undermine the morale of the troops. The strategic

Table 1.2 Typical ground attack systems

Avionics	Communications	Mission system	Weapons
	VHF	Radar	Air-to-ground missiles
Navigation/GPS	UHF	Electrooptics	
FMS	HF	ESM	Free-fall bombs
Autopilot	SHF SatCom	DASS	Laser-guided bombs
ADF	Link 16	Laser designator	Airfield denial
DME			Internal gun
TACAN		Mission recording	gun pod
TCAS		Data loader	Rockets
Landing aids		Cameras	
GPWS			
LPI RadAlt			
Air data			
Digital map			
MDP			
Displays and controls			
Head-Up display			
Helmet-mounted display			
IFF/SSR			
Avionics data bus		Mission system data bus	Weapons bus

bomber was usually a very high-flying aircraft capable of carrying a large load of bombs which were released in a 'carpet bombing' pattern. The modern aircraft may choose to fly low and fast and rely on stealth to evade enemy radar defences. Different weapons may also be employed such as Cruise missiles and joint direct attack munition (JDAM).

Lockheed F-16 (Lockheed Martin)

A-10 (US Department of Defence)

Harrier GR7 (BAE SYSTEMS)

Figure 1.7 Ground attack aircraft types.

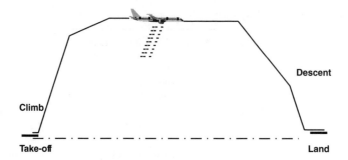

Figure 1.8   Strategic bomber mission profile.

1.4.2   Key Performance Characteristics

Strategic bomber aircraft attributes include high altitude cruise, long range and high payload capacity.

1.4.3   Crew Complement

The crew includes pilots, a navigator, an engineer and specialist mission crew. For very long missions a relief crew may be provided.

1.4.4   Systems Architecture

A typical strategic bomber platform architecture is shown in Figure 1.9. Typical strategic bomber systems are listed in Table 1.3.

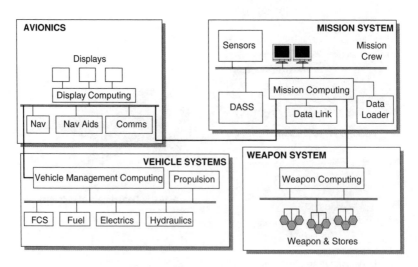

Figure 1.9   Typical strategic bomber platform architecture.

Table 1.3 Typical strategic bomber systems

Avionics	Communications	Mission system	Weapons
	VHF	Radar	Air-to-ground missiles
Navigation/GPS	UHF	Electrooptics	Free-fall bombs
FMS	HF	ESM	Laser-guided bombs
Autopilot	SHF SatCom	DASS	Airfield denial
ADF	Link 16	MAD	Cruise missiles
DME			
TACAN		Mission recording	
TCAS		Data loader	
Landing aids		Cameras	
GPWS			
LPI RadAlt			
Air data			
Digital map			
MDP			
Displays and Controls			
IFF/SSR			
Avionics data bus		Mission system data bus	Weapons bus

1.4.5 Strategic Bomber – Aircraft Types

The various types of strategic bomber aircraft are as follows (Figure 1.10):

- Boeing B-52;
- AVRO Vulcan;

B-52 (US Department of Defence)

B-1B Lancer (US Department of Defence)

B-2 Spirit (US Department of Defence)

Figure 1.10 Strategic bomber aircraft types.

- Northrop B-2;
- Tupolev Tu-22M Backfire;
- Tupolev TU-160 Blackjack;
- General Dynamics F-111.

1.5 Maritime Patrol

Over 60% of the earth's surface is covered by oceans – a natural resource that is exploited by many means: as a medium for transportation of cargo, as a source of food, as a means of deploying naval assets such as capital ships and submarines and for movement of men and materiel. It is also used for pleasure and for criminal purposes such as the smuggling of drugs, liquor, tobacco and illegal aliens. It is not surprising, therefore, that surveillance of the ocean's surface is of importance to military and paramilitary forces.

The most practical way of carrying out surveillance or reconnaissance is by air, and the flexibility of the fixed-wing aircraft with its comparatively high speed, long range and excellent detection capability from high altitude made it an excellent complement to surface vessels in carrying out naval or policing duties.

Over 90 years of development have led to the emergence of the maritime patrol aircraft (MPA) as one of the most complex of systems aircraft with a demanding role embracing a broad spectrum of tactical and strategic tasks, as well as support for civilian and humanitarian activities.

The general MPA specification that has evolved calls for the ability to transit at high speed to a distant patrol area of interest, and then to remain in that area for a long time, carrying out searches for surface, subsurface or both types of activity. Operational requirements typically ask for an ability to fly over 800 miles to an area, remain on task for over 7 h, return to base and have sufficient fuel remaining to carry out a bad weather diversion. To perform such a task requires an aircraft weighing up to 100 t, a crew of 12 or more and a suite of electronic sensors and communication systems.

1.5.1 Role Description

The typical tasks that an MPA is called upon to perform include:

Anti-surface unit warfare (ASuW)
- Reconnaissance;
- Shadowing;
- Strike against surface vessels;
- Tactical support of maritime strike aircraft;
- Over-the-horizon targeting for friendly vessels;
- Intelligence collection;
- Communications relay;
- Limited airborne early warning capability.

Anti-submarine warfare (ASW)
- Close air support to task forces and convoys;
- Open ocean searches;
- Extended tracking of submerged targets;

- Deterrence of hostile submarines;
- Cooperation with friendly submarines;
- Intelligence collection.

Search and rescue (SAR)
- Location of survivors;
- Dropping of survival equipment;
- Scene-of-action commander for rescue operations;
- Escort to rescue helicopters;
- Cooperation with rescue services;
- Escort of aircraft in difficulties.

Exclusive economic zone protection
- Oil rig surveillance;
- Fishery protection;
- Pollution detection and dispersal.

Customs and excise cooperation
- Anti-illegal immigration;
- Anti-gun running;
- Anti-terrorist operations;
- Anti-drug smuggling.

1.5.2 Anti-surface Unit Warfare (ASuW) Role

MPAs take part in all aspects of the war at sea. In the role of anti-surface unit warfare, the MPA may carry out autonomous strikes against surface targets using free-fall bombs or stand-off weapons. Alternatively, it may be used to search, identify and shadow surface forces, remaining in contact but out of range of a surface ship's weapons for long periods of time. This can be performed overtly using the integral radar of the MPA to detect, classify and track targets, or covertly using passive electronic support measure systems to detect the ship's radars while remaining outside the ship's maximum radar detection range.

Frequently it is necessary to shadow naval forces for days using a number of MPAs, each handing over the task to a relief aircraft at the end of its endurance on task. If there is then a requirement for specialist strike aircraft to carry out attacks against the ships under surveillance, then a cooperative attack is planned. The MPA can guide the attacking force accurately to suitable attack positions using its own radar, while the attacking aircraft can approach covertly under any defensive radar screen, only being detected by the air defence radar of the target at very short range. During the attack the MPA can carry out jamming of radar and communications to distract the surface ship's defensive tactics.

The MPA can also use similar tactics to cooperate with attack helicopters, or to provide over-the-horizon targeting for surface missiles launched from friendly naval ships.

1.5.3 Anti-submarine Warfare (ASW) Role

Traditionally, submarines have waged strategic warfare by effectively blockading enemy countries, preventing military supplies, reinforcements and essential food and medical supplies from arriving by sea. The counter to the submarine campaigns was to regulate shipping by organising it into strictly disciplined convoys and concentrating naval forces to protect the convoys. However, as submarines became more effective and became organised into 'wolf packs', the escort ships found themselves outperformed and sometimes out-numbered.

Furthermore, as detection ranges from improved sensors increased, surface ships did not have the speed necessary to exploit the detections and to supply a secure cordon around the enemy.

The use of the aircraft in general reconnaissance of the sea surface was a natural evolution of its role in war. However, it was not until significant performance improvements in sensors, weapons, aircraft range and endurance and communications that the MPA could play its full part in integrated close support of surface forces.

The MPA can often put itself at risk of friendly fire when joining a force. An unexpected aircraft contact in the war situation of jammed communications and strict emission control policies is often seen as a potential threat. As a result, complex joining procedures are adopted before closing within range of the defensive missile engagement zone of a friendly force.

While on task, the tactics of the MPA are likely to involve searching at low altitude using a constant radar policy to force down submarines that may have closed the force and are attempting to get periscope ranging for an attack solution. The maritime radar is optimised to detect small contacts against a backgound of reflections or 'clutter' from the sea. Although the submarine will prefer to operate submerged, there are situations in which it must expose itself above the surface. For example, there are surface-to-surface anti-ship missiles that may require the submarine to surface partially in order to fire the missile, providing opportunities for detection by the MPA radar, leading to engagement by mines, torpedo or anti-ship missile.

Sonobuoys are also used to create barriers across a perceived threat axis, allowing the MPA to listen to noises that are characteristic of different submarine types. The experienced acoustic operator can distinguish between different types and between different operating states, and can detect noises from a submarine at rest on the sea bottom. Electronic support measures are used to detect the slightest transmission from an extended communications mast, and the maritime radar can detect a fleeting extension of a communications mast or a diesel air inlet mast. A contact is confirmed by overflying the contact and using the magnetic anomaly detector to distinguish a metallic mass from a shoal of fish or other natural phenomena.

The MPA crew work closely together with their individual sensors, while the tactical commander uses fused sensor data to view the whole surface picture. The MPA works closely with other assets to detect, locate, track and prosecute an attack over many days or weeks of continuous operation. Search patterns similar to those described in the next section are routinely used to conduct an efficient open ocean search.

1.5.4 Search and Rescue (SAR) Role

Search and rescue is the public and humanitarian aspect of military maritime patrol. The extension of the lifeboat service has grown from the original requirement for the military to provide a rescue capability for military aircrew who are forced to make emergency landings in the sea. The task fell to maritime patrol aircraft, rather than a specialist aircraft, for economic reasons, particularly in peacetime when the number of available aircraft is low. The requirements for an effective SAR aircraft have much in common with a maritime patrol aircraft and include:

- Long range and endurance;
- High transit speed and long loiter time;
- All-weather operations capability;
- Precise navigation system;
- Comprehensive communications;
- Extensive sensor suite;
- Good visual platform;
- Large crew complement;
- Displays for tactical control for scene-of-action command;
- Ability to carry large quantities of air-dropped survival equipment.

A military base may keep at least one aircraft and crew on SAR standby 24 h a day, 365 days of the year. The aircraft is normally capable of taking to the air in response to a call for assistance within 2 h. As soon as the SAR aircraft is airborne, a new aircraft and crew are readied to provide cover for any subsequent calls for help.

1.5.4.1 Datum Searches

Where the distressed person or vessel has been able to pass a message to an emergency service, it is likely that a reasonably accurate datum is available and a number of datum search patterns can be used. Some example patterns are shown in Figure 1.11.

If a very accurate position is obtained and the searching aircraft is able to arrive on the scene quickly after the incident, it is possible that the cloverleaf search pattern can be used. This is often the case when military aircrew are forced to eject from their aircraft which

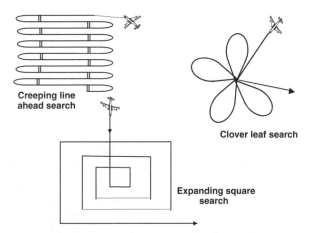

Figure 1.11   Search pattern examples.

has been tracked by a surveillance radar. The fact that the SAR aircraft has arrived shortly after the ejection means that the survivors will not have drifted far from the last known position.

However, if the SAR aircraft is delayed and the position of the datum is not so accurately defined, then there will be a need for the MPA to search out to a greater range. A more suitable search pattern would be the expanding square search, covering an area calculated by the SAR crew to enclose any inaccuracies in position or drift as a result of prevailing weather conditions. If the survivors are equipped with a personal locator beacon (PLB), then the aircraft will use its direction finding (DF) and homing systems to fly straight to the survivor's position.

1.5.4.2   Area Searches

It often happens that notification of an incident is received hours or even days after its occurrence. This usually happens when an aircraft or ship fails to make its scheduled position report. In these circumstances the SAR aircraft is likely to have to search a very large area for survivors or wreckage. Frequently the MPA would be one of a number of cooperating aircraft, helicopters, lifeboats, naval vessels and surface ships involved in the search. The procedure used for organisation, command and control are well tried and tested, and subject to international agreement and standardisation. Specific SAR radio frequencies are available, as are call signs and communication procedures that allow effective integration of civilian and military resources.

The tactics employed by an MPA to search a very large area depend on all the factors described above. If the incident involves a missing military aircraft, the MPA will normally make a number of medium- to high-altitude passes over the area allocated for search. This is firstly to establish a shipping density using radar, and secondly to determine whether any survivors are using PLBs. In the event that no PLB signals are received, the MPA will descend to an altitude suitable for visual and radar search. The type of search will frequently be creeping line ahead (CLA), with track spacing determined by the calculation $1.5 \times$ estimated visibility. Where the track spacing is very short, in poor visibility, for example, the MPA will have difficulty in maintaining the integrity of the search because of the tight turning circles at the end of each loop. A modified creeping line ahead pattern can be used to compensate for this.

1.5.4.3   Scene-of-Action Commander

One aspect of the SAR operation where a modern MPA excels is as a scene-of-action commander (SAC). A typical example of this was the Piper Alpha oil platform disaster in the North Sea. A serious explosion and fire occurred and rescue forces were mobilised. There was no shortage of ships and helicopters to carry out rescue work, but, owing to poor visibility, burning oil on the sea surface and poor communications, there was a distinct danger of the rescue craft hampering or colliding with each other. A Nimrod MR2 from RAF Kinloss was directed to the scene and was able to establish firm control of all rescue forces using radar to deconflict the various helicopters, direct firefighting ships and keep the rescue control centre fully informed of developments. Each ship and helicopter was electronically tagged and displayed on the tactical display, while the area around the rig was divided into small search boxes and allocated to specific ships or helicopters.

1.5.5 Exclusive Economic Zone Protection

1.5.5.1 Oil and Gas Rig Patrols

Many countries have a requirement to police their territorial waters, particularly those declared to be an economic exclusion zone (EEZ) and containing vital national resources such as oil, natural gas and fishing grounds. This task requires regular patrols over large areas of coastal waters by specialist aircraft cooperating with surface vessels to ensure the security of oil and gas installations which may be potential targets for terrorist action.

1.5.5.2 Anti-pollution

There is a requirement for early detection of pollution of the sea, whether by accidental discharge from ships or installations, or by illegal washing of tanks and bilges by merchant vessels. This is an ideal task for aerial surveillance with specially designed sideways looking radar (SLAR) and electrooptical devices using ultraviolet (UV) and infrared (IR) techniques to detect and measure the density and area of oil slicks at sea.

1.5.5.3 Fishery Protection

Fishing represents an increasingly important element of national economies, and there has been a growing tendency for fishing fleets to ignore international agreements for control and licensing of fishing within territorial waters. If a nation is to protect its own fishing rights, it must be capable of demonstrating a capability of detecting and apprehending any vessels fishing illegally within its EEZ. To achieve the very large-scale surveillance task effectively and in an acceptable timeframe dictates the use of an aircraft. There are obvious difficulties if an illegal fishing vessel detected by a fixed-wing aircraft needs to be arrested and brought before a court of law.

Current tactics involve very close cooperation between aircraft and surface vessels, with the aircraft locating the offender, and the ships performing the arrest. However, there is often a delay of some hours before the surface ship can transit to the offender's position, who will no doubt have got rid of any evidence of illegal fishing and possible even have sailed out of the area. It is, therefore, incumbent on the EEZ patrol aircraft to obtain sufficient evidence to allow a reliable case to be brought before an international court, if the offender escapes immediate arrest.

To have a good chance of winning a case in court, the aircraft must catch the offender in the act and obtain high-quality photography of the time and position at which each offence took place. The film needs to have a superimposed image of latitude, longitude, date and time to be used as secure evidence.

1.5.5.4 Customs and Excise Cooperation

Customs and excise operations are usually inshore, and a large military MPA may not be best suited to this kind of role. An alternative is a small twin-turboprop aircraft fitted with a minimum sensor set operating from a civilian airfield by a police or customs crew. This type of aircraft would not normally be armed, its role being surveillance and recording. However, the long-range aircraft will be called upon if a target vessel needs to be tracked over the high seas.

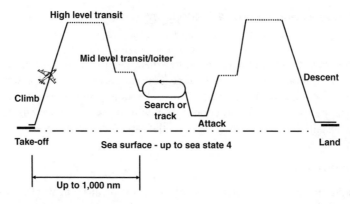

Figure 1.12   Maritime patrol aircraft mission profile.

1.5.6 Key Performance Characteristics

The key performance characteristics are:

- Long endurance;
- Long range.

1.5.7 Crew Complement

The flight deck crew consists of two pilots who may alternate the roles of flying pilot and second officer throughout a long-duration mission in order to maintain vigilance. Some types may carry an engineer who will operate the general systems and usually acts as a monitor for height. On types expected to perform very long-duration missions, for example, with air-to-air refuelling this may be in excess of 20 h, a supernumerary pilot may be carried.

The mission crew will be sized to operate the sensors and conduct the tactical mission. Crew sizes for a long-range maritime patrol and anti-submarine aircraft may exceed 10.

1.5.8 Systems Architecture

A typical maritime patrol aircraft platform architecture is shown in Figure 1.13. Typical MPA systems are listed in Table 1.4.

1.5.9 MPA Aircraft Types

The various types of MPA aircraft are as follows (Figure 1.14):

- Shackleton;
- BAE SYSTEMS Nimrod MR2;
- BAE SYSTEMS MRA4;

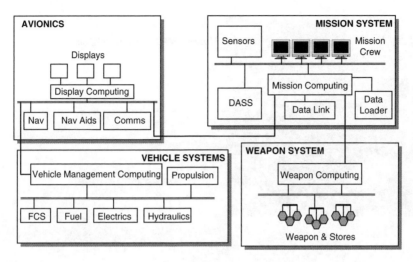

Figure 1.13 Typical maritime patrol aircraft platform architecture.

- Lockheed P-3C;
- Lockheed S-3 Viking;
- Dassault Atlantic;
- Tupolev Tu-20;
- Westland Sea King.

Table 1.4 Typical MPA systems

Avionics	Communications	Mission system	Weapons
	VHF	Maritime radar	Anti-ship missiles
Navigation/GPS	UHF	Electrooptics turret	Torpedos
FMS	HF	ESM	Free-fall bombs
Autopilot	SHF SatCom	DASS	ASR kit
ADF	Link 16	MAD	Flares
DME	Link 11	Acoustic system	Smoke markers
TACAN	Marine band	Mission recording	Sonobuoys
TCAS	Shortwave	Data loader	Mines
Landing aids		Cameras	
GPWS		Oceanographic database	
LPI RadAlt		Mission computing	
Air data		Mission crew workstations	
Digital map		Intelligence databases	
Homing			
Direction finding			
MDP			
Displays and controls			
IFF/SSR			
Avionics data bus		Mission system data bus	Weapons bus

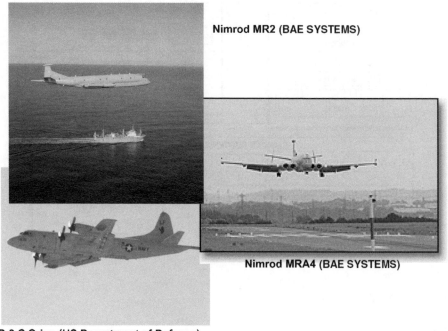

Nimrod MR2 (BAE SYSTEMS)

Nimrod MRA4 (BAE SYSTEMS)

P-3 C Orion (US Department of Defence)

Figure 1.14 Maritime patrol aircraft types.

1.6 Battlefield Surveillance

Detailed knowledge of the tactical scenario on the battlefield is of importance to military commanders and planners who need real-time intelligence of enemy and friendly force disposition, size and movement. Many commercial aircraft types have been converted to perform this role to complement specifically designed military types. The aircraft are equipped with a radar located on the upper or lower surface of the airframe that is designed to look obliquely at the ground. The aircraft flies a fixed pattern at a distance outside the range of enemy defences and detects fixed and moving contacts. These contacts are confirmed by using intelligence from other sensors or from remote intelligence databases to build up a picture of the battlefield and the disposition of enemy and friendly forces. The mission crew operate as a team to build up a surface picture, and can operate as an airborne command centre to direct operations such as air or ground strikes.

1.6.1 Role Description

A battlefield surveillance mission profile is shown in Figure 1.15.

1.6.2 Key Performance Characteristics

The key performance characteristics are high altitude, long range and a stable platform often based on a commercial airliner airframe.

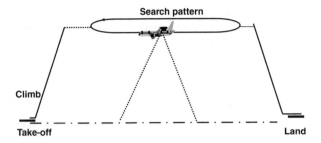

Figure 1.15   Battlefield surveillance mission profile.

1.6.3   Crew Complement

The flight deck crew consists of two pilots who may alternate the roles of flying pilot and second officer throughout a long-duration mission in order to maintain vigilance. Some types may carry an engineer who will operate the general systems and usually acts as a monitor for height. On types expected to perform very long-duration missions, for example, with air-to-air refuelling this may be in excess of 20 h, a supernumerary pilot may be carried.

The mission crew will be sized to operate the sensors and conduct the tactical mission. Crew sizes for a long-range, long-duration mission may exceed 10.

1.6.4   Systems Architecture

A typical battlefield surveillance platform architecture is shown in Figure 1.16. Typical battlefield surveillance systems are listed in Table 1.5.

1.6.5   Battlefield Surveillance Aircraft Types

The various types of battlefield surveillance aircraft are shown in Figure 1.17.

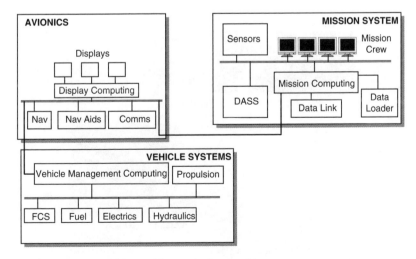

Figure 1.16   Typical battlefield surveillance platform architecture.

Table 1.5 Typical battlefield surveillance systems

Avionics	Communications	Mission system	Weapons
	VHF	Radar	
Navigation/GPS	UHF	Electrooptics	
FMS	HF	ESM	
Autopilot	SHF SatCom	DASS	
ADF	Link 16	Moving target indicator	
DME			
TACAN		Mission recording	
TCAS		Data loader	
Landing aids		Cameras	
GPWS		Mission computer	
LPI RadAlt		Mission crew workstations	
Air data		Intelligence databases	
Digital map			
MDP			
Displays and controls			
IFF/SSR			
Avionics data bus		Mission system data bus	

1.7 Airborne Early Warning

1.7.1 Role Description

Early detection and warning of airborne attack is important to give air superiority and defensive forces sufficient time to prepare a sound defence. It is also important to alert ground and naval forces of impending attack to allow for suitable defence, evasion or countermeasures action.

E-8 JSTARS (US Department of Defence)

ASTOR (Raytheon)

Figure 1.17 Battlefield surveillance aircraft types.

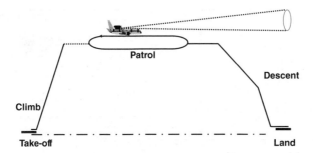

Figure 1.18 Airborne early warning mission profile.

Operating from high altitude gives the airborne early warning (AEW) aircraft an advantage of detecting hostile aircraft at longer range than surface radar, which gives vital seconds for ground defence forces.

1.7.2 Key Performance Characteristics

A long-range, long-endurance aircraft enables a patrol pattern to be set up to cover a widesector area from which attack is most likely. A radar with a 360° scan, and a capability to look down and look up, provides detection of incoming low-level and high-altitude attack. The radar will usually be integrated with an interrogator to enable friendly aircraft to be positively identified. The aircraft will also act as an airborne command post, controlling all airborne movements in the tactical area, compiling intelligence and providing near real-time displays of the tactical situation to both local forces and remote headquarters.

1.7.3 Crew Complement

The flight deck crew consists of two pilots who may alternate the roles of flying pilot and second officer throughout a long-duration mission in order to maintain vigilance. Somevtypes may carry an engineer who will operate the general systems and usually acts as a monitor for height. On types expected to perform very long-duration missions, for example, with air-to-air refuelling this may be in excess of 20 h, a supernumerary pilot may be carried.

The mission crew will be sized to operate the sensors and conduct the tactical mission. Crew sizes for a long-range, long-duration mission may exceed 10.

1.7.4 Systems Architecture

A typical airborne early warning platform architecture is shown in Figure 1.19. Typical AEW systems are listed in Table 1.6.

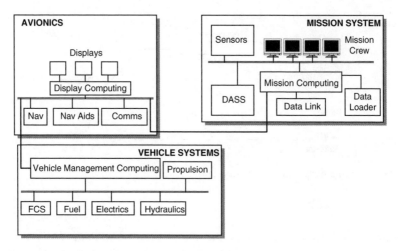

Figure 1.19 Typical airborne early warning platform architecture.

1.7.5 AEW Aircraft Types

The various types of AEW aircraft are as follows (Figure 1.20):

- Grumman E-2 Hawkeye;
- Boeing E-3 Sentry;
- Lockheed P-3 AEW;
- Tupolev Tu-126 AEW;
- Westland Sea King.

Table 1.6 Typical AEW systems

Avionics	Communications	Mission system	Weapons
	VHF	AEW radar	
Navigation/GPS	UHF		
FMS	HF	ESM	
Autopilot	SHF SatCom	DASS	
ADF	Link 16		
DME			
TACAN		Mission recording	
TCAS		Data loader	
Landing aids		Mission computer	
GPWS		Mission crew workstations	
LPI RadAlt		Intelligence databases	
Air data			
Digital map			
MDP			
Displays and controls			
IFF/SSR			
Avionics data bus		Mission system data bus	

E-3 Sentry (US Department of Defence)

P-3 AEW (Lockheed Martin)

Figure 1.20   Airborne early warning aircraft types.

1.8   Electronic Warfare

1.8.1   Role Description

Electronic warfare (EW) refers to a number of related subjects across a wide spectrum of activities:

- Electronic countermeasures (ECM);
- Electronic support measures (ESM);
- Signals intelligence (SIGINT);
- Electronic counter-countermeasures (ECCM).

1.8.2   Electronic Countermeasures

Electronic countermeasures or jamming are a commonly used form of electronic warfare used to disrupt communications or defence radars. In noise jamming, radio frequency at the same frequency as a target emitter/receiver is modulated and transmitted at the target. Depending on the transmitted power level, it is capable of denying range information to the target or degrading communications to an unacceptable level. As the jamming power increases, or the range between the jammer and the receiver reduces, the jamming can become sufficiently strong to break down the directional properties of the target antenna. In this case both range and directional information can be denied.

Deception jamming is a more subtle form of countermeasure, where the intention is to confuse the enemy as to the correct bearing, range and number of targets. It has major implications in the countering of weapon guidance systems, where the technique of range gate or velocity gate stealing can be particularly effective.

Chaff is a passive ECM application, in which the transmitter energy of a threat radar is reflected to create false targets. Chaff can be used in a distraction sense by dispensing small discrete bundles to create an impression of specific small targets to confuse a radar or seduce a missile guidance system. Large clouds of chaff can be dispensed in a confusion sense completely to obscure one's own position.

Jamming is a key tactical role on the battlefield and is often carried out by fast jets equipped with jamming equipment. These aircraft are known in the United States as Wild Weasel squadrons.

1.8.3 Electronic Support Measures

Electronic support measures (ESM) comprise the division of electronic warfare involving actions taken to intercept, locate, record and analyse radiated electromagnetic energy for the purpose of gaining tactical advantage. An important advantage of ESM when used as a sensor is that they are completely passive. They also provide the potential for detecting enemy radars at much greater ranges than the detection ranges of these radars. Signals of ESM interest are usually radar systems. However, they can also include communications, guidance or navigational emissions in the radio-frequency spectrum, as well as laser emissions and infrared radiation in the electrooptics field.

Electronic intelligence information is required for both short- and medium-term planning and also for immediate tactical use in support of offensive and defensive EW operations. ESM are primarily used to support activities such as:

1. Threat warning – the short-term or tactical activity of ESM concerned with detecting transmissions that pose a physical threat. A typical example is the use of radar warning receivers (RWR) to provide an indication of impending attack by fighter or surface-to-air missiles.
2. Target acquisition – the presence of radar systems can indicate the existence of a target, or can assist in the identification of a radar-defended target. An example of this is a maritime patrol aircraft detecting and classifying surface ships by ESM.
3. Homing – an attack radar homing passively onto land or naval base radar-defended targets.

1.8.4 Signals Intelligence (SIGINT)

Signals intelligence consists of a number of different but related activities that are usually complementary in their employment and results. SIGINT is acknowledged as being used by most military forces and governments. Security surrounds its exact operational deployment and the degree of capability available to governments. SIGINT consists of three major activities:

1. Communications intelligence (COMINT) which is achieved by the interception of communication signals of all types – telegraphy, voice or data – and obtaining intelligence on a prospective enemy's intentions, capabilities and military preparedness. Frequently, the text of messages is enciphered and cannot be read immediately. However, there is still a great deal of intelligence to be derived from signals traffic analysis and direction finding which can provide both tactical and strategic advantages. Patterns of communication can be used to identify the state of readiness and location of participants.

Sudden communication activity may indicate battle readiness or changes of plan. Silence can indicate departure of forces from an area, or deliberate radio silence prior to an attack. A database of such communications activity is useful in establishing a potential enemy's radio discipline and movements. Selective or broad band jamming can be deployed to restrict useful communications.

2. Radar intelligence (RADINT) which uses standard or special-purpose radar systems to obtain intelligence on an enemy's capabilities, deployments and intentions. These radar systems may be space, air, sea or land based. RADINT is collected by an electronic support measures (ESM) system employing a number of sensitive antennas that are able to detect radar signals in different bands. As well as detecting the signal, the ESM also establishes a precise direction of arrival of the signal. Analysis of the signal characteristics such as frequency, pulse duration, amplitude and the spacing of main power lobes and side lobes will identify a particular type of radar. Continued collection of analysed data mapped onto the types of platform carrying different radar systems enables an experienced EW operator to identify a particular ship, aircraft or land-based system type. There are claims that experienced operators can even identify an individual platform by its radar transmitter characteristics. A successfully managed EW campaign can identify radar types and their exact locations for subsequent database update or for selective jamming. This technique is known as 'fingerprinting' and it requires some very clever waveform analysis equipment.

3. ELINT Electronic Intelligence involves the interception and analysis of non-communication radio-frequency signals, usually radar, to obtain many aspects of a nation's intentions and capabilities such as technological progress, military preparedness, orders of battle, military competence, intentions etc.

1.8.5 Key Performance Characteristics

For high-altitude, long-duration intelligence gathering, a high-altitude stable platform is required. For cooperative 'Wild Weasel' support, a fast, low-level aircraft is required.

1.8.6 Crew Complement

The flight deck crew consists of two pilots who may alternate the roles of flying pilot and second officer throughout a long-duration mission in order to maintain vigilance. Some

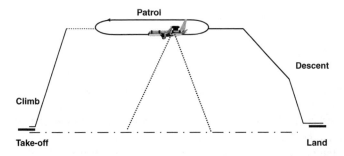

Figure 1.21 Electronic warfare mission profile.

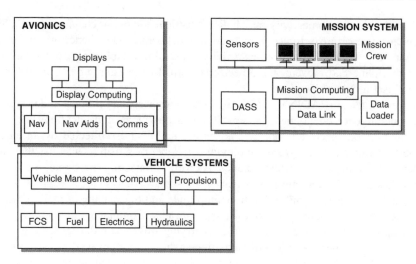

Figure 1.22   Typical electronic warfare platform architecture.

types may carry an engineer who will operate the general systems and usually acts as a monitor for height. On types expected to perform very long-duration missions, for example, with air-to-air refuelling this may be in excess of 20 h, a supernumerary pilot may be carried.

The mission crew will be sized to operate the sensors and conduct the tactical mission. Crew sizes for a long-range, long-duration mission may exceed 10. A typical wild weasel aircraft will have a one- or two-person crew.

1.8.7   Systems Architecture

A typical electronic warfare platform architecture is shown in Figure 1.22. Typical EW systems are listed in Table 1.7.

1.8.8   Example Aircraft Types

Various types of electronic warfare aircraft are shown in Figure 1.23.

1.9   Photographic Reconnaissance

1.9.1   Role Description

Photographic imagery (IMINT) can be used to confirm SIGINT intelligence by providing a high-resolution permanent image using ground-mapping cameras. Such images can be analysed by specialists and used to confirm the types identified by SIGINT. The images may provide further intelligence by providing numbers, groups and battalion identification marks, as well as the deployment of troops.

Although often obtained from satellite imaging systems, IMINT can also be collected by high- and low-altitude fixed-wing aircraft overflying the battlespace in wartime, or by

Table 1.7 Typical EW systems

Avionics	Communications	Mission system	Weapons
	VHF	Maritime radar	
Navigation/GPS	UHF	Electrooptics turret	
FMS	HF	ESM	
Autopilot	SHF SatCom	DASS	
ADF	Link 16	Active jamming	
DME			
TACAN		Mission recording	
TCAS		Data loader	
Landing aids		Mission computer	
GPWS		Mission crew workstations	
LPI RadAlt		Intelligence databases	
Air data			
Digital map			
Displays and controls			
IFF/SSR			
Avionics data bus		Mission system data bus	

overflying territory in peacetime. This activity is risky, leading to loss of aircraft from missile attack, or leading to diplomatic incidents. An example of this is the loss of the US U-2 aircraft in 1960 by surface-to-air missile over the Soviet Union, and over Cuba in 1962.

Cameras mounted in the fuselage or in a pod provide forward, rearward, downward and oblique looking views. Oblique cameras provide the opportunity to film territory without the need to conduct direct overflights – a necessity to prevent diplomatic incidents. The mission computer determines the best rate of height and speed to optimise the film frame rate and varies lens aperture to match light conditions. High-speed focal plane shutters reduce the

EA-6 Prowler (US Department of Defence)

EF-111A Raven (US Department of Defence)

Nimrod R Mk1 (Federation of American Scientists)

RC-135 (US Department of Defence)

Figure 1.23 Electronic warfare aircraft types.

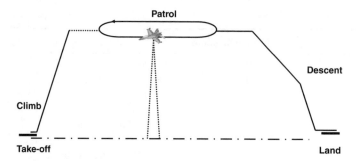

Figure 1.24 Photographic reconnaissance mission profile.

effect of vibration. A window in the airframe or pod is heated to prevent icing or condensation. Wet-film and digital cameras are often complemented by sideways looking synthetic aperture radar and infrared cameras. The reconnaissance pod fitted to Jaguar contained both wet-film and IR cameras, and the Tornado Raptor pod provides real-time high-resolution images from its CCD camera.

Low-altitude, high-speed passes over terrain by fast jets using cameras installed in underwing pods can obtain tactical information at very short notice, and this can be used in conjunction with gun and head-up display (HUD) cameras to provide an instant battlefield picture and confirmation of inflicted damage.

1.9.2 Key Performance Characteristics

Often a very high-altitude aircraft to allow enemy territory overflights beyond radar detection and missile engagement range. To enable large areas of terrain to be covered, extremely high speed is an advantage. To obtain pictures of battlefield damage, low-level high-speed flights may be used.

1.9.3 Crew Complement

Usually a single crew, although aircraft carrying out long-duration precision mapping type photography may carry camera operators and film processors.

1.9.4 Systems Architecture

A typical photographic reconnaissance platform architecture is shown in Figure 1.25. Typical photographic reconnaissance systems are listed in Table 1.8.

1.9.5 Typical Aircraft Types

Typical photographic reconnaissance aircraft types are as follows (Figure 1.26):

- Lockheed U2;
- Lockheed SR71 Blackbird;

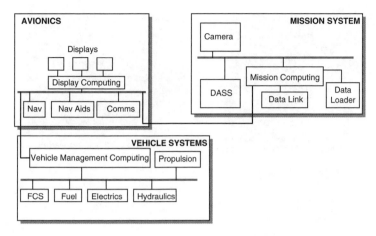

Figure 1.25 Typical photographic reconnaissance platform architecture.

- Canberra PR9;
- Jaguar GR1;
- Tornado GR4.

1.10   Air-to-Air Refuelling

1.10.1   Role Description

Military aircraft of nearly all types find it necessary to extend their range or endurance. This may be the result of the global nature of conflict which is giving rise to long-range missions

Table 1.8   Typical photo reconnaissance systems

Avionics	Communications	Mission system	Weapons
	VHF	High-resolution cameras	
Navigation/GPS	UHF	Film storage	
FMS	HF	Film processing	
Autopilot	SHF SatCom	Infrared cameras	
ADF	Link 16	Optional FLIR	
DME		Optional SAR	
TACAN			
TCAS			
Landing aids			
GPWS			
Air data			
Digital map			
MDP			
Displays and controls			
IFF/SSR			
Avionics data bus			

Lockheed SR-71 (Lockheed Martin)

Lockheed U-2 (Lockheed Martin)

Canberra PR-9 (BAE SYSTEMS)

Figure 1.26   Photographic reconnaissance aircraft types.

or long ferry flights over large oceanic distances, such as the United Kingdom to Falklands flights. This is especially true for large military aircraft which may be sent to an area to establish a presence or force projection as a diplomatic move. If there are no intermediate airfields available, or no airfields readily accessible in friendly nations, then air-to-air refuelling becomes a necessity.

Air superiority aircraft or aircraft on combat air patrol are better refuelled close to their CAP, rather than returning to base to refuel. Not only are time and fuel saved, but so is the need constantly to send replacement aircraft to ensure a continuous presence on patrol.

Incidentally, aircraft companies have extended the duration of their test missions by the use of air-to-air refuelling, extending test flights from a nominal 1 h to up to 8 h.

Most tankers in operation today are conversions of well established commercial aircraft types such as the Boeing 707, Lockheed L1011 Tristar and the BAE SYSTEMS (Vickers) VC10. Military types have also been used, such as the Boeing C-17 and the Handley Page Victor.

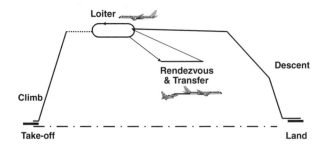

Figure 1.27   Air-to-air tanker mission profile.

There are two types of refuelling method in widespread use:

1. The probe and drogue method is widely used by most air forces. The tanker aircraft deploys and trails one or more hoses, each with a 'basket' at the end. The receiving aircraft homes on to the tanker, extends its refuelling probe and engages the probe tip in the basket. When positive contact is made, refuelling commences, the aircraft matching their speed – a tricky manoeuvre with the tanker becoming lighter and the receiver heavier as fuel is transferred. It is possible to refuel up to three receivers simultaneously.
2. Most US Air Force tanker aircraft are fitted with a boom that is deployed from the tanker. The receiver aircraft homes on to the tanker and flies close. The tanker boom operator engages the boom tip into a receptacle on the aircraft.

1.10.2 Key Performance Characteristics

Long range, capable of carrying a large fuel load and the ability to fly a stable refuelling pattern. Often a converted commercial airliner equipped with additional fuel tanks and loading equipment.

1.10.3 Crew Complement

Flight deck crew and boom operators.

1.10.4 Systems Architecture

A typical air-to-air tanker platform architecture is shown in Figure 1.28. Typical tanker systems are listed in Table 1.9.

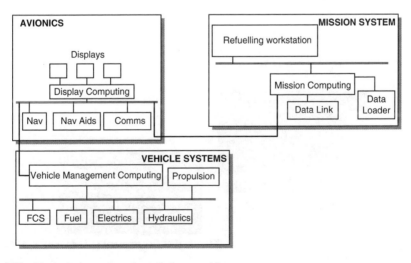

Figure 1.28 Typical air-to-air tanker platform architecture.

Table 1.9   Typical tanker systems

Avionics	Communications	Mission system	Weapons
	VHF	Radar	
Navigation/GPS	UHF		
FMS	HF	ESM	
Autopilot	SHF SatCom	DASS	
ADF	Link 16		
DME			
TACAN		Mission recording	
TCAS		Data loader	
Landing aids			
GPWS		Refuelling equipment	
LPI RadAlt			
Air data			
Digital map			
MDP			
Displays and controls			
IFF/SSR			
Avionics data bus		Mission system data bus	

1.10.5   Aircraft Types

The various types of air-to-air tanker aircraft are as follows (Figure 1.29):

- KC 135;
- KC-10 (DC-10 militarised version);
- Lockheed L-1011 Tristar;
- VC10.

VC-10 (BAE SYSTEMS)

KC-135 (US Department of Defence)

Figure 1.29   Air-to-air tanker aircraft types.

1.11 Troop/Materiel Transport

1.11.1 Role Description

The global nature of conflict and peacekeeping operations demands the movement of troops and materiel to remote theatres of operation. While the bulk of the task is performed by marine vessels, rapid deployment in order to establish a military position requires fast aerial transport. This is often seen by the military as force projection – the ability to establish a rapid presence in times of tension.

Troops can be carried with their personal arms and equipment by fixed-wing or rotary-wing aircraft, disembarking from a landed platform or by parachute. Stores, ammunition and light vehicles can be carried by the same types and either unloaded on the ground or dropped on pallets during a low-speed, very low-height transit. They may also be dropped by parachute if the terrain or the defences do not permit a low-speed pass.

Larger and heavier items of equipment such as artillery, trucks, personnel carriers or helicopters are unloaded from large short take-off and landing (STOL) transport aircraft.

Transport aircraft are used to assist in evacuations from battlefields and also to assist in humanitarian operations by carrying foodstuffs and relief supplies, as well as evacuating refugees.

1.11.2 Key Performance Characteristics

A large volume of cargo-carrying space with secure stowage and a rapid deployment ramp is required, combined with the ability to lift heavy loads from short poorly prepared strips. Long-range transit with air-to-air refuelling if necessary. Good accurate navigation and the ability to operate with mobile landing aids into new designated landing areas. The ability to operate in poor weather conditions and from ill-prepared airstrips.

1.11.3 Crew Complement

Flight deck crew of pilots and flight engineer, with crew to provide loading, unloading and paratroop deployment services.

1.11.4 Systems Architecture

A typical troop/material transport platform architecture is shown in Figure 1.31. Typical transport systems are listed in Table 1.10.

Figure 1.30 Troop/materiel transport mission profile.

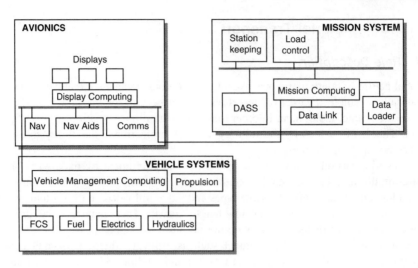

Figure 1.31 Typical troop/materiel transport platform architecture.

1.11.5 Typical Aircraft Types

Typical types of troop/materiel transport aircraft are as follows (Figure 1.32):

- Lockheed C-130;
- Boeing C-17;
- Lockheed C-5 Galaxy;
- Antonov An-124 Condor;

Table 1.10 Typical transport systems

Avionics	Communications	Mission system	Weapons
	VHF	Radar	
Navigation/GPS	UHF	Electrooptics turret	
FMS	HF	ESM	
Autopilot	SHF SatCom	DASS	
ADF	Link 16		
DME			
TACAN			
TCAS		Data loader	
Landing aids			
GPWS		Station keeping	
LPI RadAlt		Materiel deployment	
Air data		Paratroop deployment	
Digital map			
MDP			
Displays and controls			
IFF/SSR			
Avionics data bus			

Lockheed C-130 (Lockheed Martin)

C-5 Galaxy (US Department of Defence)

C-17 (US Department of Defence)

Figure 1.32   Troop/materiel transport aircraft types.

- Transall C-16;
- Fiat G-222;
- Chinook;
- Puma;

1.12 Unmanned Air Vehicles

Many aircraft have evolved to perform the roles described in the previous paragraphs, with ever-increasing performance and sophistication. Aircraft, sensors and systems have become more capable, as have the crews who perform the flying and sensor management tasks. This has led to the situation where the cost of an individual aircraft is very high – tens or even hundreds of millions of dollars per copy, and the value of the crew in terms of training time and experience is also very high. A need to operate over the battlefield without risking human life or expensive assets has led to the development of unmanned aircraft.

Unmanned air vehicles [or uninhabited aerial vehicles (UAVs)] have been used for some time as pilotless target drones. Often these were obsolete aircraft converted for use as an aerial target. Aircraft such as Meteor, Sea Vixen and Canberrra have been used in the United Kingdom and F4 Phantom in the United States. Special-purpose drones were constructed using aircraft parts or designed specifically for the task. Most UAVs today are designed specifically to fulfil a particular role. Without the need to carry a crew, the vehicle can be relatively small and simple, hence reducing cost. The reduced mass also leads to vehicles that can operate for very long duration and at high altitudes. There are examples of UAVs that can operate for over 50 h continuously at altitudes in excess of 70 000 ft.

Some UAVs are small, and careful low-observability design makes them extremely difficult to detect on radar.

An ideal role for the unmanned air vehicle is aerial observation and intelligence gathering. Optical sensors using infrared or TV are used to transmit information to a relay for visual image intelligence (IMINT) to record enemy asset deployment or bomb damage assessment, while antennas are used to obtain communications intelligence (COMINT) or signals intelligence (SIGINT). Some UAVs carry wetfilm cameras, the film being processed on return to base.

Some simple UAVs are used as decoys to entice defence radars to illuminate them as targets. This enables a force to test the alertness of the defensive screen, and also positively to identify the radar type, and hence defence weapon, in order to deploy suitable counter-measures.

All UAVs require a ground station for operators and mission planners, with associated data links for control and data communication. The human factor elements of the traditional cockpit design must be incorporated into the ground station control panels.

Some UAVs can operate autonomously for large portions of a mission without ground station intervention. It is anticipated, as technology evolves, that many of the roles and missions described in the previous paragraphs will be subsumed by role-specific UAVs. Current-generation cruise missiles are an example of a capable UAV, although they are expendable.

Some examples of UAVs are listed in Table 1.11 and shown in Figure 1.33. Unmanned combat air vehicles (UCAVs) are now emerging that are capable of carrying out offensive operations. This covers the delivery of air-launched weapons to attack ground targets, or a capability of engaging fighter aircraft and launching air-to-air missiles.

Table 1.11   Examples of unmanned air vehicles

Type	Manufacturer	Country of origin	Typical use
Jindivik		Australia	Target drone
Firebee	Teledyne Ryan	USA	Reconnaissance
Phoenix	BAE SYSTEMS	UK	Battlefield surveillance
Pioneer	Israeli Aircraft Industries	Israel	
Predator	General Atomics	USA	Surveillance, COMINT
Mirach 26	Meteor	Italy	
Seeker	Kentron	South Africa	
Banshee	Flight Refuelling	UK	Target drone
CL227 Sentinel	Canadair	Canada	Maritime surveillance
Yak 60	Yakovlev	Russia	Surveillance
TU141 Strizh	Tupolev	Russia	Surveillance
D4	Xian ASN	China	Surveillance
Skoja III	Ominpol	Czech Republic/ Hungary	Surveillance
Fox AT1/4	CAC	France	Surveillance
Brevel	STN Atlas	France/Germany	Surveillance
Global Hawk	Teledyne Ryan	USA	Surveillance
Raven	Flight Refuelling	UK	Surveillance
Storm Shadow			

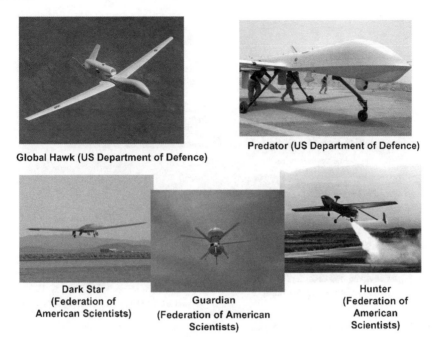

Global Hawk (US Department of Defence)

Predator (US Department of Defence)

Dark Star
(Federation of
American Scientists)

Guardian
(Federation of American
Scientists)

Hunter
(Federation of
American
Scientists)

Figure 1.33   Examples of unmanned air vehicles.

Further developments include the design of micro-UAVs that act in flocks to provide a multiple redundant sensor system capability. These devices can be programmed to mimic bird flocking behaviour so that a number of UAVs will follow a leader.

1.13 Training

Training of aircrew is an important task, from primary training through conversion to type to refresher training to maintain combat readiness. Training is accomplished by a variety of means. Initial training (*ab initio*) is usually carried out using a dynamic simulator to gain familiarisation before transferring to a single-engine light aircraft and then to a single-engine jet trainer. Conversion to an operational type again uses a simulator before transferring to a two-seat version, and then to solo. This is usually carried out on an operational conversion unit (OCU), a squadron staffed by qualified flying instructors.

Routine mission and tactics training can be performed on dynamic simulators, designed to have a high degree of fidelity in six degrees of motion and an outside world real-time picture. However, realism is an important aspect of training, and nations have designated areas of their territory for low-level or combat training. While this includes ranges in restricted areas for very high-speed flight, live weapons delivery and electronic warfare training, much flying takes place in airspace that conflicts with other users such as the general public, private flying, commercial helicopters and leisure flying of balloons, hang-gliding and microlights. Training in these circumstances is essential for combat readiness but requires delicate handling to maintain public sympathy. For this reason, many air forces have resident public liaison officers to deal with public complaints.

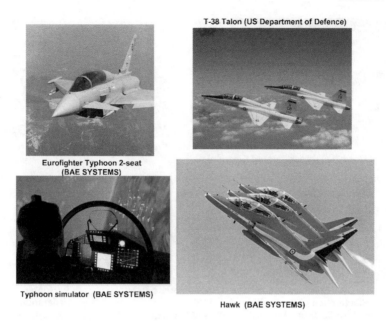

Figure 1.34 Training system aircraft types.

Simpler simulations of parts of the mission task are often undertaken in classrooms, especially for cockpit familiarisation or for mission systems crews to practise team tasks such as MPA or AEW. These are known as part task trainers. In this situation, real missions can be replayed and unexpected scenarios generated. Occasionally, flying classrooms using modified business aircraft to seat a number of trainees at workstations are used.

The training role also includes the provision of aerial targets. These may be implemented as remotely piloted drones such as converted obsolete aircraft or specially designed target vehicles. Banner targets can be towed by converted operational aircraft for air-to-air gunnery

Figure 1.35 Examples of special role types.

practice. Converted small business jets have been used to carry underwing pods equipped with radio-frequency transmitters for use in a simulated EW environment.

1.13.1 Typical Aircraft Types

Typical types of aircraft as follows:

- BAE SYSTEMS Hawk;
- Aermacchi MB339;
- Aermacchi MB326;
- Dassault/Dornier Alphajet;
- Embraer EMB-312Tucano;
- Pilatus/BAE SYSTEMS PC-9;
- Fairchild-Republic T-46;
- -endNorthrop T-38 Talon.

1.14 Special Roles

Military aircraft are often called on to perform roles beyond their original design intention. This may be for research and development of sensors and systems, development of new tactics, for intelligence gathering or for peaceful information-gathering missions. Conversion to these roles may be by major modification to the type or by adding payload internally or externally – such as under-wing pods.

Many of these special roles are, of course, so special that their existence is not revealed to the general public.

1.14.1 Examples of Special Roles

Examples of special roles are as follows:

- Gunship;
- Air ambulance;
- Arctic survey;
- Nuclear contamination detection;
- Biological/chemical sensing;
- Meteorological research;
- Covert troop deployments;
- Remotely piloted vehicle (RPV) aerial launch.

1.15 Summary

This list of roles is by no means exhaustive, the roles may also be known by other titles in different nations and different air forces. The nature of the roles and the system solutions to enable them to be fulfilled will also vary according to the perceived threat, the prevailing defence policy, local politics and the national air force inventory.

What they have in common is a comprehensive system and sensor solution that will include the following elementes:

- Communications and navigation;
- Mission sensors;
- Mission computing and data communications;
- Weapons and stores carriage and release;
- Man-machine interface.

The following chapters will describe the elements and will provide some examples of typical practical implementations.

Further Reading

Adamy, D.A. (2003) *EW 101 A First Course in Electronic Warfare*, Artech House.
Airey, T.E. and Berlin, G.L. (1985) *Fundamentals of Remote Sensing and Airphoto Interpretation*, Prentice-Hall.
Bamford, J. (2001) *Body of Secrets*, Century.
Beaver, P. (ed.) (1989) *The Encyclopaedia of Aviation*, Octopus Books Ltd.
Budiansky, S. (2004) *Air Power*, Viking.
Burberry, R.A. (1992) *VHF and UHF Antennas*, Peter Pergrinus.
Gardner, W.J.R. (1996) *Anti-submarine Warfare*, Brassey's.
Hughes-Wilson, J. (1999) *Military Intelligence Blunders*, Robinson Publishing Ltd.
Jukes, M. (2004) *Aircraft Display Systems*, Professional Engineering Publishing.
Kayton, M. and Fried, W.R. (1997) *Avionics Navigation Systems*, John Wiley.
McDaid, H. and Oliver, D. (1997) *Robot Warriors – The Top Secret History of the Pilotless Plane*, Orion Media.
Oxlee, G.J. (1997) *Aerospace Reconnaissance*, Brassey's.
Poisel, R.A. (2003) *Introduction to Communication Electronic Warfare Systems*. Artech House.
Schleher, C.D. (1978) *MTI Radar*, Artech House.
Schleher, C. (1999) *Electronic Warfare in the Information Age*, Artech House.
Skolnik, M.I. (1980) *Introduction to Radar Systems*, McGraw-Hill.
Stimson, G.W. (1998) *Introduction to Airborne Radar*, 2nd edn, SciTech Publishing Inc.
Thornborough, A.M. and Mormillo, F.B. (2002) *Iron Hand – Smashing the Enemy's Air Defences*, Patrick Stephens.
Urick, R.J. (1983) *Sound Propagation in the Sea*, Peninsula Publishers.
Urick, R.J. (1982) *Principles of Underwater Sound*, Peninsula Publishers
Van Brunt, L.B. (1995) *Applied ECM*, EW Engineering Inc.
Walton, J.D. (1970) *Radome Engineering Handbook*, Marcel Dekker.

2 Technology and Architectures

2.1 Evolution of Avionics Architectures

The introduction gave examples of how an avionics architecture may be structured and explained that, in the main, aircraft level and equipment level architectures would be addressed within the book.

The application of avionics technology to military aircraft has occurred rapidly as aircraft performance has increased. The availability of reliable turbojet engines gave a huge performance boost to both military and civil operators alike. New and powerful sensors such as multimode radars, electrooptics and other advanced sensors have provided an immense capability to modern military aircraft that enables them better to perform their roles. At the same time, the advances in digital avionics technology in the areas of processing and digital communication – by means of data buses – enabled the new systems to be integrated on a much higher scale.

This chapter addresses some of the developments and technology drivers that have led to many of today's advanced platforms; in many cases significant barriers have defeated the attainment of the original aims. These may be summarised as:

- The evolution of avionics architectures from analogue to totally integrated digital implementations (section 2.1).
- Aerospace-specific data buses – the 'electronic string' that binds avionics systems together (section 2.2).
- A description of the joint industrial avionics working group (JIAWG) architecture; originally conceived as a US triservice standard (section 2.3).
- An overview of commerical off-the-shelf (COTS) data buses which are providing a new cost-effective means of integration for the latest avionics systems (section 2.4).
- An overview of the latest developments in software real-time operating systems (RTOS); these developments are leading to increasing portability of software while also providing integrity and security partitioning (section 2.5).

Military Avionics Systems   I. Moir and A. Seabridge
© 2006 John Wiley & Sons, Ltd

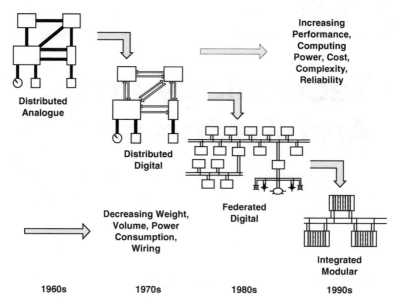

Figure 2.1 Evolution of avionics architectures.

- The increasing integration of radio-frequency subsystems (section 2.6).
- The evolution of the Pave Pace/JSF shared aperture radio-frequency architecture (section 2.7).

To utilise these improvements, the aircraft avionics systems rapidly grew in terms of capability and complexity. Technology brought improvements in terms of increased performance, computing power, complexity and reliability, although all at an increase in cost. Other benefits included a decrease in weight, volume, power consumption and wiring (Figure 2.1).

Figure 2.1 portrays how avionics architectures for modern fighter aircraft have evolved from the 1960s to the present day. The key architectural steps during this time have been:

- Distributed analogue architecture;
- Distributed digital architecture;
- Federated digital architecture;
- Integrated modular architecture; also digital.

The evolution of these different architectures has been shaped in the main by aircraft-level design drivers (Moir and Seabridge, 2004). Their capabilities and performance have been both enabled and constrained by the avionics technology building blocks available at the time. As shown in Figure 2.1, there have been changes in many characteristics throughout the period.

Prior to the 1960s, military aircraft had been manufactured in a similar way to their World War II (WWII) forebears. Avionics units were analogue in nature and interconnected via a considerable quantity of aircraft wiring. Key advances were enabled by the advent of digital computing technology in the 1960s which first found application in the architectures reaching fruition during the 1970s. The availability of digital computers that could be adopted for the rugged and demanding environment of the aerospace application brought computing power and accuracies that had not been available during the analogue era. The development of serial digital data buses greatly eased the interconnection and transfer of data between the major system units. In the early days this was achieved by means of fairly slow half-duplex (unidirectional), point-to-point digital links such as ARINC 429 and Tornado serial data link.

The arrival of microelectronics technology and the first integrated circuits (ICs) enabled digital computing techniques to be applied to many more systems around the aircraft. At the same time, more powerful data buses such as MIL-STD-1553B provided a full-duplex (bidirectional), multidrop capability at higher data rates, of up to 1 Mbit/s. This enabled the federated architectures that evolved during the 1980s, when multiple data bus architectures were developed to cater for increased data flow and system segregation requirements. At this stage the aerospace electronic components were mainly bespoke, being dedicated solutions with few, if any, applications outside aerospace.

The final advance occurred when electronic components and techniques were developed, driven mainly by the demands of the communications and IT industry which yielded a far higher capability than that which aerospace could sustain. This heralded the use of COTS technology, the use of which became more prevalent, and integrated modular avionics architectures began to follow and adapt to technology developed elsewhere.

The key attributes of each of the evolutionary stages of architectural development are described below.

2.1.1 Distributed Analogue Architecture

The distributed analogue architecture is shown in Figure 2.2. In this type of system the major units are interconnected by hardwiring; no data buses are employed. This results in a huge amount of aircraft wiring and the system is extremely difficult to modify if change is necessary. This wiring is associated with power supplies, sensor excitation, sensor signal voltage and system discrete mode selection and status signals. These characteristics are evident in those aircraft conceived and designed throughout the 1950s and 1960s, many of which are still in service today.

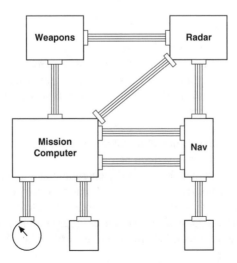

Figure 2.2 Distributed analogue architecture.

These systems have dedicated subsystems, controls and displays. The displays are electro-mechanical and often extremely intricate in their operation, requiring instrument maker skills for assembly and repair.

The use of analogue computing techniques does not provide the accuracy and stability offered by the later digital systems. Analogue systems are prone to bias or drift, and these characteristics are often more pronounced when the aircraft and equipment are subject to a hot or cold soak over a prolonged operating period. The only means of signalling rotary position in an analogue system is by means of synchro angular transmission systems. The older analogue aircraft – termed classic in the industry – therefore contain a huge quantity of synchros and other systems to transmit heading, attitude and other rotary parameters. Pallet (1992) is an excellent source of information on many of the older analogue techniques.

The older equipment is very bulky and heavy and tends to be unreliable as there are many moving parts. This is not a criticism; the designers of the time did their best with the technology available, and many very elegant engineering solutions can be found in this type of equipment. Futhermore, the skills required to maintain some of the intricate instruments and sensors are gradually becoming scarcer, and consequently the cost of repair continues to rise even assuming spare parts are available. Many educational and training institutions no longer teach at this technology level, giving rise to a knowledge gap, which in turn has implications for organisations wishing to refurbish or maintain legacy aircraft and products.

As has already been mentioned, these systems are very difficult to modify, which leads to significant problems when new equipment such as a flight management system has to be retrofitted to a classic aircraft. This is required when military aircraft are upgraded to comply with modern Air Traffic Control (ATC) procedures or a global air transport system (GATM) which are described in Chapter 7.

Typical aircraft in this category are: Boeing 707, VC10, BAC 1-11, DC-9 and early Boeing 737s. Many of these types are still flying, and some such as the VC-10 and the KC-135 and E-3/E-4/E-6 (Boeing 707 derivatives) are fulfilling military roles. They will continue to do so for several years, but gradually their numbers are dwindling as aircraft structural problems are manifested and the increasing cost of maintaining the older systems takes its toll.

2.1.2 Distributed Digital Architecture

The maturity of digital computing devices suitable for airborne use led to the adoption of digital computers, allowing greater speed of computation, greater accuracy and removal of bias and drift problems. The digital computers as installed on these early systems were a far cry from today, being heavy, slow in computing terms, housing very limited memory and being difficult to reprogram – requiring removal from the aircraft in order that modifications could be embodied.

A simplified version of the distributed digital architecture is shown in Figure 2.3. The key characteristics of this system are described below.

Major functional units contained their own digital computer and memory. In the early days of military applications, memory was comprised of magnetic core elements which were very heavy and which in some cases could only be reprogrammed off-aircraft in a maintenance shop. This combined with the lack of experience in programming real-time computers with limited memory and the almost total lack of effective software development tools resulted in heavy maintenance penalties.

As electrically reprogrammable memory became available, this was used in preference to magnetic memory. A significant development accompanying the emergence of digital

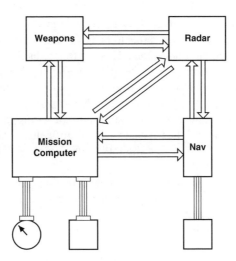

Figure 2.3 Distributed digital architecture.

processing was the adoption of serial half-duplex (unidirectional) digital data buses – ARINC 429 (civil aircraft) and Tornado serial (UK military) – which allowed important system data to be passed in digital form between the major processing centres on the aircraft. Although slow by today's standards (110 kbit/s for ARINC 429 and 64 kbit/s for Tornado serial), the introduction of these data buses represented a major step forward, giving navigation and weapon-aiming systems major performance improvements by adopting this technology.

At this stage, individual system components and equipment were still dedicated in function, although clearly the ability to transfer data between the units had significantly improved. The adoption of data buses, particularly ARINC 429, spawned a series of ARINC standards which standardised the digital interfaces for different types of equipment. The uptake of this standardisation led manufacturers producing inertial navigation systems (INS) to prepare standard interfaces for these systems. This eventually led to the standardisation between systems of different manufacturers, potentially easing the prospect of system modification or upgrade. The ARINC data bus is still important in military systems since many civil platforms adopted for military use rely upon the bus for baseline avionics system integration. The Boeing 737 [maritime patrol and airborne early warning and command system (AWACS)], Boeing 767 (AWACS and tanker) and A330 (tanker) are all recent examples. The Nimrod MRA4 uses an Airbus-based architecture for the avionics and flight deck display system which was based on ARINC 429. This is a modern example of a new military aircraft successfully blending a commercial system with military standard systems.

Displays in the cockpit were dedicated to their function as for the analogue architecture already described. The displays were still the intricate electromechanical devices used previously, with the accompanying problems. In later implementations the displays become multifunctional and multicolour, and the following display systems were developed in the civil field:

- Electronic flight instrument system (EFIS);
- Engine indication and crew alerting system (EICAS) – Boeing and others;
- Electronic checkout and maintenance (ECAM) – Airbus.

Data buses offered a great deal of flexibility in the way that signals were transferred from unit to unit. They also allowed architectures to be constructed with a considerable reduction in interunit wiring and multipin connectors. This led to a reduction in weight and cost, and also eased the task of introducing large and inflexible wiring harnesses into the airframe. This, in turn, led to reductions in the non-recurring cost of producing harness drawings, and the recurring cost of manufacturing and installing harnesses.

Although data buses did remove a great deal of aircraft wiring, the question of adding an additional unit to the system at a later stage was still difficult. In ARINC 429 implementations, data buses were replicated so that the failure of a single link between equipment did not render the system inoperable.

Overall the adoption of even the early digital technology brought great advantages in system accuracy and performance, although the development and maintenance of these early digital systems was far from easy.

Aircraft of this system vintage are:

- Military – Jaguar, Nimrod MR2, Tornado and Sea Harrier;
- Civil – Boeing 737 and 767 and Bombardier Global Express; these aircraft are relevant as many military platforms in the tanker, AWACs and intelligence gathering roles use these baseline civilian platforms.

2.1.3 Federated Digital Architecture

The next development – the federated digital architecture – is shown in Figure 2.4. The federated architecture – from now on all architectures described are digital – relied principally upon the availability of the extremely widely used MIL-STD-1553B data bus. Originally conceived by the US Air Force Wright Patterson development laboratories, as they were called at the time, it evolved through two iterations from a basic standard, finally ending up with the 1553B standard, for which there are also UK Def-Stan equivalents (UK Def Stan 00–18 series).

The adoption of the 1553B data bus standard offered significant advantages and some drawbacks. One advantage was that this was a standard that could be applied across all North

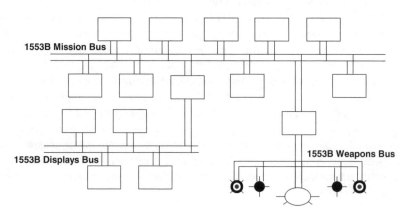

Figure 2.4 Federated avionics architecture.

Atlantic Treaty Organisation (NATO) members, offering a data bus standard across a huge military market, and beyond. The 1553B data bus has been an exceptionally successful application, and the resulting vast electronic component market meant that prices of data bus interface devices could be reduced as the volume could be maintained. It also turned out, as had been the case with earlier data bus implementations, that the interface devices and hence the data buses were far more reliable that anyone could have reasonably expected. Consequently, the resulting system architectures were more robust and reliable than preceding architectures and more reliable than the designers had expected.

Federated architectures generally use dedicated 1553B-interfaced line replaceable units (LRUs) and subsystems, but the wide availability of so much system data meant that significant advances could be made in the displays and other aircraft systems such as utilities or aircraft systems where avionics technology had not previously been applied.

Although the higher data rates were most welcome – approximately 10 times that of the civil ARINC 429 and about 15 times that of the earlier Tornado serial data link – this standard was a victim of its own success in another way. The full-duplex (bidirectional), multidrop protocol meant that it was rapidly seized upon as being a huge advance in terms of digital data transfer (which it was). However, system designers soon began to realise that in a practical system perhaps only 10–12 of the 31 possible remote terminals (RTs) could generally be used owing to data bus loading considerations. At the time of the introduction of 1553B, it was the policy of government procurement agencies to insist that, at system entry into service for a military system, only 50% of the available bandwidth could be utilised to allow growth for future system expansion. Similar capacity constraints applied to processor throughput and memory. Therefore, system designers were prevented from using the last ounce of system capability either in terms of data transfer or computing capability. This led to the use of subsystem dedicated data buses, as shown in Figure 2.4, in which each major subsystem such as avionics, general systems and mission systems had its own bus, complete with a dual-redundant bus controller.

It was also recognised that it was not necessary to have every single piece of data bus equipment talking to every other across the aircraft. Indeed, there were sound systems reasons for partitioning subsystems traffic by data bus to enable all similar task-oriented systems to interchange information on the same bus. The provision of interbus bridges or links between different buses allowed data to be exchanged between functional subsystems. Therefore, during the late 1980s/early 1990s, many multibus architectures similar to the one portrayed in Figure 2.4 were evolved. With minor variations, this architecture is representative of most military avionics systems flying today, including the F-16 mid-life update, SAAB Gripen and Boeing AH-64 C/D.

The civil aircraft community was less eager to adapt to the federal approach, having collectively invested heavily in the single-source–multiple sink ARINC 429 standard that was already widely established and proving its worth in the civil fleets. Furthermore, this group did not like some of the detailed implementation/protocol issues associated with 1553B and accordingly decided to derive a new civil standard that eventually became ARINC 629. To date, ARINC 629 is not envisaged as having a military application.

MIL-STD-1553B utilises a 'command–response' protocol that requires a central control entity called a bus controller (BC), and the civil community voiced concerns regarding this centralised control philosophy. The civil-oriented ARINC 629 is a 2 Mbit/s system that uses a collision avoidance protocol providing each terminal with its own time slot during which it may transmit data on to the bus. This represents a distributed control approach. To be fair to both parties in the debate, they operate in differing environments. Military systems are subject to continuous modification as the Armed Forces need to respond to a continually evolving threat scenario requiring new or improved sensors or weapons. In general, the civil operating environment is more stable and requires far fewer system modifications.

ARINC 629 has only been employed on the Boeing 777 aircraft where it is used in a federated architecture. The pace of aerospace and the gestation time required for technology developments to achieve maturity probably mean that the Boeing 777 will be the sole user of the ARINC 629 implementation.

Along with the developing maturity of electronic memory ICs, in particular non-volatile memory, the federated architecture enabled software reprogramming in the various system LRUs and systems via the aircraft-level data buses. This is a significant improvement in maintainability terms upon the constraints that previously applied. For military systems it confers the ability to reprogram essential mission equipment on a mission-by-mission basis. For the civil market it also allows operational improvements/updates to be speedily incorporated.

The more highly integrated federated system provides a huge data capture capability by virtue of extensive high-bandwidth fibre-optic networks.

2.1.4   Integrated Modular Architecture

The commercial pressures of the aerospace industry have resulted in other solutions being examined for military aircraft avionics systems. The most impressive is the wholesale embracement of integrated modular architectures, as evidenced by US Air Force initiatives such as the Pave Pillar and JIAWG architectures which will be described later.

The resulting architectures use open standards, ruggedized commercial technology to provide the data bus interconnections between the major aircraft systems and integrated computing resources (Figure 2.5).

2.1.5   Open Architecture Issues

As the use of digital technology became increasingly widespread across the aircraft throughout the 1970s, a number of issues became apparent. The use of digital technology introduced new and difficult issues that the system designer and end-user had to embrace in order successfully to develop and support digital avionics systems.

The use of computers and, later, microprocessors, the functionality of which lay in the software applications that were downloaded, introduced a new and far-reaching discipline: that of software tools and software languages. Early processors were slow and cumbersome without the benefit of structured and standardised tools, instruction sets and languages. The memory available for the operational software was strictly limited and was a major constraint upon how the program was able to fulfil its task. A further compounding factor is that the avionics processor is required to perform its operational tasks in real time. At that point in the evolution of computer systems, such operational 'real-time' design experience as was available was centred on the use of mainframe applications where size and memory are less of an impediment. This stimulated a separate offshoot of the computing community that had new issues to address and precious few tools to help in its endeavour.

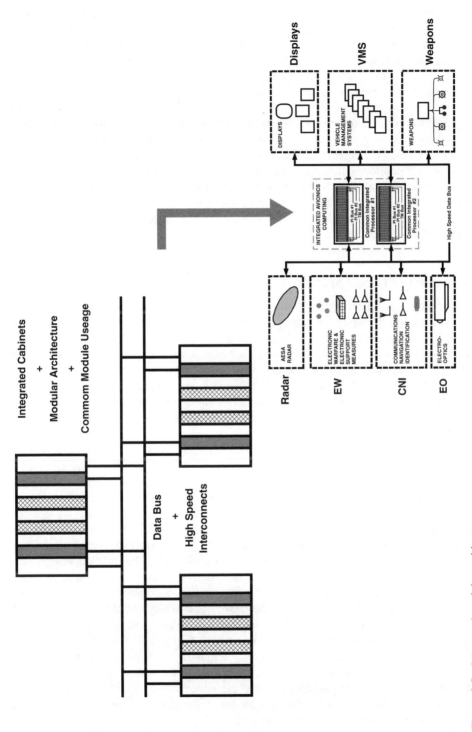

Figure 2.5 Integrated modular architecture.

Another factor had an impact upon the adoption of digital technology. In the main the aerospace community is and was conservative and generally prefers to adopt bespoke solutions more suited to its unique operational tasks, particularly where high integrity and safety are often primary considerations. Furthermore, airborne applications, particularly those on military aircraft designed for worldwide use, have to survive severe operational environments compared with a mainframe computer operating in a comfortable air-conditioned environment. As a consequence, the development timescales, particularly of large or multinational projects requiring extensive software designs, became much longer than those of corresponding commercial projects. Therefore, while the military aerospace designers conservatively used technology that was proven and that would work within their severe environments, the commercial computer world was able to adopt more flexible approaches. As both worlds developed independently, the military tended to become a much smaller player in terms of component development and procurement while the commercial fields of IT and telecommunications became the major driving force. In particular, in this environment, the commercial computing world were early to adopt open architectures, whereas the military community did so much later.

2.1.6 Impact of Digital Systems

Although early implementations of digital technology followed the distributed digital architectures shown in Figure 2.3, the availability of dedicated specifications such as MIL-STD-1553B permitted the aircraft wide interconnection of functional aircraft and avionic functions. While somewhat limited in capacity at 1 Mbit/s, the standard proved to be reliable and the multidrop, full-duplex architecture was central to the adoption of the federated digital architecture described above. Digital avionics technology was used not only to implement navigation and weapon-aiming functions but also became embedded in sensors, communications, flight and engine control, stores management systems, displays and utility control functions.

As functionality and the use of digital technology increased, so the proliferation of processor types and software languages also increased. Therefore, while an aircraft system was integrated in the sense that the data buses were able readily to interchange data between major systems and subsystems, the individual lower-level equipment implementations were beginning to diverge.

One factor that became a major issue was the non-availability of a standardised software language. In many cases the time constraints imposed by the real-time application meant that the program had to be written in assembler or even machine code. The ability to modify and support such implementations once the original design team had dispersed became impossible. Reliable and detailed software documentation was generally sparse or nonexistent, with comment fields to explain the design philosophy. As the scope of digital technology expanded, the need for widespread and standardised software tools, languages and procedures became recognised. The military community addressed these issues by developing more military specifications to enforce standardisation and impose disciplines to improve the visibility, documentation and supportability of software application programs.

Therefore, while the widespread application of digital technology in distributed digital and federated digital architectures brought considerable performance advantages, there was a significant downside that needed to be addressed.

2.1.7   Response of the Services to Digital Avionics Systems Issues

The problems resulting from widespread application of digital technology were addressed – at the risk of considerable oversimplification – by the adoption of the following measures:

- Adoption of federated multibus architectures using a range of line replaceable units (LRUs), commonly known as 'black boxes' to the layman;
- Standardisation of processors;
- Standardisation of high-order language (HOL).

These measures were accompanied with a number of standards imposing defined hardware and software development methodologies.

The adoption of the federated bus architecture was generally a success inspite of the limited bandwidth of MIL-STD-1553B, and many so-called fourth-generation fighter aircraft are flying and enjoying the benefits of that approach. The embedded dual-redundant nature of the data bus standard coupled with the inherent reliability of the data bus interface devices has provided a robust and durable solution. The real problems existed at a lower level in the subsystems and LRUs that comprised the overall avionics system.

The US military took the approach of standardising upon a common processor in the hope that major benefits would result in terms of lower costs and enhanced supportability. It was hoped that the use of a standard machine used force-wide would provide these benefits across a range of platforms, and the standard adopted by the US Air Force was the MIL-STD-1750A limited instruction set processor. As often happens as a result of ambitious standardisation initiatives, there is a tendency to 'round up' the performance of the common device to meet the more demanding (and more risky) applications; this tends to result in a 'fit-all' product that is overly complex for many mundane applications. The standardisation focus within the US Navy was the AN/AYK-14 computer which suffered similarly. Therefore, while the Air Force and Navy were both pursuing standardisation initiatives, their chosen solutions differed.

The US Air Force also attempted to standardise the software language, and during the mid-1980s the JOVIAL language was commonly specified. However, by the 1970s, the US Department of Defense (DoD) was using more than 2000 languages for its mission-critical programming. Most of these were languages that were developed for one specific application. Finally, in 1975, the DoD formed the US Department of Defense High-Order Language Working Group (HOLWG) to find a solution to what was often called the 'software crisis'.

The HOLWG group members decided that they needed to create a language that could be used for just about anything, whether it be systems programming, artificial intelligence or, most important of all, real-time programming and embedded systems. Rather than create this new language themselves, they decided to hold a contest. Coincidentally, all of the competing teams created Pascal-based languages. In the end, the winner was the French-led CII Honeywell-Bull. Eventually, the language was christened 'Ada', in honour of Lady Ada Lovelace, daughter of the poet Lord Byron and assistant to mathematician Charles Babbage, who invented the analytical machine. Lady Ada is often considered to be the world's first programmer, and more details of her life can be found in Woolley (1999).

These initiatives were essentially aimed at producing a military off-the-shelf (MOTS) suite of products whereby the military system designer could select known products for the implementation of the design. The idea was that these military products would be supported by the commercial industry computer base, suitably funded, to provide the military community with the technology and devices that they desired. The drawback to this approach was that the commercial community was moving ahead faster than the military could keep pace, not only with regard to devices and software but also in the adoption of open system architectures. The military were therefore forced to consider open architecture modular integrated cabinet approaches that the commercial world was developing with success.

The adoption of modular architectures was not totally new to the services: the US Air Force in particular had initiatives under way throughout the 1970s and 1980s. The integrated communication navigation and identification architecture (ICNIA) was a serious attempt to use common radio-frequency (RF) building blocks to address the proliferation of RF equipment. A later initiative called Pave Pillar also addressed modular architecture issues. This programme questioned the black box approach to avionics. Pave Pillar architecture physically comprised a number of building blocks called common modules. Each module contained the circuitry to perform a complete digital processing function including interface control and health diagnosis. The common modules were developed from a limited very high-speed integrated circuit (VHSIC) chip set. A number of common module types could then be built up from a small family of VHSIC chips. Modules could then, in turn, be built up to form the basis of any of several avionic subsystems. Some uncommon modules are still required for the odd specific function; however, reduction in the types of spare required as a result of common module usage provided a significant cost improvement. Common modules result in increased production runs for specific modules, which reduces initial purchase cost, while the associated reduction in spares reduces maintenance and support costs. A modular concept allows maintenance engineers to remove and replace components while allowing system designers to adapt the avionics suite to new requirements, and at reduced risk. This concept eventually developed into the integrated modular avionics architecture discussed later in this chapter.

By the late 1980s the Joint Integrated Avionics Working Group (JIAWG) had adopted an approach that was mandated by the US Congress to be used on the three major aircraft developments. These were the US Air Force advanced tactical fighter (ATF), now the F-22 Raptor; the US Navy advanced tactical aircraft (ATA), or A-12 Avenger, which was cancelled in 1990; and the US Army LHX helicopter which became the RAH-66 Comanche and which was cancelled in early 2004. The JIAWG architecture as implemented on the F-22 and later IMA/open architectures will be described later in this chapter. Meanwhile in Europe, the Allied Standards Avionics Architecture Council (ASAAC) developed a set of hardware and software architecture standards around the IMA/open architecture concept.

2.1.8 Need to Embrace COTS

As it became clear that the military aerospace community was a follower rather than a leader in terms of component technologies and architectures, a number of key issues had to be addressed if COTS were to embraced. These included:

- The ability to obtain devices suitable for operation in specified military operating temperature ranges: $-55°C$ to $+125°C$; other environmental issues such as vibration and humidity would also be important;

- Achieving support for the military requirements of integrity/safety, security and certification: issues to which the donor technology is not necessarily exposed in the commercial field;
- Determining the heritage of the hardware device or software program – integrity of development tools, documentation, design traceability and assurance;
- Lack of control of the service authorities over the design standards used;
- The relative short life span and volatility of commercial products, leading to huge lifetime support issues.

The federated architecture already described has proved to be successful but not without a number of drawbacks: the relatively limited bandwidth of MIL-STD-1553B has already been mentioned. In such a system, units are loosely coupled, and this may lead to non-optimum use of data owing to latency issues and overall system performance degradation. While system upgrades may be relatively easily accomplished at the data bus level simply by adding another remote terminal, this is only part of the story. Any change in data transfer leads to changes in the bus controller transaction tables. More importantly, system-wide upgrades usually affect a number of different units, each of which may have its own issues relating to the ease (or difficulty) of modification. Furthermore, if the units involved in the upgrade are provided by different vendors, as is often the case, the modification or upgrade process becomes even more complicated as all vendors need to be managed throughout the programme.

The fundamental advantage offered by an integrated modular avionics (IMA) approach is that, from the outset, the system is conceived using standard building blocks that may be used throughout the aircraft level system. Therefore, common processor modules, common memory modules and, where possible, common input/output modules offer the means of rapidly conceiving and constructing quite extensive system architectures. This approach reduces risk during the development phase, as well as offering significant supportability advantages. The IMA philosophy readily adapts to redundancy implementation in a most cost-effective manner so that economies of scale are easily achieved.

The adoption of COTS-based IMA architectures provides another significant advantage, that is, rapid prototyping. As the baseline modules are off-the-shelf produced in a commercial format, these may be readily purchased in order that a candidate architecture may be built and prototyped using mature commercial boards. Previously, prototyping had to wait until early development hardware was available for all the contributing subsystems; this hardware tended to be immature, and often featured development bugs.

A true comparison of the benefits to be attained by utilising COTS can only be reached after reviewing all the areas where the commercial technology offers benefits. In virtually every respect, COTS offers advantages:

1. *Data buses and networks*. Evolution from the initial 1 Mbit/s MIL-STD-1553B data bus to high-speed COTS 1 Gbit/s, plus solutions such as the scalable coherent interface (SCI), the asynchronous transfer mode (ATM), fibre channel technology and the gigabit Ethernet (GE).
2. *Software*. The adoption of a multilayer software model that segregates the application software from the functional hardware, thereby allowing software to be portable across a range of differing and evolving hardware implementations.
3. *Processing*:

    - *Signal processing*. This offers significant advances in some of the more challenging tasks in the area of radar, electronic warfare (EW) and electroopics (EO); many technological advances have resulted from the development of the Internet.

- *Data processing*. There is often a significant task involved in processing data for transmission or display. The pace at which commercial office IT processors are moving in terms of clock speed far outweighs any advance that could have been achieved using military drivers alone.

These issues and the competing technologies are discussed in detail by Wilcock *et al.* (2001).

In summary, there are many advantages that the military community may gain by the adoption of COTS technology. Apart from the obvious advantage of cost, much greater capability and functionality benefits result, and COTS solutions have been vigorously sought since around the mid-1990s. While not totally without drawbacks, COTS has been adopted for many of the new build or upgrade programmes launched in the last decade.

A few of the many examples of COTS technology in military systems are as follows:

1. The US JSTARS aircraft is fitted with standard DEC Alpha operator workstations.
2. The Israeli Python 4 air-to-air missile avionics system is powered by an Intel 486 processor.
3. The Chinese F-7MG has been fitted with a Garmin GPS 150 receiver (as available at most good high-street electronics dealers!).

2.2. Aerospace-specific Data Buses

As has already been described, digital data buses have been one of the main enablers in the use of digital electronics in aircraft avionics systems. Early data buses were single-source, single-sink (point-to-point), half-duplex (unidirectional) buses with relatively low data rates of the order of a few tens to 100 kbit/s data rates. The next generation of data buses as typified by MIL-STD-1553B were multiple-source, multiple-sink (bidirectional) buses with data rates of around 1–10 Mbit/s. Later fibre channel buses achieve 1 Gbit/s data rates, with the prospect of expanding to several Gbit/s.

Figure 2.6 depicts a comparative illustration of the data bus transmission rates of buses used onboard military avionics platforms.

The dedicated aerospace data buses used within the military aerospace community are:

- Tornado serial data bus;
- ARINC 429;
- MIL-STD-1553B and derivatives;
- STANAG 3910.

Other bus standards such as the JIAWG high-speed data bus (HSDB), the IEEE 1394b and fibre channel buses are all commercial standards that have been adopted for military use.

2.2.1 Tornado Serial

The Tornado serial data bus – more correctly referred to by its PAN standard (Panavia standard) – was the first to be used on a UK fighter aircraft. The bus was adopted for the Tornado avionics system and also used on the Sea Harrier integrated head-up display/

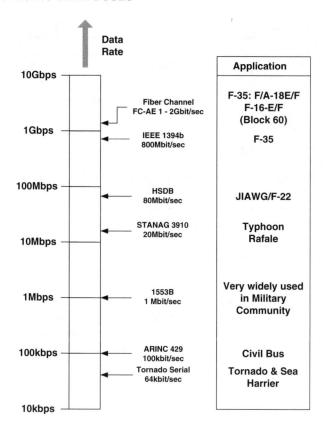

Figure 2.6   Comparative data bus transmission rates.

weapon-aiming computer (HUD/WAC) system. This is a half-duplex serial bus operating at a rate of 64 kbit/s and is used to pass data between the avionics main computer and other sensors, computers and displays within the Tornado nav attack and weapon-aiming system, as shown in Figure 2.7.

The bus comprised four wires implemented as a twisted screened quad format. The lines carried clock and complement and data and complement respectively.

The Tornado system architecture utilising this bus is depicted in the lower part of Figure 2.7. It shows the main computer interfacing via the Tornado serial bus with major avionics subsystems:

- Doppler radar;
- Radar (ground-mapping radar (GMR) and terrain-following radar (TFR));
- Laser range finder/marked target receiver (MTR);
- Attitude Sources – inertial navigation system (INS) and secondary attitude and heading reference system (SAHRS);
- Autopilot/flight director system (AFDS);
- Stores management system (SMS);
- Front cockpit:

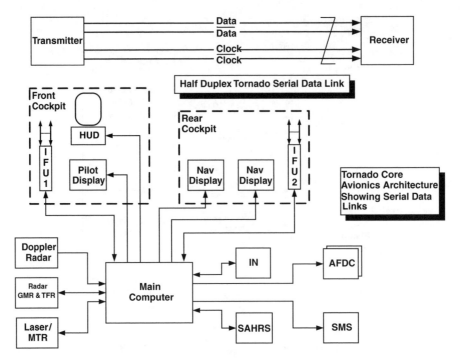

Figure 2.7   Tornado serial data bus and application.

- head-up display (HUD),
- pilot's navigation display,
- interface unit 1 (IFU 1) for front cockpit I/O;

- Rear cockpit:

  - navigator's displays (2),
  - interface unit 2 (IFU 2) for rear cockpit I/O.

This system was designed in the early 1970s and entered service in 1980.

2.2.2   ARINC 429

ARINC 429 is a single-source, multiple-sink, half-duplex bus that operates at two transmission rates; most commonly the higher rate of 100 kbit/s is used. Although the data bus has its origins in the civil marketplace, it is also used extensively on civil platforms that have been adopted for military use, such as the Boeing 737, Boeing 767 and A330. High-performance business jets such as the Bombardier Global Express and Gulfstream GV that are frequently modified as electronic intelligence (ELINT) or reconnaissance platforms also employ A429.

The characteristics of ARINC 429 were agreed among the airlines in 1977/78, and it was first used throughout the B757/B767 and Airbus A300 and A310 aircraft. ARINC, short for Aeronautical Radio Inc., is a corporation in the United States whose stockholders comprise US and foreign airlines and aircraft manufacturers. As such it is a powerful organisation

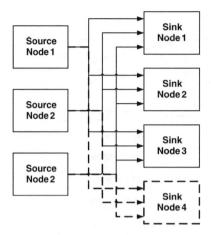

Figure 2.8 A429 topology and the effect of adding units.

central to the specification of equipment standards for known and perceived technical requirements.

The ARINC 429 (A429) bus operates in a single-source–multiple sink mode so that a source may transmit to a number of different terminals or sinks, each of which may receive the data message. However, if any of the sink equipment needs to reply, then each piece of equipment will require its own transmitter and a separate physical bus to do so, and cannot reply down the same wire pair. This half-duplex mode of operation has certain disadvantages. If it is desired to add additional equipment as shown in Figure 2.8, a new set of buses may be required – up to a maximum of eight new buses in this example if each new link needs to operate in bidirectional mode.

The physical implementation of the A429 data bus is a screened, twisted wire pair with the screen earthed at both ends and at all intermediate breaks. The transmitting element shown on the left in Figure 2.9 is embedded in the source equipment and may interface with up to

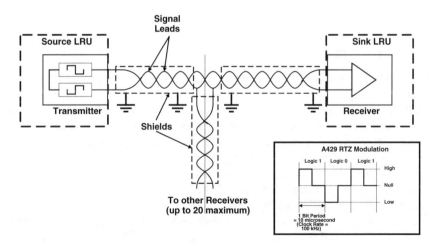

Figure 2.9 A429 data bus and encoding format.

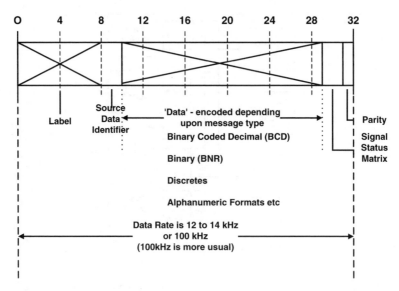

Figure 2.10 A429 data word format.

20 receiving terminals in the sink equipment. Information may be transmitted at a low rate of 12–14 kbit/s or a higher rate of 100 kbit/s; the higher rate is by far the most commonly used. The modulation technique is bipolar return to zero (RTZ), as shown in the box in the figure. The RTZ modulation technique has three signal levels: high, null and low. A logic state 1 is represented by a high state returning to zero; a logic state 0 is represented by a low state returning to null. Information is transmitted down the bus as 32 bit words, as shown in Figure 2.10.

The standard embraces many fixed labels and formats, so that a particular type of equipment always transmits data in a particular way. This standardisation has the advantage that all manufacturers of particular equipment know what data to expect. Where necessary, additions to the standard may also be implemented. Further reading for A429 may be found in Moir and Seabridge (2004).

2.2.3 MIL-STD-1553B

MIL-STD-1553B has evolved since the original publication of MIL-STD-1553 in 1973. The standard has developed through 1553A standard issued in 1975 to the present 1553B standard issued in September 1978. The basic layout of a MIL-STD-1553B data bus is shown in Figure 2.11. The data bus comprises a screened twisted wire pair along which data combined with clock information are passed. The standard generally supports multiple-redundant operation with dual-redundant operation being by far the most common configuration actually used. This allows physical separation of the data buses within the aircraft, therby permitting a degree of battle damage resistance.

Control of the bus is performed by a bus controller (BC) which communicates with a number of remote terminals (RTs) (up to a maximum of 31) via the data bus. RTs only perform the data bus related functions and interface with the host (user) equipment they support. In early systems the RT comprised one or more circuit cards, whereas nowadays it is

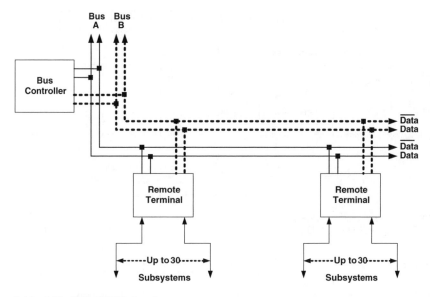

Figure 2.11   MIL-STD-1553B data bus.

usually an embedded chip or hybrid module within the host equipment. Data are transmitted at 1 MHz using a self-clocked Manchester biphase digital format. The transmission of data in true and complement form down a twisted screened pair offers an error detection capability. Words may be formatted as data words, command words or status words, as shown in Figure 2.12. Data words encompass a 16 bit digital word, while the command and status

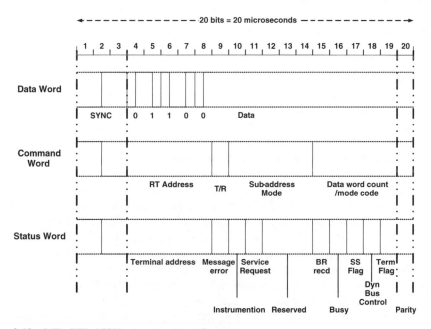

Figure 2.12   MIL-STD-1553B data bus word formats.

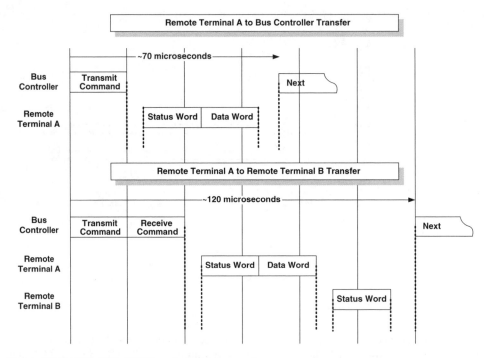

Figure 2.13 MIL-STD-1553B typical data transactions.

words are associated with the data bus transmission protocol. Command and status words are compartmented to include various address, subaddress and control functions, as shown in Figure 2.12.

MIL-STD-1553B is a command–response system in which transmissions are conducted under the control of a single bus controller at any one time; although only one bus controller is shown in these examples, a practical system will employ two bus controllers to provide control redundancy.

Two typical transactions are shown in Figure 2.13. In a simple transfer of data from RT A to the BC, the BC sends a transmit command to RT A, which replies after a short interval known as the response time with a status word, followed immediately by one or more data words up to a maximum of 32 data words. In the example shown in the upper part of the figure, transfer of one data word from RT A to the BC will take approximately 70 µs (depending upon the exact value of the response time plus propagation time down the bus cable). For the direct transfer of data between two RTs as shown from RT A to RT B, the BC sends a receive command to RT B followed by a transmit command to RT A. RT A will send its status word plus the data (up to a maximum of 32 words) to RT B which then responds by sending its status word to the BC, thereby concluding the transaction. In the simple RT to RT transaction shown in Figure 2.13, the total elapsed time is around 120 µs for the transmission of a single data word, which appears to be rather expensive on account of the overhead of having to transmit two command words and two status words as well. However, if the maximum number of data words had been transmitted (32), the same overhead of two command and two status words would represent a much lower percentage of the overall message time. For further reading, see MIL-STD-1553B (1986).

MIL-STD-1553B has proved to be a very reliable and robust data bus and is very well established as a legacy system. Attempts have been made to increase the data rate which is the only major shortcoming. A modification of 1553 called 1553 enhanced bit rate (EBR) running at 10 Mbit/s has been adopted for bomb carriage on the JSF/F-35 using the miniature munitions/store interface (MM/SI). Other vendors have run laboratory demonstrators at 100 Mbit/s and above, and a feasibility program has been initiated to demonstrate 1553 bit rates of 100 Mbit/s with the aim of extending data rates to 500 Mbit/s. This possible derivative is termed enhanced 1553 (EB-1553), and the US Air Force recently hosted a workshop to investigate the possibilities. In October 2003 the Society of Automotive Engineers (SAE) formed an 'Enhanced Performance 1553' task group to address the prospect of launching applications with throughputs of 200–500 Mbit/s using existing cables and couplers. A further standard using 1553 is MIL-STD-1760 – a standard weapons interface which is described in Chapter 9.

2.2.4 STANAG 3910

The evolution of STANAG 3910 was motivated by a desire to increase the data rate above the 1 Mbit/s rate provided by MIL-STD-1553B. The basic architecture is shown in Figure 2.14. The high-speed fibre-optic data terminals pass data at 20 Mbit/s and are connected using a star coupler. Control is exercised by MIL-STD-1553B using electrical connections. The encoding method is Manchester biphase, as for 1553, and data transactions are controlled by means of a bus controller, as is also the case for 1553.

The use of fibre optics passing data at 20 Mbit/s offers a significant improvement over 1553. Furthermore, the ability to transfer messages of up to 132 blocks of 32 words (a total of 4096 data words) is a huge advance over the 32 word blocks permissible in 1553. A total of 31 nodes (terminals) may be addressed, which is the same as 1553.

There are four possible implementations of STANAG 3910 (Table 2.1). The standard also makes provision for the high-speed channel to be implemented as an optical transmissive star coupler, a reflective star coupler or a linear Tee coupled optical bus. Eurofighter Typhoon utilises the type A network with an optical reflective star coupler in a federated architecture.

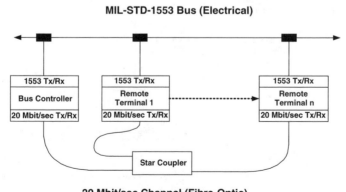

Figure 2.14   STANAG 3910 architecture.

Table 2.1   STANG 3910 implementations

Type A	Low-speed channel	1553B data bus
	High-speed channel	Fibre-optic data bus
Type B	High-speed channel	Fibre-optic equivalent of 1553B
	Low-speed channel	Physically separate fibre-optic data bus
Type C	High-speed channel	Fibre-optic equivalent of 1553B
	Low-speed channel	Wavelength division multiplexed with low-speed channel
Type D	High-speed channel	1553B data bus
	Low-speed channel	Physically separate wire data bus

Note: Both the low-speed and high-speed buses use Manchester biphase encoding.

The STANAG bus is used for the avionics bus and the attack bus while standard MIL-STD-1553B is used for the weapons bus. The Typhoon architecture is discussed in Chapter 9 of this book.

The similarity of transactions to 1553 may be seen by referring to the remote terminal (RT) to bus controller (BC) transaction on the low-speed bus as shown in Figure 2.15. The BC issues a standard 1553 command word followed by a high-speed (HS) action word (taking the place of a 1553 data word). After a suitable interval the RT issues a 1553 status word that completes the transaction. After an intermessage gap, the next transaction is initiated. This cues a high-speed message frame on the high-speed bus which enables the transmission of up to 132 blocks of 32 data words, up to a maximum of 4096 words, as has already been stated. A full description of all the STANAG data transactions may be found in Wooley (1999) which is also a useful source on many other data buses that may be used in avionics applications.

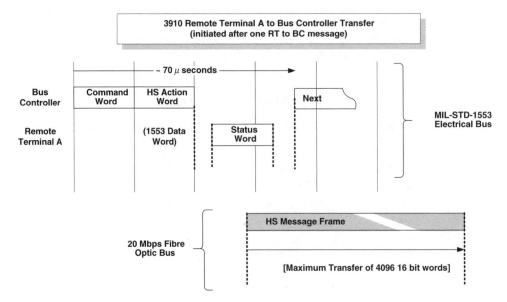

Figure 2.15   STANAG 3910 RT to BC data transaction.

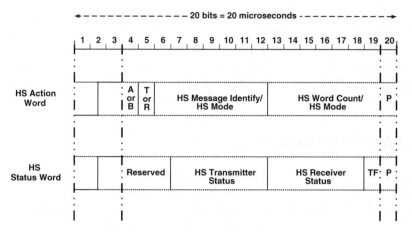

Figure 2.16 STANAG 3910 HS word formats (low-speed bus).

The formats of the low-speed bus HS action word and HS status word are shown in Figure 2.16. It can be seen that the general format is similar to the 1553 words, except that these protocol words have positive-going synchronisation pulses as they are replacing the standard 1553 data word [which also has a positive-going synchronisation pulse (Figure 2.12)]. Also, the word content, instead of containing a 16 bit data word, contains message fields that relate to the high-speed bus message content or transmitter/receiver status. As for other 1553 words, the final bit (bit 20) is reserved for parity.

The format of the STANAG 3910 high-speed bus data structure is shown in Figure 2.17. The data content is preceded by 56 bits associated with the message protocol and followed

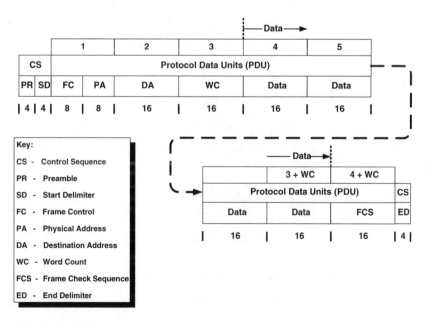

Figure 2.17 STANAG 3910 data structure.

by a further 20 bits, giving a total overhead of 76 bits in all. Therefore, a maximum 4096 data word transfer will take $(56 + 4096 \times 16 + 20) = 65\,612$ bits. At a the high-speed bus data rate of 20 Mbit/s, the total elapsed time is 3280.6 µs.

The application of STANAG 3910 is confined to two European fighter aircraft programmes. Eurofighter Typhoon uses a variant called the EFA bus, which is a type A implementation, while the French Rafale employs type D, using the electrical high-speed bus version.

2.3 JIAWG Architecture

The Joint Integrated Avionics Working Group (JIAWG) was a body charged with developing an avionics architecture based upon the Pave Pillar principles of a modular integrated architecture. The resulting advanced avionics architecture (A^3) was mandated by the US Congress to apply to the following major projects of the late 1980s:

US Air Force	Advanced tactical fighter/F-22 Raptor
US Navy	Advanced tactical aircraft (ATA)/A-12
	(cancelled in the early 1990s)
US Army	LH helicopter/RAH-66 Comanche
	(cancelled in early 2004)

2.3.1 Generic JIAWG Architecture

A generic avionics architecture embracing the main features of the JIAWG architecture is illustrated in Figure 2.18. The major functional elements are:

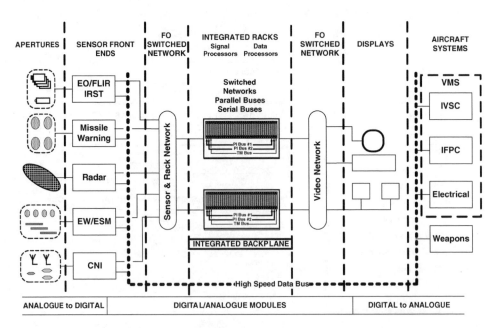

Figure 2.18 Generic JIAWG avionics architecture.

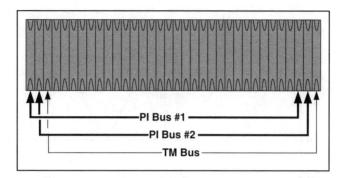

Figure 2.19   Generic JIAWG integrated cabinet architecture.

- Radio-frequency (RF) apertures and sensor front ends associated with electrooptic (EO), missile warning, radar, electronic warfare/electronic support measures (EW/ESM), and communications, navigation and identification (CNI) systems;
- A fibre-optic switched network handling incoming preprocessed sensor data;
- Integrated avionic racks encompassing signal and data processing and interconnected using switched networks, parallel buses and serial buses;
- A fibre-optic switched network handling video data destined for displays;
- Aircraft and weapons systems;
- A high-speed data bus (HSDB) (fibre-optic bus) interconnecting the avionics major systems to the integrated cabinets.

The integrated cabinet architecture selected for the JIAWG architecture is portrayed in generic form in Figure 2.19. The main processor is called the common integrated processor (CIP) and two cabinets are provided in the F-22 architecture with space provision for a third. In fact the term processor is a misnomer as the CIP function contains a cluster of processors (up to seven different types) working together to perform the aircraft-level signal processing and data processing tasks. It has been reported that the combined throughput of the CIP is of the order of 700 million operations per second (MOPS). The F-22 CIP is the responsibility of Hughes.

The F-22 cabinet has space to accommodate up to 66 SEM-E size modules, but in fact only 47 and 44 modules are used respectively in CIP 1 and CIP 2; the remaining 19/22 modules are growth. The density of the packaging and the high power density mean that liquid cooling is used throughout for the cabinets.

To achieve the necessary communications to interchange data with the donor and recipient subsystems, the cabinet interfaces internally and externally using the following data buses:

- A high-speed data bus (HSDB) – a fibre-optic (FO) bus connecting the cabinet with other aircraft subsystems, as shown in Figure 2.18;
- A dual-redundant parallel interface bus (PI bus) (backplane bus) used to interconnect the modules within the rack;
- A serial test and maintenance bus (TM bus) for test and diagnosis.

The FO buses and interconnects are provided by the Harris Corporation.

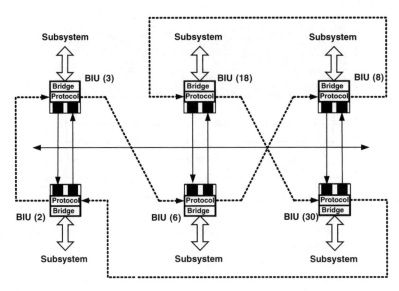

Figure 2.20   FO linear token passing topology (HSDB).

2.3.2   High-speed Data Bus

The high-speed data bus is a linear token ring FO bus operating at 80 Mbit/s. The protocol is in accordance with SAE AS4074.1, for a linear token passing multiplexed data bus. The key characteristics of the HSDB are:

- Manchester biphase encoding;
- Message length up to 4096 16 bit words;
- Ability to service up to 128 terminals.

Figure 2.20 shows a FO liner token passing topology typical of that used for the HSDB.

The SAE linear token passing bus (LTPB) employs a broadcast technique whereby all the terminals can receive all the traffic passing on the bus. However, the transactions are structured such that only those stations whose address matches the destination address within the message header can copy the message. A token passing path is superimposed upon the linear transmission media, and it is this that provides the control logic for each bus interface unit (BIU).

A token is passed around the ring from the lowest address to the highest and then to the lowest again, forming a continuous loop. Stations that have data to transmit claim the token and send the message. The philosophy used is that known as a 'brother's keeper' where every terminal is responsible for providing a valid token to pass to the next terminal. Therefore, in the example given in Figure 2.20, the token and therefore control of the bus would be passed as follows:

$$\text{BIU(2)} \gg \text{BIU(3)} \gg \text{BIU(6)} \gg \text{BIU(8)} \gg \text{BIU(18)} \gg \text{BIU(30)} \gg \text{BIU(2)}$$

and the sequence would repeat.

The media access control (MAC) protocol adopted for the SAE TTPB is one in which timers are used to bound the time a station is allowed to pass data on the bus. As well as having bounded time constraints, each BIU has four message priority levels such that the highest-priority messages are transmitted first and those of lowest priority last. In the event that lower-priority messages cannot be passed owing to insufficient time, these messages can be deferred until the next period.

AIR4288, AIR4271, AS4290 and RTCA DO-178B in the references section of this chapter provide further information upon the operation, use and validation of SAE AS4074.

Unfortunately, this data bus standard, although adopted by the SAE, has not attracted the general support and take-up that the system designers had hoped for. Consequently, it will need to be replaced with a more recent alternative such as the fibre channel (FC) bus very early in the service life of the aircraft. Before its recent cancellation, the RAH-66 Comanche had already adopted a fibre channel implementation in place of the HSDB.

2.3.3 PI Bus

The PI bus is a fault-tolerant parallel backplane bus very similar in structure to VME. The bus operates using a 32 bit wide bus which can be expanded to 64 bit wide. The transfer rate is 50 Mbit/s and the JIAWG architecture employs a dual-redundant implementation. The PI bus is supported by the SAE 4710 and STANAG 3997 standards.

2.3.4 TM Bus

The TM bus is a serial bus comprising five wires that supports test and diagnostics according to the IEEE Std 1149.5-1995 which allows a master controller to interface with up to 250 slave units. Depending upon the precise implementation of the TM bus, diagnosis may be achieved at the board (module) or chip (component) level.

2.3.5 Obsolescence Issues

While the JIAWG architecture was a valiant attempt to achieve cross-service avionics standardisation, it has not succeeded owing to a number of factors, none of which could have been foreseen. At the outset, each of the services was expecting to procure several hundred ATFs, ATAs and LH helicopters. The Cold War had not ended and the USSR/Warsaw Pact was seen as the prime threat. The Navy ATA/A-12 was cancelled early for other reasons, but the remaining ATF/F-22 and LH/RAH-66 programmes had to contend with a post-Cold War and post-Desert Storm world. Inevitably, some of the momentum was lost once the USSR disintegrated and contempory weapons systems performed so well in the Iraq conflict. Finally, it would have been a visionary who could have forecast the true scale and pace of the impact of commercial technology upon military platforms.

Before cancellation of the RAH-66 Comanche, the following problems had been recognised with the JIAWG architecture:

1. The JIAWG-driven processor (Intel 80960 MX) product line was closed down and the processor changed to the i86.
2. The JIAWG-driven backplane (PI bus) product line was closed down.

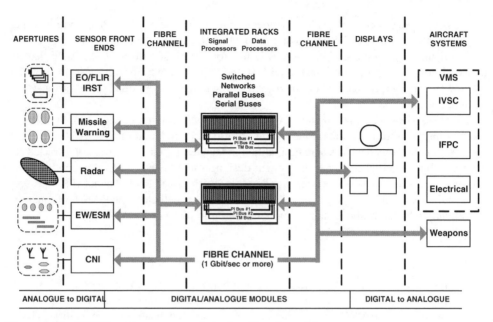

Figure 2.21 JIAWG architecture upgraded using fibre channel buses.

3. The JIAWG-driven test and maintenance (TM bus) is non-standard.
4. The JIAWG/SAE-driven 80 Mbit/s HSDB failed to achieve commercial or military acceptance.
5. The JIAWG-driven SEM-E module format is too small for some COTS technology.

These problems will probably be addressed on the F-22 by the adoption of a fibre channel (FC) for the aircraft-level and major subsystem interconnects, permitting the existing original architecture to be 'grown' in bandwidth while addressing the obsolescence issues. Figure 2.21 shows how the baseline JIAWG generic architecture could be modified to accept FC buses in place of the original HSDB, PI and TM buses. The overall impression is of simplification, as the original 80 Mbit/s linear token ring HSDB is replaced by FC buses capable of carrying 1 Gbit/s or more. The higher bandwidth provided is more suited to 'data stream' the sensor data to the signal and data processing areas within the integrated cabinets. Similar architectural upgrades have been examined for the Apache AH-64 C/D (which predates the JIAWG architecture) and RAH-66/Comanche helicopters.

Developments on more recent projects have provided a forward path for the integration of portable legacy software and new processor types without the need to invest further development work on the operational software packages. This approach is outlined later in the chapter.

2.4 COTS Data Buses

The lack of progress in sponsoring a sustainable high-speed military data bus has led systems designers towards developments in the commercial field, namely COTS. There are a number of possibilities, including the asynchronous transfer mode (ATM) and the fibre distributed data interface (FDDI), but those viewed with the most favour at the time of writing are:

- The fibre channel bus;
- IEEE 1394 firewire.

2.4.1 Fibre Channel

The fibre channel (FC) is a high-throughput, low-latency, packet-switched or connection-oriented network technology; in aerospace applications the latter configuration is usually employed. Data rates are presently available up to 1 Gbit/s, although 10 Gbit/s versions are under development. A whole series of standards are evolving, with one dedicated to avionics applications – the avionics environment project (FC-AE).

The FC bus is designed to operate in an open environment that can accommodate multiple commercial protocols such as the small computer system interface (SCSI) and transport control protocol/Internet protocol (TCP/IP). For avionics applications the FC-AE standard specifies a number of upper-level protocols (ULPs) that align more directly with aerospace applications. These are:

- FC-AE-1553;
- FC-AE-ASM (anonymous subscriber messaging);
- FC-AE-RDMA (remote direct memory access);
- FC-AE-LP (lightweight protocol).

FC-1553 is very useful as it allows the MIL-STD-1553 protocol to be mapped on to the high-bandwidth FC network, creating a low-overhead highly deterministic protocol with a high bandwidth capability. FC-AE-1553 allows a large number of nodes to communicate: increasing from 32 for the baseline 1553 implementation to 2^{24}, while the number of possible subaddresses increases from 32 to 2^{32}. Likewise, the maximum word count increases from 32 to 2^{32}. While these features offer an enormous increase in capability, a further benefit is that FC-AE-1553 provides a bridge between legacy 1553 networks and the much higher-bandwidth FC networks. Therefore, an upgrade introducing an FC network to provide additional bandwidth in certain parts of an avionics architecture can be readily achieved while maintaining those parts that do not require a bandwidth increase intact. The FC-AE-ASM, RDMA and LP options are lightweight protocols that can be variously adopted for specific avionics applications, depending upon the exact requirement.

SCSI is a widely used commercial mass storage standard that allows FC nodes easily to access SCSI-controlled disc space, while TCP/IP is a widely used networking protocol that allows modern computer peripherals and software to communicate. This allows the avionics system designer to utilise commercially available packages. The compatibility of SCSI and TCP/IP with the FC-AE specifications also allows rapid prototyping early in the development cycle.

The FC standards define a range of topologies, the basic four of which are shown in Figure 2.22. These are as follows:

1. Point-to-point communication is used to provide a dedicated link between two computers. This is the least expensive option but one that may find application connecting a radar or EO sensor with its associated processor.

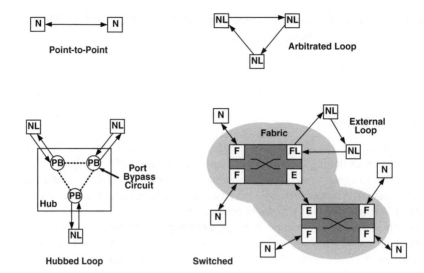

Figure 2.22   Fibre channel topologies.

2. Arbitrated loop is a ring topology providing a low-cost solution to connect up to 128 nodes on a shared bandwidth network, where each node is able to gain control of the loop and establish a connection with another node and generate traffic. Arbitrated loops are limited in their ability to support simultaneous operation and do not include fault tolerance to enable the network to withstand node or media failure. The available bandwidth is shared between the network nodes.
3. Hubbed loop is a variation of the arbitrated loop in which the node connections are made by means of a central hub. In the event of failure, a port bypass circuit enables the failed or inoperative node to be bypassed and allows traffic to continue between the remaining nodes.
4. Switched fabric is a network capable of providing multiple simultaneous full-bandwidth transfers by utilising a 'fabric' of one or more interconnected switches: this provides the highest level of performance. An offshoot of the switched fabric topology is the provision of arbitrated loops (called a public loop) enabling connection between low-bandwidth nodes (connected to the public loop) and high-bandwidth nodes connected directly to the fabric. Switched fabric represents the most powerful but also the most expensive option.

To enable these topologies, a number of different node types are defined and used, as shown in Figure 2.22. These node types are as follows:

1. *Node port (N_Port)*. This provides a means of transporting data to and from another port. It is used in point-to-point and switched topologies.
2. *Fabric port (F_Port)*. This provides the access for an N_Port to the switched fabric.
3. *Loop port (L_Port)*. This port is similar to an N_Port but has additional functionality to provide the arbitrated loop and hubbed loop functions. It can also be used to connect loops and the switched fabric.
4. *Expansion port (E_Port)*. The E_Port is used to provide interconnection between multiple switches within the fabric.

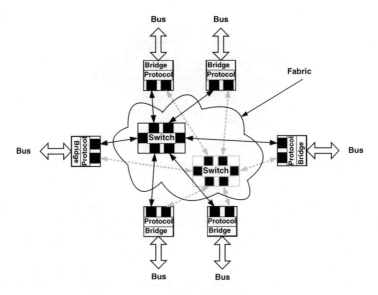

Figure 2.23 Redundant switched fabric topology with bridges.

The switched fabric topology offers a high-capacity interconnect between a number of nodes which can in turn can link to their own shared or dedicated networks (Figure 2.23). Figure 2.23 depicts a six-port configuration that is dual redundant, the lower 'ghosted' network being a replica of the upper network. Although functionally this appears as a network in the figure, in practical terms it will probably be packaged as a module which can be located in an integrated modular cabinet with the other buses or systems being hosted within the same cabinet.

2.4.2 Fibre Channel Options

There are many ways in which the FC may be implemented and products are being offered by a range of vendors. At the time of writing, four main implementations appear to be attracting the most interest:

- StarFabric;
- Rapid IO;
- Infiniband;
- 100/1000 Gbit/s Ethernet.

Of these, StarFabric has been selected for applications on the joint strike fighter (F-35). Figure 2.24 shows an eight-port StarFabric switched fabric incorporating Quad PC processing.

2.4.3 IEEE 1394 Firewire

IEEE 1394 firewire is a widely used data bus scaleable in its original form from 50 to 400 Mbit/s. It has an extremely wide market capture, being commonly used in the electronic domestic consumer market: video cameras, etc. This marketplace has also paved the way for IEEE 1394 to be widely applied in civil aircraft in-flight entertainment (IFE) systems.

Figure 2.24   Eight-port StarFabric switched fabric module.

A significant disadvantage of the baseline standard is that it utilises a daisy-chain architecture to connect the network devices together. Like the arbitrated loop configuration already described, it is therefore unable to tolerate failures by the nodes or the transmission media and is not an attractive option for avionics applications.

Later versions developed under the IEEE 1394b standard are able to work in a network form up to 800 Mbit/s, and there are reports that this standard has been adopted for interconnecting portions of the vehicle management system (VMS) on the F-35.

2.5   Real-time Operating Systems

As for a home computer or laptop, the operating system (OS) of an avionics system and its major subsystems is key to correct operation and achievement of performance goals. The OS needs to be reliable and robust, not prone to malfunctions or crashes. On an aircraft there are many functions that may be flight critical or mission critical, and therefore the software must be reliable and safe. Furthermore, the aircraft systems perform their functions in real time, and therefore the operational programs depend upon a real-time operating system (RTOS) to provide the means for them to execute their programs.

For many years individual companies developed their own operating systems which were used in their own products, often associated directly with a particular processor or computer application. Such dedicated or proprietary systems are still being developed today, but increasingly designers are seeking commercial solutions that are created, maintained and supported by technology specialists. This releases high-grade software resources to work on more specific application software that executes the aircraft or weapons systems functions. An RTOS will be smaller, more modular and more focused in application than the commercial OS used on laptops and PCs. In particular, for high-integrity real-time applications the RTOS should be kept as compact as possible, featuring only essential functions. It also needs to provide a guaranteed level of service, always responding within specified time constraints.

2.5.1   Key Attributes

The key attributes expected of a COTS RTOS are:

1. *Versatility.* The RTOS must be applicable to multiple systems such that a minimum number of RTOSs are required across the aircraft.

2. *Safety.* The system must be capable of being partitioned such that software integrity and therefore aircraft safety are maintained.
3. *Security.* The RTOS must be capable of separating multiple levels of classified and sensitive data.
4. *Supportability.* The OS must be maintained and enhanced throughout its operational life.

2.5.2 Safety

The software developed for high-integrity or safety-critical systems used in aerospace applications is subject to intense scrutiny. All processes associated with developing the software are clearly identified, and extensive plans and documentation are produced to ensure that the correct validation and verification processes are undertaken. Within the civil aerospace community, the specification is issued by the Radio Technical Commission for Aerospace (RTCA). This body has issued DO-178B, entitled 'Software considerations in airborne systems and equipment certification', which was developed by the avionics industry to establish software considerations for developers, installers and users when aircraft equipment design is implemented using microcomputer techniques. DO-178B is recognised 'as an acceptable means to secure FAA approval of digital computer software', RTCA DO-178B. In Europe an equivalent specification is issued by the European Organisation for Civil Aviation Equipment (EUROCAE) under EUROCAE ED-12B. A joint RTCA/EURO-CAE committee is expected to commence work on a third revision standard, DO-178C.

DO-178B acknowledges that not all computer failures or software malfunctions affect an aircraft to the same degree. A malfunction in the aircraft flight control system is clearly more hazardous than the failure of a reading light. Accordingly, software is divided into five different categories, level A through to E, where level A represents the highest level of approval and level E the lowest.

During the initial aircraft design, all those failures that can cause the various levels of failure severity are identified and used to modify the aircraft systems design accordingly. Therefore, long before an aircraft is built, all these conditions are identified and appropriate design steps taken and quality of design assured. This process helps to define the system architecture, the number of control and power channels, the level of redundancy, etc. It also specifies a design assurance level according to what the effects of a failure might be; these design assurance levels are reflected in the RTOS software certification levels (Table 2.2).

There are five main categories of failure severity. The most serious is a catastrophic failure which would result in the loss of the aircraft and passengers. The probability of such an event occurring is specified as extremely improbable, and in analytical or qualitative terms it is directed that a catastrophic failure should occur less than 1×10^{-9} per flight hour. That is less than once per 1000 million flying hours. Other less significant failures are 'hazardous', 'major', 'minor' and 'no-effect'; in each case the level of risk is reduced and the probability of the event occurring is correspondingly increased. Therefore, a minor failure – perhaps the failure of a reading light – can be expected to be reasonably probable, with the event occurring less than 1×10^{-3} per flight hour or less than once every 1000 flying hours. A brief summary of the applicable failure severities is shown in Table 2.3.

DO-178B is not mandated for military aircraft use, indeed its use is not mandated within the civil community. However, as it is a recognised method of successfully achieving

Table 2.2 RTOS software certification levels

Software certification level	Definition
A	Software whose anomalous behaviour would cause or contribute to a failure of a system function resulting in a **catastrophic failure** condition for the aircraft
B	Software whose anomalous behaviour would cause or contribute to a failure of a system function resulting in a **hazardous failure** condition for the aircraft
C	Software whose anomalous behaviour would cause or contribute to a failure of a system function resulting in a **major failure** condition for the aircraft
D	Software whose anomalous behaviour would cause or contribute to a failure of a system function resulting in a **minor failure** condition for the aircraft
E	Software whose anomalous behaviour would cause or contribute to a failure of a system function resulting in a **no-effect failure** condition for the aircraft

certification, the industry effectively uses it as a de facto requirement. There are several reasons why the military avionics community should adopt the standard:

1. It should do so where dual-use software exists, for example flight management system (FMS) software developed for a civil application and adopted for a military program.
2. The adoption of 'commercial best practices' is encouraged, and the fact that DO-178B is the accepted best practice in the civil community offers some reassurance.
3. There is a need to adopt DO-178B in areas where military aircraft have to operate in situations governed by the civil regulations. FMS software designed to satisfy global air transport management (GATM) requirements will also be subject to communications, navigation, surveillance/air transport management (CNS/ATM) regulations imposed by the civil authorities.
4. The standard should be adopted to provide robust software products capable of future reuse.

2.5.3 Software Partitioning

In recent years there has been a tendency for software to be co-hosted on shared processor assets, and this has been given added impetus by integrated modular avionics architectures.

Table 2.3 Summary of the applicable failure severities

Failure severity	Probability	Analytical
Catastrophic	Extremely improbable	Less than 1×10^{-9} per flight hour
Hazardous	Extremely remote	Less than 1×10^{-7} per flight hour
Major	Remote	Less than 1×10^{-5} per flight hour
Minor	Reasonably improbable	Less than 1×10^{-3} per flight hour

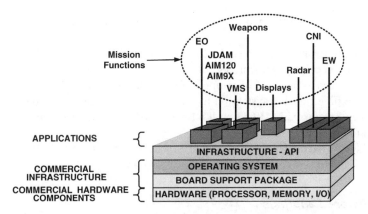

Figure 2.25 Software and hardware integration.

In federated architectures of the type already described earlier in this chapter there is little opportunity to rationalise processor usage in this way.

However, systems integrators have over the past 10 years embraced the modular concept and have confronted the new software issues that result. Multitasking of processors requires software partitioning such that the software applications cannot interfere with each other. In particular, there is a pressing need to ensure that high-integrity applications cannot be adversely affected by lower integrity functions co-hosted on the same processor.

ARINC 653 is an industry specification – again originating from the civil community – that defines partitioning of software and addresses these issues. Software partitioning is an important issue in its own right, but this methodology has helped to address other important issues that have proved very difficult to overcome in the past, including the obsolescence of hardware. This has always been a problem but has been greatly accentuated by the adoption of COTS technology with its rapid hardware development cycles and consequent early obsolescence.

The philosophy adopted is typified by the tiered approach illustrated in Figure 2.25. These tiers or layers provide the following:

1. At the top level, an ARINC 653 compliant software infrastructure or application program-ming interface (API) provides the partitioning of weapons systems functions that may have been developed in a previous program (legacy software) or may be new software developed specifically for the platform. These will generally, but not exclusively, be military-specific applications (the dual use of software has already been discussed).
2. The API infrastructure interfaces with the RTOS which will generally be a commercial package. In many cases the RTOS will be DO-178B compliant.
3. A board-level (i.e. sub-LRU-level) support package will provide the necessary software support to enable the COTS hardware to interface with the commercial RTOS and API layers.
4. The hardware layer based upon COTS technology will be the most dynamic and rapidly varying component of this implementation as the rapid evolution of processor and FO/FC network technology continues. The fact that the most rapidly varying hardware content is decoupled from the application software means that hardware obsolescence can be contained and technology advances enjoyed while the investment in software applications and system functionality is protected.

Recent applications of partitioned and certifiable RTOS have included:

- Green Hills software with level A, INTEGRITY-178B RTOS on the Sikorsky S-92 helicopter avionics management system (AMS);
- Lynux Works Lynx-OS level operating system in association with Rockwell Collins on the adaptive flight display system on the Bombardier Challenger 300 business jet;
- Wind River systems with AE653 on the Boeing 767 tanker transport and C-130 avionics modification program (AMP);
- CsLEOS RTOS developed by BAE SYSTEMS and certified to DO-178B level A for a fly-by-wire flight control system upgrade to the Sikorsky S-92 helicopter.

2.5.4 Software Languages

An added complication to the portability of software application packages relates to the software language used. At an early stage in the use of digital technology it was recognised that the software burden in terms of initial programming effort and lifetime support was in many ways more difficult to quantify and manage than the development of the aerospace-specific hardware. By the mid-1970s the US services were in crisis owing to the vast proliferation in software languages: reportedly in 1976 there were more than 450, some as dialects of standard languages but generally all of which were low in interoperability and reliability and high in maintenance costs. This led to the development of Ada as a high-order language.

By the late 1970s/early 1980s the US services and others specified Ada as the preferred language, and Ada 83 was later upgraded to Ada 95 with additional features and capability. As the number of compilers available, available expertise and development tools increased, Ada became the language of preference in the late 1980s and early 1990s, representing a large investment for the Pentagon and their supplier base.

During the 1990s the armed services withdrew their support for Ada, and new languages such as C and C++ were permitted. There have been and are on-going technical and business debates of great intensity about the merits and demerits of Ada versus these commercial languages. It is not the place of this publication to pass judgement on either approach. However, from the avionics systems designer's viewpoint, the issue is how to integrate legacy software packages that might largely be written in Ada. In extreme cases, software may need to be recoded in order to be compatible with the new environment. The benefits accruing from the adoption of COTS in terms of increased performance and longevity therefore have to weighed against these issues.

2.5.5 Security

As well as the need to partition software functions for reasons of integrity, a more recent requirement is the need to partition for reasons of security. The US authorities now want the RTOS in some applications to be able to operate at multiple levels while maintaining security between them. The evolution of network-centric warfare doctrines means that the same processor that is handling flight management functions in the civil air traffic domain may also have to handle highly classified target data, weapons engagement profiles, rules of engagement, etc.

The rules for protecting sensitive data have evolved over the past three decades but originally resides in a publication issued in the United States in 1983 and known as the 'orange book'. More recently, an international IT security project called 'Common Criteria'

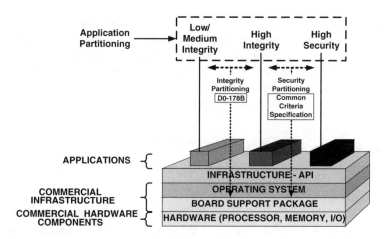

Figure 2.26 Integrity and security partitioning.

has replaced the original publication. 'Common Criteria' specifies seven evaluation assurance levels (EALs) that correspond to earlier classifications. The highest level, EAL-7, requires the system to be multilevel secure, and is able to separate three or more levels of data while processing them on shared hardware.

Usually, the necessary assurance levels are achieved using a combination of hardware [perhaps using the memory management unit (MMU) of the processor] and software means by using the core of the RTOS or 'microkernel'. This concept, referred to as multiple independent levels of security (MILS), permits different levels of classified data to be accommodated. The MILS approach is under active consideration for a number of programs including C-130 AMP, F-22, F-35, the global positioning system (GPS) and the joint tactical radio system (JTRS). MILS can co-exist with ARINC 653/DO-178B implementations, as shown in Figure 2.26.

2.6 RF Integration

Earlier in this chapter the integration of aircraft at the top level using FC networks was examined. Another area where the aircraft can benefit from integration is in the area of the radio-frequency (RF) subsystems. Modern fighters are fitted with a plethora of RF systems, some of which are listed below:

- Radar;
- Electronic warfare (EW);
- Identification friend or foe (IFF);
- Radar warning in several RF bands;
- Navigation aids: TACAN, ILS, MLS, GPS;
- Communications: VHF, UHF, HF SatCom joint tactical information distribution system (JTIDS), Link 11; secure radio.

These systems each have their own antenna, RF sections, signal and data processing: the result is a vast collection of non-standard, heavy and sometimes unreliable hardware modules.

It has been recognised for some time that this is an area ripe for rationalisation and functional integration using a common suite of modules. In the past, initiatives such as integrated communications, navigation and identification architecture (ICNIA) and the integrated electronic warfare suite (INEWS) have attempted to address this problem. For whatever reason, these failed to attain the benefits sought, but advances in RF processing technology are now offering the prospect of fuller integration.

2.6.1 Primary Radar Evolution

2.6.1.1 Independent Systems of the 1950s

Integration of the radar, communications and navigation functions has been occurring over the past four decades, albeit at a slow pace. Figure 2.27 shows a 1950s era system with stand-alone radar, communications and navigation functions. By the standards of today the radar was quite rudimentary, with airborne search and tracking modes. Ground-mapping capabilities were incidental and a pulse Doppler facility with a look-down capability was not yet available. This system was analogue in nature and quite similar to the distributed analogue system described earlier. Systems were interconnected using a great deal of dedicated wiring; and computer and displays would be dedicated to the relevant system. Typical aircraft in this category would be the North American F-86D, Lockheed F-94 Starfire and Northrop F-89 Scorpion.

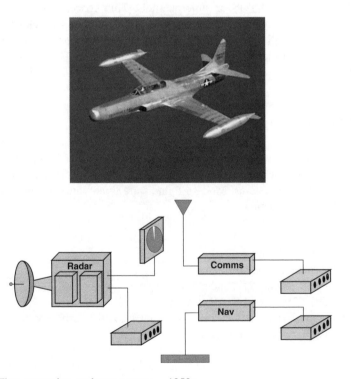

Figure 2.27   First-generation analogue system – 1950s.

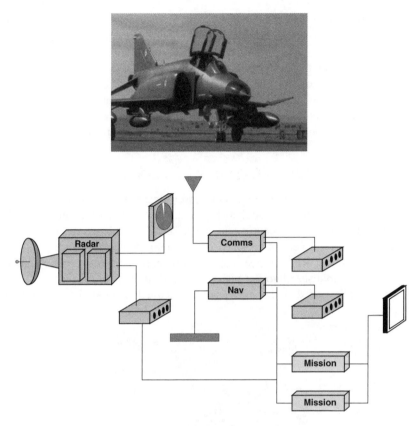

Figure 2.28 Integrated system – 1960s and 1970s.

2.6.1.2 Integrated Systems of the 1960s and 1970s

The next stage is typical of systems entering service during the 1960s, of which the McDonnell Douglas F/A-4 Phantom is a good example (Figure 2.28). In this system the radar, communications and navigation systems are integrated into a mission system with a display providing mission rather than subsystem data. Although these systems were largely analogue, later variants did introduce some digital subsystems. Most system interconnection was still very much undertaken by hardwiring as data buses were only just becoming mature at this stage. These systems introduced further complexity brought about by additional-functionality pulse Doppler radars and inertial navigation systems (INS) and by systems integration. The addition of other, new, more capable navigation aids and integrated radar warning radar (RWR) suites added a further twist of complexity. This integration had to be achieved without the advantage of mature and reliable data buses to enable a federated system to be constructed. These aircraft were very challenging to maintain, suffering from unreliable equipment, high power consumption, vast amounts of wiring and LRUs that were often buried deep in the aircraft as space was very much at a premium. The wiring-intensive nature of the system meant that modifications were always difficult and very expensive to embody. Nevertheless, in spite of the maintenance penalties, these systems brought a step change in aircraft functional capability.

During the 1980s the availability of mature and cost-effective data buses such as 1553 eased the integration task and removed much of the interconnecting wiring, leading to the multibus federated system seen in the F-16 Fighting Falcon and F/A-18 Hornet in the United States and in the Eurofighter Typhoon, SAAB Gripen and Dassault Rafale in Europe. The AH-64 Apache was one of the first helicopters to use a multiple 1553 bus network to integrate its weapons system. At this stage the need for standardisation and modularisation of hardware and software were recognised as the initial adoption of digital technology brought with it many teething problems as well as performance improvements.

2.6.1.3 Integrated Modular Architecture of the 1990s

The 1990s federated architecture built upon the lessons learned from the previous generation, and modular implementations were sought (Figure 2.29). The JIAWG architecture adopted a modular approach but ran into component obsolescence problems, as has already been described. However, in this generation of avionics systems, the performance explosion came in the form of additional RF and electrooptic (EO) sensors. By now, radar antennas had evolved from parabolic dishes into flat plates and used limited 'beam-shaping' techniques, but the antenna still needed to be mechanically scanned. Digital signal processing had

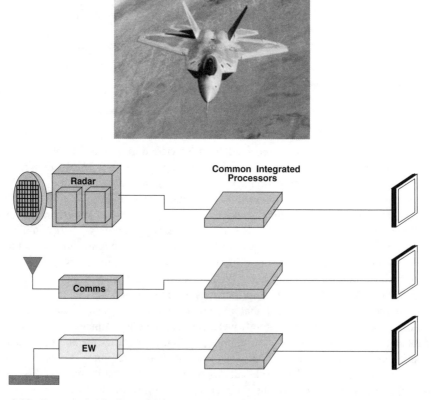

Figure 2.29   Integrated avionics architecture.

evolved to offer true multimode functionality, i.e. the ability to use the same radar for airborne intercepts, ground mapping and missile guidance, for example. Later radars such as those fitted in the F-15E, A/F-18E/F upgrade and block 60 F-16 upgrade (F-16E/F) included an active electronically scanned array (AESA) in place of the flat plate antenna. This antenna is fixed, and radar beams are shaped and steered entirely electronically, without the need for any moving parts. This also brought increased range performance, and highly reliable multimode radar capable of operating in several modes simultaneously. Common integrated processor cabinets provided a modular computing resource for all the mission functions.

However, as will be seen, the RF content of the avionics system was increasing and the need for true integration of the RF system, sharing receiver, signal processor and transmitter resources, became more pressing. Consequently, technology studies preceding the joint strike fighter (JSF) were conducted to consider the route map for future development. The joint advanced strike technology (JAST) program embraced a number of study and technology demonstrator programs to identify technologies suitable for the JSF and reduce risk where possible.

The JAST architecture highlighted the adoption of a number of technologies to aid systems integration, but perhaps the most innovative was the concept of using shared apertures and antenna, as shown in Figure 2.30, as well as a modular approach to the rest of the RF architecture. Much of the material gained from the JAST studies, combined with experience gained from the JIAWG and the F-22 Raptor, will be embodied in the F-35 joint strike fighter program. For more information, see the Joint Advanced Strike Technology Program Avionics Architecture Definition of 8 August and 9 August 1994.

2.6.2 JIAWG RF Subsystem Integration

The RF integration on the F-22 using the JIAWG architecture is indicative of the conservative state of the art 10 years ago when the F-22 program entered the engineering

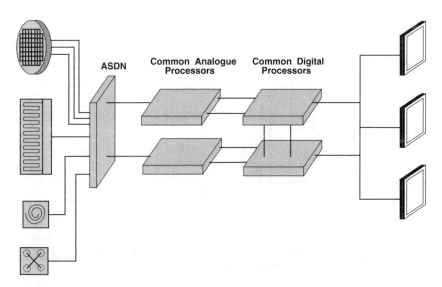

Figure 2.30   2000 + integrated architecture with shared apertures.

manufacturing and development (EMD) phase. The aim was to have a highly capable radar, EW and CNI suite that would enable the aircraft to survive and prosecute its mission in a high-threat environment, and to do so using stealth techniques. The resulting architecture and equipment suite, driven particularly by the need to carry all sensors and weapons internally to preserve stealth, led to a more complex system than other contemporary 'fourth-generation' fighters which, although being very capable in their own right, could not satisfy all the requirements of the F-22 mission. These aircraft do not have as comprehensive an avionics system, often needing to add additional 'podded' sensors for specific roles and carry weapons externally.

The major RF subsystems on the F-22 are regarded as separate entities containing dedicated RF processing, although signal processing and mission data processing is integrated within the CIPs. These subsystems are:

- Active electronically scanned array (AESA) radar;
- Electronic warfare (EW) suite;
- Communications, navigation and identification (CNI) equipment.

Figure 2.31 depicts the F-22/JIAWG architecture, with these sensor subsystems and centralised processing functions highlighted.

Table 2.4 has been compiled with the help of the Joint Advanced Strike Technology Program Avionics Architecture Definition of 8 August 1994 and 9 August 1994. It is representative of the JIAWG/F-22 implementation and serves the purpose of demonstrating the RF architecture integration advances over the past decade or so.

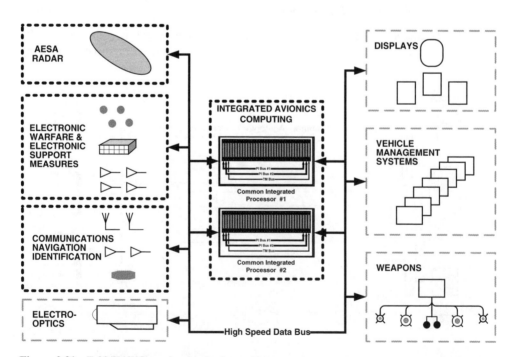

Figure 2.31   F-22/JIAWG top-level avionics architecture.

Table 2.4 Summary of JIAWG apertures

Number	System	Type	Elements	Band (GHz)	Location	Functions
1	Radar	AESA	2000	8–12	FWD	Radar, active and passive targeting
2–7	EW	Spiral	1	2–18	FWD Port and STBD AFT PORT and STBD TOP/BOT	Radar warning receiver (RWR)
8–13	EW	LP	1	2–6	WATERLINE	ECM TX
14–19	EW	LP	1	6–18	WATERLINE	ECM TX
20–21	EW	Spiral	1	2–18	TOP/BOT	ECM RX
22–33	EW	Spiral	1	2–18	6 PORT 6 STBD	Situational awareness (SA); Fwd sector – two arrays, each uses 1 RWR element: Az and El DF
34–35	CNI	Slot	1	5	FWD-BOT AFT-BOT	Microwave landing system (MLS)
36–43	EW	Spiral	1	0.5–2	4 PORT 4 STDB	SA; fwd sector – two arrays, each uses RWR element: Az and El DF
44–45	CNI	Linear array	8	1–1.1	FWD PORT FWD STDB	IFF interrogater
46–47	*CNI*	*Slot*	*1*	*1–1–1*	*TOP/BOT*	*IFF transponder*
48	*CNI*	*Slot*	*1*	*0.9–1.2*	*TOP/BOT*	*TACAN/JTIDS*
49–52	*CNI*	*Slot*	*4*	*1.2–1.5*	*TOP*	*GPS*
53	*CNI*	*Slot*	*2*	*0.1–0.33*	*BOT*	*ILS glideslope ILS localiser*
54	*CNI*	*Slot*	*1*	*0.076*	*BOT*	*ILS marker*
55	*CNI*	*Slot*	*1*	*0.2–0.4*	*TOP*	*UHF-SatCom*
56	CNI	LP	1	15	FWD BOT	Special communications
57–59	CNI	AESA	100	10	FWD PORT WING FWD STBD WING TAIL	Common high-band data link (CHBDL)
60–62	*CNI*	*AESA*	*64*	*Class*	*FWD PORT WING FWD STBD WING TAIL*	*Cooperative engagement capability (CEC)*
63–64	*CNI*	*CNI*	*Ferrite*	*0.002–0.03*	*PORT/STBD*	*HF Comm; Link 11*

Note: CNI omnidirectional antenna in italics are not shown in Figure 2.31.

Table 2.4 identifies all the apertures – 64 in all – and gives a brief summary of the subsystem, type of antenna, number of active elements, frequency band (GHz) and location, and a brief summary of the associated function. A diagrammatic portrayal of these apertures – ignoring the omnidirectional antenna for clarity – is shown in Figure 2.32.

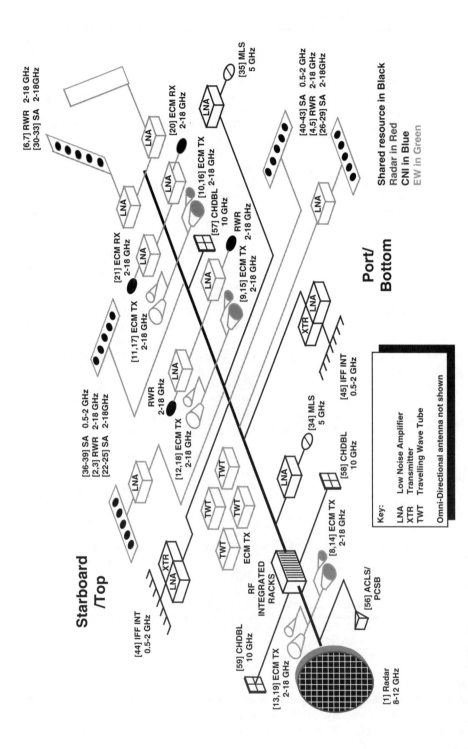

Figure 2.32 JIAWG RF aperture architecture.

The capabilities that this assembly of apertures offer, together with the appropriate electronics units, are summarised below:

1. The AESA radar provides active and passive targeting. This includes ~2000 active transmit/receive (TR) elements. The AESA radar is addressed in Chapter 4.
2. The electronic warfare (EW) suite uses arrays to provide:

   - Radar warning and situational awareness (SA) around the aircraft using a combination of spiral antennas over the 0.5–18 GHz frequency bands. A total of 24 spiral antennas are located around the aircraft to provide full spherical coverage for any RF source and provide direction-finding capabilities for the detected signals.
   - Jamming directed at the aircraft is sensed by a further two spiral antennas. The aircraft can, if desired, respond by transmitting electronic countermeasures in the 2–12 GHz bands using a total of 12 log periodic (LP) antennas located on the water-line.

   The principles of EW are described in Chapter 6.
3. The CNI functions are addressed by a total of 21 slot, linear array, LP, ferrite and phased array antennas that provide the aircraft CNI capability:

   - microwave landing system (MLS) – 2;
   - IFF interrogator – 2;
   - TACAN/JITDS – 2;
   - GPS – 1 array with 4 elements;
   - ILS glideslope, localiser and marker – 2;
   - UHF SatCom – 1;
   - Common high-band data link (CHBDL) – 3;
   - Cooperative engagement capability (data link) – 3;
   - HF communications and Link 11 – 2.

   The principles of CNI equipment are described in Chapter 7.
4. The electronics units associated with these apertures are:

   - Rf integrated racks to provide the RF amplification, detection and signal demodulation for incoming signals and the modulation and power amplification for outgoing signals. In this architecture the RF 'front end' is dedicated to each subsystem.
   - Low-noise amplifiers (LNA) located throughout the aircraft to amplify signals before transmission to the RF racks.
   - Travelling wave tube (TWT) amplifiers to provide power for EW countermeasure transmissions and IFF transmitters for the IFF interrogator.

Figure 2.33 does not portray the full picture. In order to secure the best unimpeded fields of view, many of the apertures need to be located on or near the aircraft extremities. To illustrate this fact, Figure 2.33 shows the provision of the CNI apertures (upper aspect) for the F-22. This gives a true impression of the complexity of installing a highly capable system while abiding by the constraining factors that are imposed by the need for the aircraft to remain stealthy commensurate with its mission.

The situation with regard to the EW apertures is equally if not more complex. Figure 2.34 indicates a further aspect of distributing apertures in this manner. Many apertures need

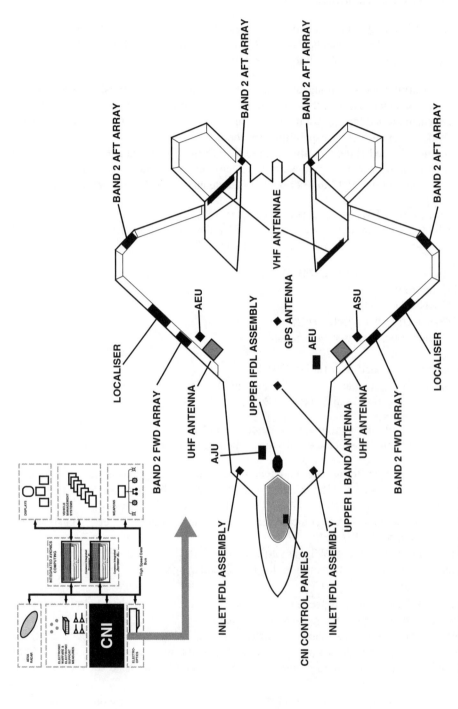

Figure 2.33 F-22 CNI apertures (upper aspect).

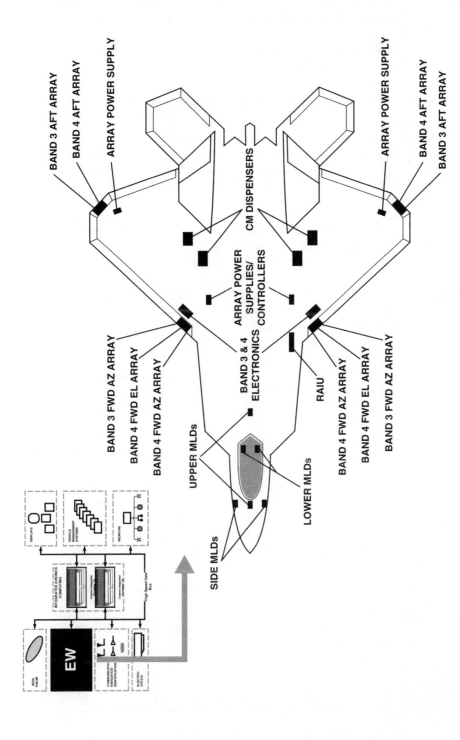

Figure 2.34 F-22 EW apertures.

dedicated power supplies provided locally and the provision of LNAs for many functions. The RF signals that are gathered by the arrays have to feed through the aircraft to the integrated RF racks using coaxial wiring to avoid undue signal attenuation. The fact that the EW suite also embodies wide-band transmitters to provide radiated electronic counter-measures is a further complicating factor.

2.7 Pave Pace/F-35 Shared Aperture Architecture

One of the objectives in developing the Pave Pillar architectures was to address the RF functional area and seek rationalisation of the receiver and demodulation and modulation and amplification/transmitting functions. In the JIAWG architecture above these are handled on a subsystem basis, and the aim of Pave Pace is to provide an integrated RF sensor system – sometimes called an integrated sensor system (ISS). The sharing of resources between the functional system can enable significant savings in cost, weight, volume and reliability. Studies have quantified these savings by comparing a third-generation ISS (JIAWG) with a fourth-generation version (Pave Pace) as shown in Table 2.5.

The basis of a fourth-generation RF ISS is shown in Figure 2.35. The primary arrays may typically comprise a large active array: multiarm spiral arrays (MASAs), slot arrays and multiturn loops (MTLs). These arrays are connected via an RF interconnect to a collection of receive frequency converters that convert the signal to intermediate frequency (IF). The IF receive signals are fed through an IF interconnect to the receiver modules. After detection, the baseband in-phase (I) and quadrature (Q) components are fed through the fibre-optic interconnect to the integrated core processing.

For transmission the reverse occurs, signals are passed to the multifunction modulators and through a separate IF interconnect to the transmit frequency converters. After modulation and power amplification the output signals are passed via the RF interconnect to the appropriate array(s). It is the sharing of these functions within a common RF host that enables the major savings to be made. In the comparison given above it is estimated that about a third was achieved by the use of more advanced components and improved packaging while the remainder came from the integration process.

The JAST documentation already mentioned produced the Pave Pace equivalent of the JIAWG RF architecture shown in Figure 2.32 and expanded in Table 2.4 above.

Figure 2.36 shows the rationalisation of apertures that can be gained from a fourth-generation integrated system, totalling 22 apertures as opposed to 64 apertures for its predecessor. Table 2.6 lists and groups these apertures by function.

Table 2.5 Comparison of third- and fourth-generation RF integrated sensor systems

	Fourth-generation RF ISS	Third-generation RF ISS
Cost* ($)	3.9 million	7.9 million
Weight (lb)	500	1245
Volume (ft³)	8	16
Reliability (h)	225	142

* 1989 US dollars.
Source: Reference 2 in JAST report.

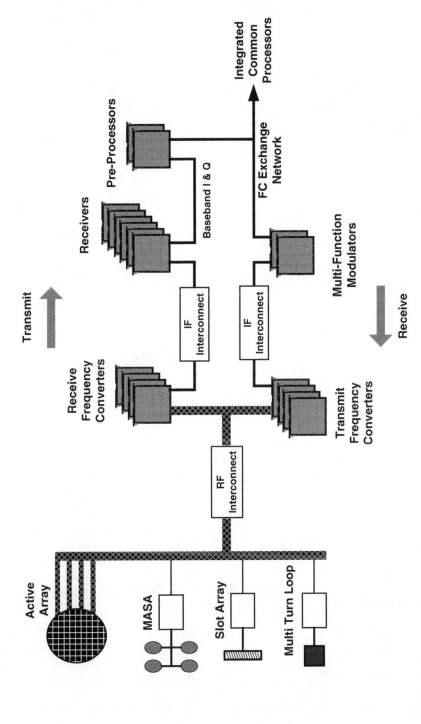

Figure 2.35 Pave Pace integrated RF architecture.

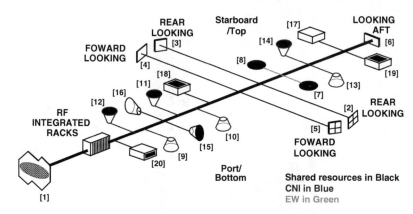

Figure 2.36 Pave Pace – shared aperture RF architecture.

Table 2.6 Pave Pace aperture listing

Number	System	Type	Elements	Band (GHz)	Location	Functions
1	Radar, EW, CNI	WBSA	3000	6–12	FWD	A/A and A/G radar, RWR, ECM, SA, passive targeting, CHBDL, weapon data link
2,3	EW, CNI	WBSA	200	6–18	PORT WING – AFT STBD WING – AFT	RWR, ECM, SA,passive targeting, CHBDL
4–6	EW, CNI	WBSA	64	2–6	PORT WING – FWD STBD WING – FWD TAIL – AFT	RWR, ECM, SA, data link, MLS
7–8	EW	Spiral	1	2–18	TOP/BOT	RWR, ECM receive
9–12	CNI	MASA	8-ARM	0.2–2	2 TOP 2 BOT	UHF radio, GPS, have quick, glideslope, JTIDS, TACAN, IFF transponder, TCAS (functions spread among four apertures to match coverage and functional mix)
13–14	CNI	MASA	8-ARM	0.2–2	TOP/BOT	IFF interrogator
15–16	Radar, EW	MASA	8-ARM	0.2–2	PORT/STBD	RWR, SA, ECM, SAS
17–18	EW, CNI	MTL	1	0.03–0.2	TOP/BOT	VHF radio, SINCGARS, self-protect
19–20	EW, CNI	MTL	1	0.03–0.2	BOT	VOR, localiser, marker beacon, self-protect, SINCGARS, VHF radio
21–22	*CNI*	*MTL*	*1*	*0.002– 0.03*	*PORT/STBD*	*HF Comm, Link 11*

Note: CNI HF Comms/Link 11 antenna are not shown in Figure 2.36.

It can be seen that there is considerable rationalisation compared with the previous architecture, particularly in terms of rationalising functions: Radar + EW, EW + CNI, etc. The AESA has been replaced by a wide-band synthetic array (WBSA) (number 1 in Table 2.6) containing ~3000 elements that services radar, EW and CNI functions. Five similar but smaller arrays provide forward (numbers 4 and 5), rear (numbers 2 and 3) and aft (number 6) coverage for EW and CNI usage. A variety of spiral, MASA and MTL antennas provide the entire gamut of EW and CNI equipment coverage as described in Table 2.6.

References

ARINC Specification 429: Mk 33 digital information transfer system, Aeronautical Radio, Inc., 1977.

ARINC 653, Avionics application software standard interface, Aeronautical Radio Inc., January 1997.

AIR4271 – Handbook of system data communication.

AIR4288 – Linear token passing multiple data bus user's handbook.

AS4290 – Validation test plan for AS4074.

Guide to digital interface standards for military avionic applications, Avionic Systems Standardisation Committee, ASSC/110/6/2, Issue 2, September 2003.

Joint Advanced Strike Technology Program, avionics architecture definition – issues/decisions/rationale document, Version 1, 8 August 1994.

Joint Advanced Strike Technology Program, avionics architecture definition – appendices, Version 1, 9 August 1994.

MIL-STD-1553B digital time division command/response multiplex data bus, Notice 2, 8 September 1986.

Moir, I. and Seabridge, A.G. (2004) *Design and Development of Aircraft Systems – An Introduction*, Professional Engineering Publishing, ISBN 1-86058-437-3.

Pallet, E.H.J. (1992) *Aircraft Instruments and Integrated System*, Longman, ISBN 0-582-08627-2.

RTCA DO-178B, Software considerations in airborne systems and equipment certification.

UK Def Stan 18-00 Series.

Wilcock, G., Totten, T. and Wilson, R. (2001) The application of COTS technology in future modular avionics systems. Electronics and Communications Engineering Journal, August 2001.

Wooley, B. (1999). *The Bride of Science – Romance, Reason and Ada Lovelace*, Macmillan.

3 Basic Radar Systems

3.1 Basic Principles of Radar

The original concept of radar was demonstrated by laboratory experiments carried out by Heinrich Hertz in the 1880s. The term RADAR stands for Radio Aid to Detection And Ranging. Hertz demonstrated that radio waves had the same properties as light (apart from the difference in frequency). He also showed that the radio waves could be reflected from a metal object and could be refracted by a dielectric prism mimicking the behaviour of light.

The concept of radar was known and was being investigated in the 1930s by a number of nations, and the British introduced a ground-based early warning system called Chain Home. In the late 1930s, as part of the world's first integrated air defence, this system has been credited with the winning of the Battle of Britain in 1940. The invention of the magnetron in 1940 gave the ability to produce power at higher frequencies and allowed radar to be adopted for airborne use. The first application was to airborne interception (AI) radars fitted to fighter aircraft to improve the air defence of Great Britain when used in conjunction with the Chain Home system. By the end of WWII, rudimentary ground-mapping radars had also been introduced under the dubious name of H_2S. *Echoes of War* (Lovell, 1992) gives a fascinating account of the development of radar during the war. Since that time, radar has evolved to become the primary sensor on military aircraft and is widely used in civil aviation as a weather radar able to warn the flight crew of impending heavy precipitation or turbulence. For further information, see *Pilot's Handbook – Honeywell Radar RDR-4B*.

Since that time, enormous advances have been made in airborne radars. Fighter aircraft carry multimode radars with advanced pulse Doppler (PD), track-while-scan (TWS) and synthetic aperture (SA) modes that impart an awesome capability. Larger aircraft with an airborne early warning (AEW), such as the E-3, carry large surveillance radars aloft with aerial dishes in excess of 20 ft in diameter. At the other end of the scale, attack helicopters such as the AH-64 C/D Longbow Apache deploy near-millimetric radars in a 'doughnut' on top of the rotor, measuring no more than 3 ft across (Figure 3.1).

Military Avionics Systems I. Moir and A. Seabridge
© 2006 John Wiley & Sons, Ltd

Figure 3.1   Contrasting airborne radar applications. (I. Moir)

The performance and application of radar is highly dependent upon the frequency of operation. Figure 3.2 shows the range of electromagnetic (em) applications used in modern military avionics systems. The applications may be grouped into three categories in ascending order of frequency:

1. Communications and Navaids, more correctly referred to as Communications, Navigation and Identification (CNI), operating in the band from 100 kHz to just over 1 GHz. CNI systems are addressed in Chapter 7.
2. Airborne radar from ∼400 MHz to a little under 100 GHz. This is the subject of this chapter and of Chapter 4.
3. Electrooptics (EO) including visible light in the band from a little over 10 000 GHz (10 THz) extending to just over 1 000 000 GHz (1000 THz). The frequency numbers are so high at this end of the spectrum that wavelength tends to be used instead. The EO band encompasses visible light, infrared (IR) and laser systems which are described in Chapter 5.

Focusing on the airborne radar systems that are the subject of this and the next chapter, these cover the frequency range from ∼400 MHz to 94 GHz, as shown in Figure 3.3. This illustrates some of the major areas of the spectrum as used by airborne platforms. In ascending order of frequency, typical applications are:

- E-2C Hawkeye US Navy surveillance radar operating at ∼400 MHz;
- US Air Force E-3 airborne warning aircraft command system (AWACS) employing a surveillance radar operating at ∼3 GHz;
- Radar altimeters operating at ∼4 GHz, commonly used on civil and military aircraft;
- Fighter aircraft operating in the 10–18 GHz range;
- US Army AH-64 C/D Apache attack helicopter with Longbow radar (AH-64 D variant) operating at ∼35 GHz;
- Active radar-guided, air-launched or ground-launched antiarmour missiles: either Hellfire or Brimstone operating at ∼94 GHz.

The entire frequency range used by radar and other radio applications is categorised by the letter identification scheme shown in Table 3.1. However, only those frequencies assigned by

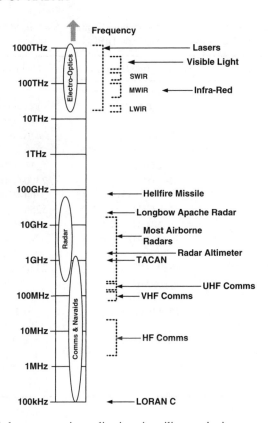

Figure 3.2 Range of electromagnetic applications in military avionics.

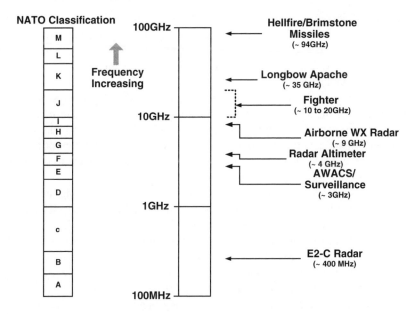

Figure 3.3 Airborne radar frequency coverage.

Table 3.1 Designation of radar bands [source: Skolnik, M.I. (1980) *Introduction to Radar Systems*, McGraw-Hill]

Band designator[a]	Nominal frequency range	ITU assignment
HF	3–30 MHz	
VHF	30–300 MHz	138–144 MHz
		216–225 MHz
UHF	300–1000 MHz	420–450 MHz
		850–942 MHz
L	1–2 GHz	1215–1400 MHz
S	2–4 GHz	2300–2500 MHz
		2700–3700 MHz
C	4–8 GHz	5250–5925 MHz
X	8–12 GHz	8500–10 680 MHz
Ku	12–18 GHz	13.4–14.0 GHz
		15.7–17.7 GHz
K	18–27 GHz	24.05–24.25 GHz
Ka	27–40 GHz	33.4–36 GHz
V	40–75 GHz	59–64 GHz
W	75–110 GHz	76–81 GHz
		92–100 GHz
mm	110–300 GHz	126–142 GHz
		144–149 GHz
		231–235 GHz
		238–248 GHz

[a]IEEE Std 521–1984.

the International Telecommunications Union (ITU) are available for use. This categorisation does not mandate the use of a particular band or frequency but merely indicates that it is available to be used. Other factors decide which band to be used in a particular application: most notable are the effects of atmospheric absorption and the size of antenna that the platform can reasonably accommodate.

The effect of atmospheric absorption is a constraint depending upon physics that is totally outside the control of the designer. Physical antenna size is to some extent under the control of the designer, although the platform dimensions will be determined by factors relating to its airborne performance, range and so on. As for many systems, the design of a radar system is subject to many considerations and trade-offs as the designer attempts to reconcile all the relevant drivers to obtain an optimum solution.

The effects of atmospheric absorption are shown in Figure 3.4. The diagram illustrates the loss in dB per kilometre across the frequency spectrum from 1 to 300 GHz. This curve varies at various altitudes – the particular characteristic shown is for sea level. A 10 dB loss is equivalent to a tenfold loss of signal, so the loss per kilometre at 60 GHz is almost a 1000 times worse than the loss at around 80 GHz. These peaks of atmospheric absorption occur at the resonant frequency of various molecules in the atmosphere: H_2O at 22 and 185 GHz and O_2 at 60 and 120 GHz, with the resonance at 60 GHz being particularly severe.

Also shown on the diagram are four key frequency bands used by some of the weapons systems of today:

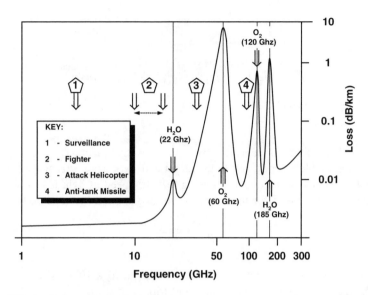

Figure 3.4   Effects of atmospheric attenuation.

- Surveillance radar operating at ~3 GHz;
- Fighter radar radiating from 10 to 18 GHz;
- Attack helicopter operating at 35 GHz;
- Anti-armour missile transmitting at 94 GHz.

It can be seen that the atmospheric absorption effects have a significant impact upon the portions of the spectrum that the radar designer can reasonably utilise.

The basic principle used by radar is portrayed in Figure 3.5. The energy radiating from a radar transmitter propagates in a similar fashion to the way ripples spread from an object dropped in water. If the radiated energy strikes an object – such as an aircraft – a small

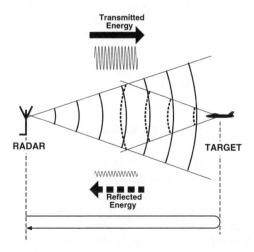

Figure 3.5   Basic principles of radar.

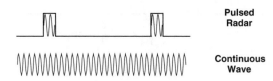

Figure 3.6   Pulse and continuous wave.\

proportion of that energy is reflected back towards the radar. The transmitted energy effectively has a double journey: out to the target and back again. Radar uses this principle to measure the distance to the target; knowing that the speed of light is $\approx 3 \times 10^8$ m/s, and by measuring the time taken for the reflection to arrive, makes it possible to calculate the target range:

$$R = \frac{c \times \Delta t}{2}$$

where R is the range of the target, c is the speed of light (3×10^8 m/s) and Δt is the time taken for the radar energy to perform the round trip.

Radar energy may also be transmitted in a number of ways. Figure 3.6 shows two situations; one where the RF energy is sent in pulses and the other where RF energy is radiated continuously – also known as a continuous wave.

Pulsed radar transmission is useful when information is required regarding the range of a target. Clearly, by transmitting a pulse of radar energy it is easy to measure when the reflected pulse returns and hence determine the target range using the formula given above.

Using a continuous wave transmission allows the closing (or receding) velocity of the target to be determined. This is achieved by using the Doppler effect. The Doppler effect is one by which the frequency of radiation is affected if a target is moving in the radial direction between radar and target (see Figure 3.7 which depicts a point radiating source travelling with a velocity from left to right).

If a radiating (or reflecting) target is receding from an observer, the frequency will appear to reduce as far as the observer is concerned. Conversely, if the target is approaching then the frequency will appear to increase. The classic illustration is of a train approaching, passing and receding from a stationary observer: as the train approaches, the sound pitch will be higher than when it recedes. As will be seen, the Doppler effect is a very useful property that is extensively used in various radar applications.

3.2   Radar Antenna Characteristics

In Figure 3.5 it is implied that the radar energy is directed in some manner in the direction of the target, and this is in fact the case. In order to achieve such directed energy, early systems used parabolic reflectors in which radiated energy is directed towards the reflector from a radiating horn at the focal point. Reflected energy returning from the target is 'gathered' by the reflector and concentrated at the focal point where the horn feed is situated. In this way the radiated energy is directed towards the target and the reflected energy is gathered from it, as illustrated in Figure 3.8.

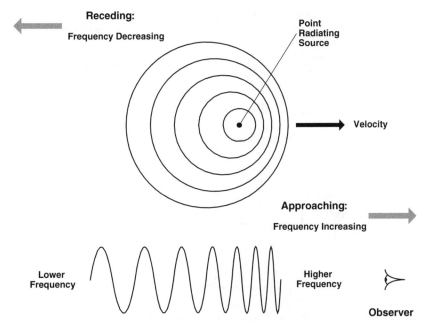

Figure 3.7 Doppler effect due to motion.

In recent years, radars have adopted the planar array shown in the lower part of Figure 3.8. Whereas the parabolic reflector achieves beam shaping by means of its physical parabolic shape, the planar array achieves a similar effect by careful phasing within the RF feeds at the rear of the planar array. This principle is described in more detail in Chapter 4.

This radar antenna directional property is extremely important to the radar as it focuses the energy into a beam on transmission and effectively 'gathers' the reflected energy during reception. This directional property enhances the operation of the radar and is known as the antenna 'gain'. The gain depends upon the size of the antenna and the frequency of the radiated energy.

The beamwidth of an antenna is actually quite a complex function, for, during the formation of the main beam, which is at the heart of the performance of the radar (owing to the antenna gain), a number of sublobes are also created which are distinctly unhelpful.

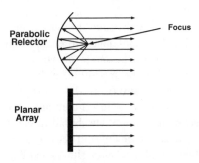

Figure 3.8 Parabolic reflector and planar array.

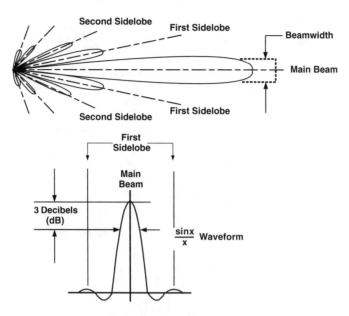

Figure 3.9   Antenna pattern showing main beam and sidelobes.

These sublobes are known as sidelobes. Whereas the main beam is central to the performance of the radar, the sidelobes detract from the radar performance since they effectively waste energy and have other adverse effects.

As shown in Figure 3.9, the upper part of the diagram portrays the main beam and sidelobes generated by a typical directional antenna. The area of interest to the radar is the main beam. The sidelobes are characterised as the first sidelobe, the second sidelobe, etc., and for the antenna pattern shown there are five sidelobes, each becoming progressively weaker the further off-boresight (the radar centre-line) they are. Not only do the sidelobes waste energy during transmission by directing energy away from the target, they also allow stray and unwanted energy to enter the antenna and therefore the radar receiver during reception. Stray energy may be noise or 'clutter' produced by spurious reflections from the ground, alternatively it may be energy being transmitted by an enemy jammer who is attempting to confuse the radar. The actual beam pattern is determined by a specific mathematical relationship known as a $\sin x/x$ waveform. The beamwidth of the main beam is defined as the point at which the signal strength of the $\sin x/x$ waveform has dropped to 3 dB below (−3 dB) the peak value. In numerical terms this relates to a signal level $1/\sqrt{2}$ below the peak signal which equates to 0.707 of the peak level.

The beamwidth varies according to the mode in which the radar is operating and the information it is trying to gather. The beamwidth also does not necessarily have to be the same in both axes (azimuth and elevation), as will be described. For an air-to-air mode the beamwidth will be narrow and be equal in azimuth and elevation, whereas for a ground-mapping mode it will be narrow in azimuth and broad in elevation.

The term decibel relates to a measurement unit that is used extensively within the radar community to describe relative signal levels in a short-hand logarithmic form according to a base of 10. In radar calculations, dynamic range between two signals may be several orders

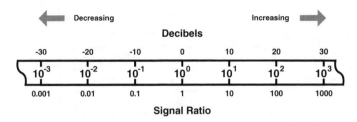

Figure 3.10 Decibels compared with numerical format.

of magnitude, and continuously using many noughts is tiresome and confusing. The principle is shown in Figure 3.10.

It can be seen that, as the signal strength increases to the right, so the decibels increase. At a positive signal ratio of 10^3 or 1000, the decibel figure is 30 dB [$\log_{10}(1000)$ is 30], whereas for a decreasing signal ratio, a negative ratio of 10^{-3} or 1/1000, the decibel figure is -30 dB [$\log_{10}(1/1000)$ is -30]. It is therefore easier to say that in a good antenna design the first sidelobe is 30 dB down (or -30 dB) on the main lobe signal than saying that the signal ratio is 1/1000th or 10^{-3} below the main lobe (this is a realistic figure in practice). Therefore, within radar terminology the decibel notation is used liberally to describe gains or losses when considering system performance.

3.3 Major Radar Modes

Examining the operation of some basic radar modes helps to understand how beamwidth and other factors such as pulse width, scan patterns, dwell time and pulse repetition frequency (PRF) are important to radar operation. The main modes that are described are:

- Air-to-air search;
- Air-to-air tracking;
- Air-to-air track-while-scan (TWS);
- Ground mapping.

3.3.1 Air-to-Air Search

One of the functions of a fighter aircraft is to be able to search large volumes of air space to detect targets. Many scan patterns are able to accomplish this function, but perhaps the most common is the four-bar scan shown in Figure 3.11. This scan comprises four bars stacked in elevation, and the radar mechanically scans from side to side in azimuth while following the four-bar pattern. The pattern shown begins in the top left-hand corner and finishes in the bottom left-hand corner before recommencing another cycle. The scan might typically cover ±30° in azimuth centred about the aircraft centre-line and about 10–12° in elevation. Alternatively, sector scans may be used, say ±10° skewed left or right off the centre-line if that is where the targets are located. The beamwidth in the air-to-air search mode will probably be ∼3°, and the scan bars will usually to be positioned one beamwidth or 3 dB apart to ensure that no target falls between bars. The search pattern is organised such that a target may be illuminated several times during each pass, as indicated by the overlapping

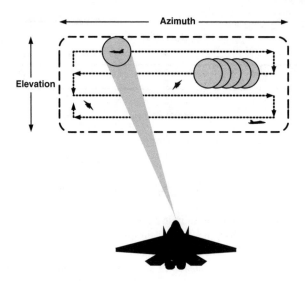

Figure 3.11   Typical air-to-air search pattern.

antenna coverage shown in the figure. This allows the target to be detected with certainty and avoids nuisance detections or false alarms.

More importantly, integration techniques (summing the return from several successive pulses) allow the signal return to be enhanced and therefore the ability to detect the target in noise or clutter to be significantly improved. For similar reasons, integration techniques can enhance the range at which the target is initially detected. In earlier-generation radars, all the targets detected would be shown on the radar display and could lead to a confusing picture, particularly when trying to separate friend from foe. More modern radars with digital processors are able to categorise multiple targets more easily, thereby simplifying the engagement procedure.

It should be recognised that, when the radar is operating in this air-to-air search mode, enemy targets fitted with a radar warning receiver (RWR) or other detection equipment will know that they are being illuminated or 'painted' by the searching radar. Furthermore, by categorising radar signal parameters such as radiated frequency, pulse width and PRF, the enemy target will be able to identify what type of radar and what aircraft type is being encountered. The RWR will also give a bearing to the illuminating radar, and the 'blip/scan' ratio will indicate whereabouts in the radar scan pattern it is located. It will also be obvious that the radar energy only has to travel a single path to reach the potential target, whereas the energy has to travel out and back to the radar to produce a return. This means that the target aircraft will be receiving a much stronger signal than the radar. Therefore, while air-to-air search is a useful mode, radar operators should also be aware that at the same time they are also giving potentially useful information to their adversary.

3.3.2   Air-to-Air Tracking

On occasions the radar may need to obtain more pertinent data regarding the target, perhaps in order to prepare to launch an air-to-air missile. To attain this more specific target

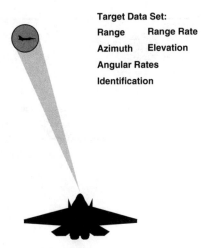

Figure 3.12   Air-to-air tracking.

information, the radar needs to 'lock on' to the target. When this occurs the scan pattern changes and the radar antenna tracks the target in azimuth and elevation. The target is also locked in terms of range using a range gate. The radar is now able to track the precise movements of the target. In some tracking modes the PRF may be switched to higher frequency to increase the target data update rate (Figure 3.12).

The target dataset will include the following data:

- Range;
- Azimuth;
- Azimuth rate;
- Target identification;
- Range rate;
- Elevation;
- Elevation rate;
- Target classification.

The accompanying changes in the radar characteristics detected by the potential target following lock-on is a warning that the engagement is becoming more serious. At this point the target may attempt evasive tactics – deploy countermeasures or chaff or jam the target radar.

The means by which the radar achieves target angle tracking and target range is described later in the chapter.

3.3.3   Air-to-Air Track-While-Scan

The disadvantage of locking on to the target and thereby signifying engagement intentions has already been described, but the advent of digital signal/data processing has enabled an

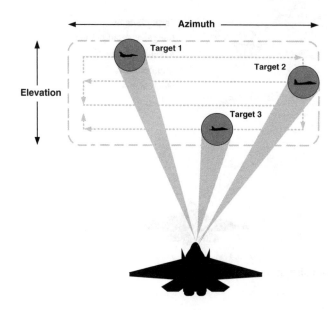

Figure 3.13 Air-to-air TWS.

elegant solution to be developed. Track-while-scan automates the process of deciding which target to engage (Figure 3.13). As TWS is under way, the radar processor progressively builds up a history of the flight path of targets within the scan. If successive measurements disagree, then the track is rejected; if the data agree, then the track is maintained. Gates are initiated that assign angular information, range and range rate to each track and predict where the target will be at the time of the next observation. If the track is stable, then the forecast gates will become more accurate and statistical filters will establish that the predicted fit is good. Techniques are used to arbitrate when gates overlap or where more than one target appears in the same gate perimeter.

The advantages of TWS are as follows:

1. Accurate digitised tracking data are established on each track within the antenna scan pattern without alerting potential targets that they are being tracked.
2. The automation process allows many targets to be tracked accurately and independently.
3. A typical radar using TWS will be able to track 20 or more targets in three-dimensional space.

3.3.4 Ground Mapping

From the early days of radar it was known that the radar could be used to map the terrain ahead of the aircraft. Using the different reflective characteristics of land, water, buildings, etc., it was possible to paint a representative map of the terrain ahead of the aircraft where major features could be identified. With the application of digital processing and advanced signal processing techniques, the ability to resolve smaller features increased and high-resolution mapping became possible.

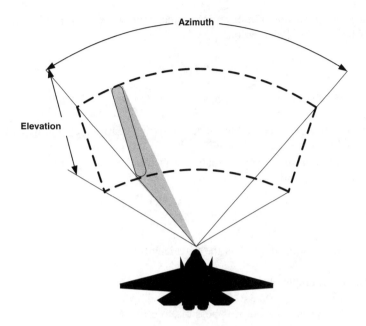

Figure 3.14   Ground mapping.

While using the ground-mapping mode, the antenna sweeps from side to side as shown in Figure 3.14. The area illuminated by the mapping beam equates to the dotted boundary shown in the figure.

Whereas the air-to-air modes use a narrow pencil beam, a fan beam is used for ground mapping. That is, a beam where one dimension is narrow — 2 or 3° — while the other is relatively broad, say 10 to 15°. The figure shows that the ground-mapping beam is narrow in azimuth and wide in elevation; this represents the optimum shape for the mapping function.

This describes the operation of a basic ground-mapping mode. In the past 20 years, improved capability and flexibility have been achieved by the use of digital computing in the radar data processor (RDP) and presignal processor (PSP). By the use of fast Fourier transform (FFT) for signal processing and aircraft motion compensation, increasingly sophisticated radar modes have been developed. These include:

- Doppler beam sharpening (DBS);
- Synthetic aperture radar (SAR);
- Inverse Synthetic Aperture Radar (ISAR).

These modes will be described in Chapter 4.

3.4   Antenna Directional Properties

Earlier it was stated that the directional properties of an antenna were determined by the radiated frequency and the size of the antenna. There are simple formulae that help to estimate the beamwidth and gain of an antenna if these parameters are known.

The frequency and wavelength of an electromagnetic wave are related to each other and the speed of light, c, by the equation

$$c = f \times \lambda$$

where c is the speed of light $(3 \times 10^8 \, \text{m/s})$, f is the frequency (Hz) and λ is the wavelength (m).

Therefore, for the airborne fighter radar described earlier, operating at a frequency of 10 GHz $(10^{10} \, \text{Hz})$, $\lambda = c/f = 3 \times 10^8 / 10^{10} = 3/100$ or 0.03 m or 3 cm.

Another formula is a ready approximation to determine the beamwidth θ of an antenna knowing the frequency and the antenna size:

$$\theta \approx \frac{65 \times \lambda}{D}$$

where θ is the beamwidth (deg), λ is the wavelength (m) and D is the antenna dimension (m).

Using again the example of the airborne fighter radar, and assuming an antenna dimension of 0.6 m ($\sim$24 in), $\theta \approx 65 \times 0.03/0.6 \approx 3.5°$.

Using a similar approximation, it is possible to estimate the antenna gain:

$$G_D \approx \frac{4 \times \pi}{\theta_B \times \varphi_B}$$

where G_D is the antenna gain, θ_B is the beamwidth (rad) in one axis and φ_B is the beamwidth (rad) in the orthogonal axis (one radian $\approx 57.3°$).

Again using the fighter radar example, $G_D = 4 \times \pi \times (57.3)^2 / (3.5)^2 \approx 3368$. This gives an idea of the advantage that the antenna gain confers. Expressed in decibels, the antenna has a gain of $\log_{10}(3368)$ or around 35 dB.

3.5   Pulsed Radar Architecture

The basic principles of radar operation have already been outlined. The detailed operation of pulse radar is described in this section. A top-level diagram of a pulsed radar system is shown in Figure 3.15.

3.5.1   Pulsed Radar Components

The diagram shows the major elements which are:

- Modulator;
- Transmitter;
- Antenna;
- Receiver;
- Video processor.

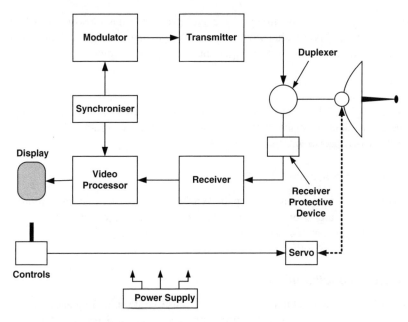

Figure 3.15   Top-level pulsed radar architecture.

3.5.1.1   Modulator

The modulator determines the pulse shape and the nature of the radar modulation. Although pulsed transmission is the most elementary form of radar operation, the modulation in a modern multimode radar may take many forms depending upon the nature of information being sought. The operation of the modulator is controlled by the synchroniser which dictates when a pulse should be initiated. The modulator uses the superheterodyne ('super-het') principle of modulation to superimpose the modulating signal upon the high-frequency carrier to provide a composite waveform.

3.5.1.2   Transmitter

The transmitter amplifies the modulated carrier signal and feeds it to the antenna via a duplexer. This serves the function of directing the transmitter energy to the antenna waveguide system to be fed by the antenna elements for transmission into the atmosphere. It also routes the reflected target energy to the receiver.

3.5.1.3   Antenna

The antenna, as has been described, directs the radar energy towards the target and receives the reflected energy from the target. Along with the target echo, a substantial amount of clutter from ground returns is also received. The antenna beam is focused according to the shape of the antenna and the nature of the beam required. Unwanted radar energy enters through the antenna sidelobes as well as the main beam. The antenna also receives noise from a variety of external sources that can help mask the true target signal.

Returning energy is passed through a receiver protective device which blocks the large amounts of transmitted power that would cause severe damage to the receiver, but also at the appropriate time allows the reflected target energy to pass through.

3.5.1.4 Receiver

The receiver amplifies the reflected target signal and performs the demodulation process to extract the target data from the surrounding noise, and the resulting target video data are passed to the video processor.

3.5.1.5 Video Processor

The video processor is also controlled by the synchroniser in order that transmitted pulse and target return pulses are coordinated and that a range measurement may be made. The resulting data are coordinated and displayed on the radar display.

3.5.2 Pulsed Modulation

The nature of the pulse modulation in terms of pulse width and frequency of repetition is highly interactive with a number of important radar characteristics and has a significant impact upon the performance of the radar. The basic parameters of a pulsed radar signal are described in Figure 3.16.

In pulsed radar operation, the carrier frequency is modulated by the envelope of a single rectangular pulse; in this case the pulse embraces a fixed carrier frequency. As will subsequently be discovered, in sophisticated radar operations there are advanced forms of modulation/transmission in which the pulse is not rectangular nor the carrier fixed in frequency. The pulse width is denoted by the symbol τ and is usually fairly narrow, perhaps

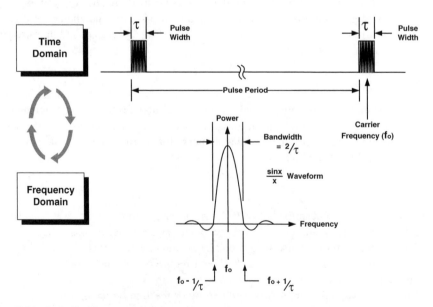

Figure 3.16 Pulsed radar transmission.

~1 µs in an air-to-air mode. After a time interval called the pulse period, a second pulse is transmitted and the sequence is repeated. The rate at which the pulses are repeated is called the pulse repetition frequency (PRF), and both the pulse width τ and the PRF are key radar parameters.

It will be noted on the diagram that the terms 'time domain' and 'frequency domain' are mentioned. The time domain is familiar in everyday life as it is the domain in which we live; the frequency domain is more abstract but is of great importance to the radar designer. In fact, the time and frequency domains are interdependent and interwoven, and this has a significant impact upon the operation of radar systems. What happens in the time domain affects the frequency domain, and vice versa.

The rectangular pulse τ results in a response in the frequency domain that has a $\sin x/x$ response, the same generic response that determines the pattern of the antenna main beam and sidelobes. However, in this case the response is occurring on an axis relating to frequency rather than angle off-boresight as is the case in the antenna pattern. When the incoming pulse is received, it results in a frequency response of received power portrayed by the $\sin x/x$ response and centred upon the radiated frequency f_0. The practical limits of the main $\sin x/x$ response are $\pm f_1$ centred on f_o, that is, $f_0 + 1/\tau$; $f_0 - 1/\tau$ (f_0 being the carrier frequency and τ the pulse width), and this determines the bandwidth required of the receiver in order to be able to pass all the components of the target return. Therefore, for a 1 µs pulse the receiver bandwidth would need to be $2/\tau = 2/(1 \times 10^{-6}) = 2 \times 10^6$ or 2 MHz (see the lower part of Figure 3.16). The narrower the transmitted pulse, the wider will be the bandwidth; the converse also applies.

In basic pulsed radar operation the pulse width also determines the range resolution of which the radar is capable. The radar can only resolve to half the pulse width, as at a lower interval than this part of the pulse has been reflected while part has not yet reached the target. A pulse of 1 µs duration will be approximately 1000 ft long (as light travels at $3.3 \times 3 \times 10^8$ or $\sim 10^9$ ft/s and the duration of the pulse is 1×10^{-6}; distance = velocity × time). Therefore, a 1 µs pulse will be able to resolve the target range to no less than 500 ft. In fact, by using more complex modulation and demodulation methods called pulse compression, it is possible to achieve much better resolution than this; pulse compression is discussed in Chapter 4.

The pulse period – and therefore PRF – also has an impact upon the radar design that affects target ambiguity, as shown in Figure 3.17. The figure shows an aircraft illuminating two targets and compares the effect of the returns from these targets for two different pulse periods. In pulse period T_1 ($PRF_1 = 1/T_1$), the returns from both targets are received before the successive pulse is transmitted, and the range is unambiguous. In the case of the shorter pulse period T_2 ($PRF_2 = 1/T_2$), the return from the most distant target occurs after the transmission of the successive pulse and to the radar appears as a relatively close target within the second period. In this case the target range is ambiguous and misleading. The selection of PRF is one of the most difficult choices the radar designer has to make, and some of the effects of range ambiguity are discussed in more detail in Chapter 4.

The fact that the radar only transmits for a portion of the time means that the average power is relative low. The average power is given by the following expression:

$$P_{av} = \frac{P_{peak} \times \tau}{T}$$

where P_{peak} is the peak power, τ is the pulse width and T is the pulse period.

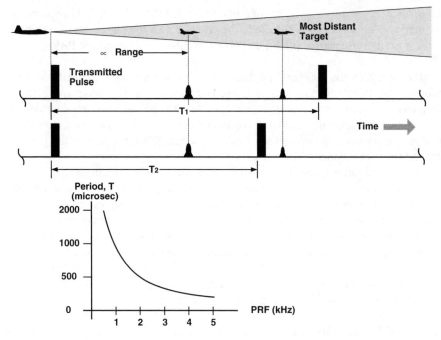

Figure 3.17 Effect of pulse period on target ambiguity.

For a peak power of 10 kW, a pulse width of 1 μs and a pulse period of 250 μs, we have

$$P_{av} = \frac{10^4 \times 1 \times 10^6}{10^6 \times 250} = 40 \text{ W}$$

3.5.3 Receiver Characteristics

In order to be able to detect the target, the radar receiver has to able to discriminate from unwanted effects. The main adverse affects are as follows:

1. Noise is either internally generated or radar transmitter induced or externally sourced. Noise is random and can only be minimised by good design.
2. Clutter due to unwanted returns from the ground and other sources is usually more systematic and can be countered by filtering and processing techniques.

3.5.3.1 Noise

The sources of noise that can affect the ability of the radar receiver to detect a target signal are shown in Figure 3.18. The total system noise includes noise from the following sources:

1. Antenna noise T_a. The antenna noise includes all those sources of noise that are external to the radar, including radiation from the sun, terrain, emissions from man-made objects

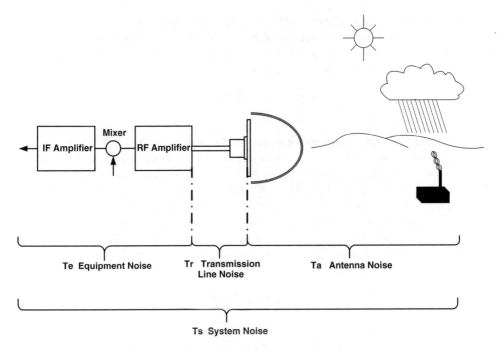

Figure 3.18 Sources of noise affecting radar signal.

and the weather. Noise from jamming may also be included in this category. The radome and the antenna itself may also generate noise. While external noise will be most troublesome when it enters the system via the antenna main beam, it should also be remembered that noise can also enter via the antenna sidelobes.

2. Transmission line noise T_r. This includes noise originating within the waveguide couplers, duplexer and the receiver protection device.

3. Equipment noise T_e. The equipment noise is generated within the receiver itself and is the most difficult to counter.

The total system noise, T_s, is the sum of these individual components:

$$T_s = T_a + T_r + T_e$$

The problem with the noise in a receiver is that, once present, it is there to stay. Signal amplification in subsequent stages will only amplify the noise as well as the signal and accentuate the problem of target detection. One technique commonly used is to insert a low-noise amplifier (LNA) at the front of the receiver to amplify the signal proportionately more than the noise. LNAs are also commonly used where antennas (or apertures) are mounted remotely throughout the airframe and where transmission losses might be relatively high.

The receiver noise is defined as noise per unit of receiver gain:

$$\text{Receiver noise} = \frac{\text{noise at output of receiver}}{\text{receiver gain}}$$

The receiver gain can be easily measured using laboratory techniques.

The receiver noise may be characterised by a figure of merit or noise figure F_n. This is defined as the ratio of the noise figure of the actual (imperfect) receiver to the hypothetical ideal receiver providing equal gain. Therefore:

$$F_n = \frac{\text{noise output of actual receiver}}{\text{noise output of ideal receiver}}$$

An ideal receiver would produce no noise; the only noise that would exist would be that from external sources. This external noise can be represented as though resulting from thermal agitation in a conductor (resistor) since the two have similar spectral characteristics. Therefore, in the derivation of F_n, for both ideal and actual receivers, the thermal noise can be portrayed as the voltage across a resistor. Thermal noise is governed by the random motion of the free electrons within the conductor and is uniformly spread across the entire spectrum. This motion is determined by the absolute temperature of the notional resistor, denoted by T_0. Also, the noise depends upon the receiver bandwidth B. Thus, to derive the mean noise power for an ideal receiver, the expression

$$\text{mean noise power} = k \times T_0 \times B \quad (\text{W})$$

may be used for an ideal receiver, where k is Boltzmann's constant $= 1.38 \times 10^{-23}$ W s/K, T_0 is the absolute temperature of the resistor representing the external noise (K) and B is the receiver bandwidth (Hz).

The external noise is the same for both receivers, and by convention T_0 is taken to be 290 K which is close to room temperature. Where the external noise is small by comparison with that generated by the receiver, as is usually the case, the mean noise figure for an actual receiver may be determined by the following:

$$\text{Mean noise power} = F_n \times k \times T_0 \times B \quad (\text{W})$$

As was shown earlier, the total noise may represented by T_s where the mean noise power (all sources) $= k \times T_s \times B$.

The nature of the modulation used also has an impact upon receiver noise. This is shown in Figure 3.19. The figure shows the simple comparison of narrow and broad rectangular pulse modulation. It was shown earlier in Figure 3.16 that the bandwidth needed to accommodate all the frequency components of a rectangular pulse was governed by the $\sin x/x$ waveform, and that the theoretical bandwidth was $2/\tau$. The narrow (sharper) pulse τ_1 needs a greater bandwidth than the broader pulse τ_2. The narrow pulse gives an improved range resolution and, for a given pulse period (PRF), a reduced mean power, so it can be seen that there are performance trade-offs to consider that affect bandwidth and hence receiver noise.

In practical systems a compromise is allowed and generally a bandwidth of $1/\tau$ is regarded as sufficient. Therefore, it is common practice to narrow the IF bandpass filter until it is $1/\tau$ wide, just wide enough to pass the bulk of the target-related energy but reject the unwanted noise. This design is called a matched filter, and the mean noise energy per pulse is kT_0/τ.

In the Doppler radars addressed in Chapter 4 the Doppler filters downstream of the IF filter are much finer, and greater noise and clutter rejection result.

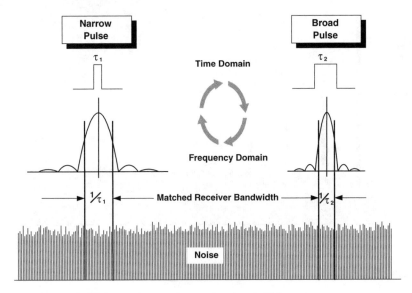

Figure 3.19 Effect of different pulses on the receiver bandwidth.

The detection and extraction of a target echo from a background of noise depends upon the four factors outlined below:

- The average power radiated in the direction of the target;
- The proportion of the radiated energy reflected back in the direction of the radar;
- The proportion of power recaptured by the radar antenna;
- The length of time the antenna beam is trained upon the target.

Average power is determined by the relationship of the peak power, P_{peak}, transmitted by the radar and the modulation characteristics of pulse width, τ, and pulse period, T, as shown in the previous section. The antenna gain, G_{D}, also increases the power density related to the beamwidth(s) and beam geometry.

As the radiated signal is directed towards the target, it spreads out an increasing area, proportional to R^2, where R is the range from the radar. This means that the power density reduces by a factor of $1/R^2$ as the energy is propagated in the direction of the target.

A fraction of the energy incident upon the target will be reflected back in the direction of the radar. In the simplest form the target may be considered to be a simple sphere with a specific cross-sectional area, denoted by the symbol σ and specified in square metres. The reality is much more complicated than that, and other factors such as reflectivity and directivity play a great part, as will be seen in the discussion on low observability or stealth in Chapter 4.

As the energy is reflected back to the target, the $1/R^2$ effect applies in terms of the reduction in received power density. The impact of this effect means that the energy received at the radar has been reduced by a total factor of $1/R^4$ in its outward and return path to and from the target. This has an impact upon the ability of the target signal to be detected above the noise, as shown in Figure 3.20. The figure shows how the returning signal (not to scale)

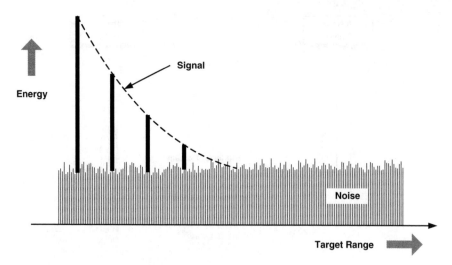

Figure 3.20 Effect of range upon the target echo.

decreases with increasing range to the point where the signal is not detectable against the noise background.

As will be seen, the equation governing the strength of the return signal is a fourth-power law, and this means that the receiver has to accommodate a very large dynamic excursion in terms of variation in target signal strength as the range varies. In certain modes this is addressed by a technique called sensitivity time control (STC) in which the receiver gain is reduced at very short ranges and increased progressively during the range sweep. This technique is sometimes referred to as swept gain and to some extent mitigates the problem of extremely high signal returns at short range.

Another technique is often used to counter this effect and prevent the receiver amplifiers from saturating: if the receivers saturate, then both signal and noise will merge as the amplifiers clip both noise and target signal returns. In this case, automatic gain control (AGC), as the name suggests, automatically reduces amplifier gain to prevent saturation occurring.

The actual detection of the target signal is determined by the setting of a target detection threshold as shown in Figure 3.21. This shows two targets, A and B, against a background of noise on a time axis: A and B are obviously at different ranges from the radar. The figure shows the importance of setting the target threshold correctly with respect to the mean noise level. If the threshold is set low, then it may be anticipated that more targets may be detected. However, as the diagram shows, setting a low target threshold has the accompanying risk of detecting a spurious target – called a false alarm. For the low threshold setting shown, the radar would detect three targets: genuine targets A and B and the false alarm.

Conversely, there are problems with setting the threshold too high to avoid false alarms. In this case the return from genuine target A is lost and only target B is detected.

One of the major factors affecting target detection was antenna time on the target. So far, only the detection of a target using a single pulse has been considered. In fact, as the radar beam sweeps through the target, a number of successive pulses will illuminate the target in a short period of time. Most radars have the capability of integrating the detected output over a number of pulses, and this has significant advantages, as can be seen from Figure 3.22.

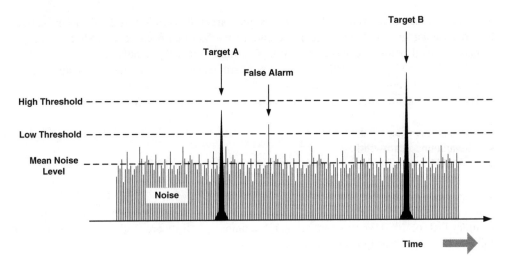

Figure 3.21 Receiver threshold setting.

Noise is generally random in terms of amplitude and phase. The target return is more systematic and repetitive in nature, at least over a range of successive pulses in an antenna scan. The effect of integrating noise over a series of pulses is to end up with noise at more or less the mean noise level before integration. The converse is true for a real target return. The target return is aggregated during the integration process and the result is a much stronger target return. The figure shows that as an example – integration over 12 pulses produces an

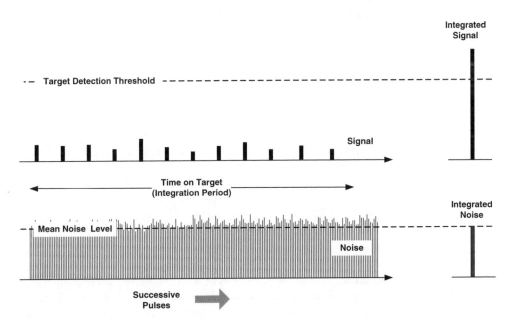

Figure 3.22 Effect of integration over several pulses.

integrated signal that comfortably exceeds the target detection threshold, whereas the integrated noise does not. This occurs in spite of the fact that each of the individual target signals are well below the target detection threshold and without pulse integration would each be subsumed by noise. This shows the powerful capability of 'extracting' a signal from noise using integration techniques.

The actual antenna time on target depends upon a combination of three factors:

- The antenna scan or slew rate;
- The antenna beamwidth;
- The PRF.

Taking some simple figures by way of illustration, if the antenna scan rate is 60 deg/s and the 3 dB beamwidth is 3°, then the antenna will dwell upon a target for 1/20th of a second. If a medium PRF of 1000 Hz is assumed, then the antenna will theoretically have a total of 50 'hits' on the target during every pass across the target.

3.5.3.2 Clutter

The effects of clutter, particularly from ground returns or precipitation, can cause large amounts of unwanted signal being returned to the receiver. Clutter can enter the receiver channel through the main beam or via the sidelobes. It can depend upon the nature of the terrain, terrain geometry and the aspect (depression angle) of the antenna boresight. If the clutter is from the water, then it may depend upon sea state (the height of the waves or the smoothness of the water surface). In some ways, clutter may be systematic in terms of the effect that it has on the radar, and in these cases it is easier to counter or filter.

Moving targets, or targets with a significant radial velocity with respect to the radar, may have a Doppler shift component that may enable the target to be distinguished against a stationery background. The use of Doppler filters and target velocity techniques is described in Chapter 4.

3.5.4 Radar Range Equation

The foregoing discussion leads us to the equation that is the most powerful and commonly used when examining the performance of radar systems, that is, the radar range equation. The radar range equation takes many forms depending upon those factors that need to be taken into account and the type of transmission being considered. In the simplest form, the maximum range for a single radar pulse is determined by the following equation:

$$R = \sqrt[4]{\frac{P_{\text{peak}} \times G \times \sigma \times \tau}{(4\pi)^2 \times S_{\text{min}}}}$$

where R is the radar range (m), P_{peak} is the peak power (W), G is the antenna gain (m^2) (this may also be expressed in decibels for ease of calculation, as explained earlier), σ is the target cross-sectional area (m^2), τ is the transmitted pulse width (s), and S_{min} is the minimum detectable signal energy (W-s). This equation does not take account of pulse integration.

There are some interesting observations to make regarding this formula:

1. Peak power P_{peak}. As the peak power only affects the radar range by the inverse fourth power, doubling the peak power of the radar only increases the range by the inverse fourth power of $2 \approx 1.19$ or 19%.
2. Antenna gain G. If the antenna is circular, doubling the size of the antenna will increase the gain of the antenna by 4, and the overall range by a factor 1/2 or by about 71%. However, commensurate with the antenna gain, the beamwidth would halve, which may make target acquisition more difficult. Dwell time might also have to increase to improve target integration. Altering the wavelength of the radiated transmission would have an effect upon radar range as the range alters by the inverse square of the wavelength. Decreasing wavelength or increasing frequency can therefore increase the range. The atmospheric absorption outlined in Figure 3.4 earlier in the chapter will be an important factor, as at certain parts of the spectrum absorption rates are punitive, more than cancelling out any benefit that increasing radiated frequency may confer.
3. Target cross-sectional area σ. Reducing the target cross-sectional area by a factor of 60 dB (equivalent to 1×10^{-6}) by using extensive low observability (LO) techniques reduces the range by a factor of $\sim$30.
4. Pulse width τ. Maintaining mean power but decreasing the pulse width increases the peak power but also increases the receiver bandwidth, allowing more noise into the receiver.
5. Minimum detectable signal S_{min}. Decreasing the minimum detectable signal increases the radar range, but the risk of false alarms may increase.

As more factors are taken into account, so more trade-offs need to be made. However, as will be seen later, the adoption of sophisticated modulation and signal processing techniques can gain significant performance enhancements in modern digital radars.

3.6 Doppler Radar

In the early part of the chapter the Doppler effect was described, that is, the effect upon radiated frequency when a moving source approaches or recedes from an observer. The same effect occurs when radar energy is reflected by ground clutter, except that the Doppler frequency shift is doubled as the radio energy has to travel out and back to the radar. Normally, ground returns are a nuisance as far as the radar is concerned, and all means are used to reject the ground clutter. However, there is one radar application where the ground clutter Doppler frequency shift is utilised, and that is the Doppler radar, sometimes called the Doppler navigator. A typical configuration for a Doppler radar is shown in Figure 3.23.

The Doppler radar comprises three or four narrow, continuous wave radar beams angled down from the horizontal and skewed to the left and right of the centre-line. The three-beam layout shown in the figure is called a lambda configuration for obvious reasons. The diagram shows a situation where the aircraft is flying straight ahead, i.e. heading equals track and there is no angle of drift because of crosswind. The forward beams 2 and 3 will experience a positive Doppler shift as the ground is advancing towards the aircraft. The Doppler shift Δf is proportional to $2V/\lambda$, where V is the aircraft forward velocity and λ is the wavelength of the radiated frequency. The aft beam 1 will experience a negative Doppler shift proportional

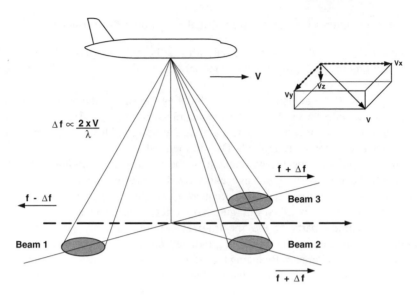

Figure 3.23   Doppler radar.

to $2V/\lambda$ as the ground is receding from the aircraft. There are several scaling factors including direction cosines associated with the beam, but subtracting forward and aft beams yields a signal proportional to $4V/\lambda$. Therefore, by manipulating and scaling the Doppler shifted returns from all three beams, the aircraft horizontal velocity with respect to the ground (i.e. ground speed), V_x, may be calculated.

   If the aircraft is drifting left or right owing to a cross-wind, then, by using the cross-track Doppler shift components and a similar manipulation process, the cross-track velocity, V_y, may be calculated. The vertical velocity component, V_z, may also be calculated. The vector sum of V_x, V_y and V_z enables the total aircraft velocity, V, to be established. Doppler radars do have one disadvantage: if the terrain is very flat with a low reflectivity coefficient, then insufficient energy may be reflected back to the radar and the Doppler shift cannot be measured. Such effects can be achieved when travelling over very calm water or ice-covered expanses of water. Doppler radars were very commonly used before inertial navigation systems (INS) became the norm about 30 years ago; more recently, INS has been augmented by the satellite-based global positioning system (GPS). The initial avionics configuration of Tornado included a Doppler radar, and they are still frequently used on helicopters as air data becomes very unreliable at low airspeeds. For further data on Doppler radars, see Kayton and Freid (1997).

   As will be explained in Chapter 4, sophisticated radars in use today combine the use of pulse techniques and Doppler to produce pulsed Doppler (PD) modes of operation.

3.7   Other Uses of Radar

3.7.1   Frequency Modulation Ranging

The use of pulsed radar techniques has hitherto been described to measure target range. However, frequency modulation may also be used to determine range as depicted in Figure 3.24.

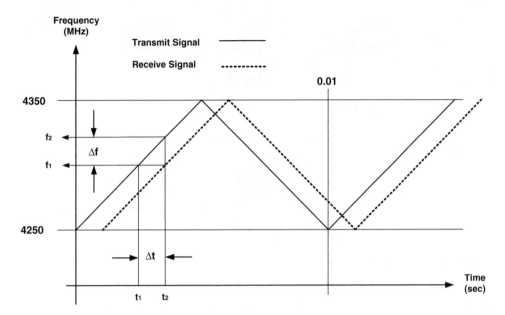

Figure 3.24   Frequency modulation ranging.

The transmitted signal consists of a triangular wave modulation, as shown, that sweeps across the frequency spectrum, completing one cycle in 0.01 s in the example given. The received frequency will lag the transmitted frequency by an amount Δf owing to the time taken to complete the out and return journey. The example shows a measurement taken when the reflected received frequency, f_1, is compared with the current frequency at the transmitter, f_2, with the difference in frequency being Δf. The associated time difference signal, Δt, is proportional to the range of the target.

The figures shown on the diagram relate to the use of this technique in a radar altimeter, where the radar returns are used to calculate the instantaneous altitude of the aircraft above the terrain over which the aircraft is flying. In this example, the transmitter is sweeping in a linear manner over a frequency range of 4250–4350 MHz in 0.01 s. The use of radar altimeters is described in Chapter 7.

3.7.2   Terrain-following Radar

Whereas the radar altimeter is useful in informing pilots where they are in relation to the terrain underneath the aircraft, it does not tell them where the terrain is in front of the aircraft. To do this, the pilot needs to use a terrain avoidance (TA) mode or, better still, a dedicated terrain-following radar (TFR). The TA function can be crudely achieved by using a normal pulsed radar in a single-bar scan mode with a fixed depression angle. This will tell the pilot where he is in relation to the terrain ahead of the aircraft, but it is not a sophisticated mode and does not readily lend itself to coupling into the autopilot (Figure 3.25).

The TFR is a dedicated radar coupling into a dedicated functional system and autopilot that allows the pilot much greater performance and flexibility when penetrating at low level at night. The TFR scans the terrain ahead of the aircraft and receives ground returns that are

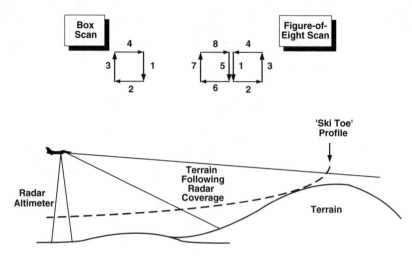

Figure 3.25   Terrain-following radar operation.

used for guidance. Normally, a simple box scan is used where the active sweeps are those in the vertical direction (sections 1 and 3). In some circumstances a figure-of-eight scan is used which provides broader lateral coverage than the simple box scan. The TFR therefore builds up a range/elevation picture of the terrain ahead of the aircraft and calculates an imaginary 'ski-toe' profile that reaches out ahead of the aircraft. This profile is calculated taking into account such factors as aircraft speed, manoeuvrability, etc., and provides an envelope within which the aircraft will not be able to avoid the terrain ahead. The system is configured so that, whenever the terrain ahead broaches the ski-toe envelope, the aircraft pitches up to rectify the situation. Similarly, if the terrain drops away in front of the aircraft, the aircraft pitches down until just operating outside the profile. The system operates just like the toe of a ski, moving up or down to follow the terrain ahead of the aircraft but always ensuring the aircraft can safely manoeuvre.

The measurements from the radar altimeter are also fed into the terrain-following system which calculates the 'most nose-up command' provided by either TFR or radar altimeter. This has the advantage of providing the pilot with an additional altitude safety buffer directly beneath the aircraft as the TFR is looking several miles ahead.

The TFR/radar altimeter commands may be coupled into the autopilot to provide an auto-TF mode while the aircraft is approaching the target area, thereby enabling the aircraft to fly at low level automatically while the pilot performs other mission-related tasks. The TFR may be an embedded system forming part of the aircraft primary radar, alternatively it may be provided in a pod that is loaded on to the aircraft. The AN/AAQ-13 LANTIRN navigation pod fitted to F-15 and F-16 aircraft performs a TFR function for these aircraft.

3.7.3   Continuous Wave Illumination

On some weapons systems a continuous wave (CW) illumination mode is provided. This mode is used when aircraft are fitted with semi-active air-to-air missiles; that is, missiles that can receive incoming RF energy and once fired can track and engage the target. As the

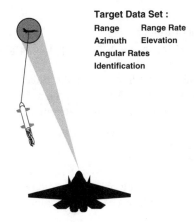

Figure 3.26 CW illumination.

missiles are unable to transmit, the aircraft radar has to provide the target illumination and it does this by using a CW illuminator co-boresighted with the aircraft radar antenna. Therefore, when the aircraft radar is locked on to the aircraft it can simultaneously illuminate the target (Figure 3.26). The disadvantage of this technique is that the aircraft radar has to remain locked on to the target and transmitting CW illumination until the engagement is complete. In high-density air-to-air combat this may not always be possible.

3.7.4 Multimode Operation

Modern radars such as those on the F-15E and F-22 have the capability of operating simultaneously in a number of modes, an example of which is shown in Figure 3.27. In this hypothetical example, three simultaneous modes are depicted:

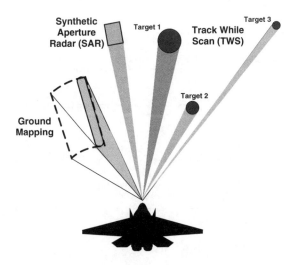

Figure 3.27 Simultaneous multimode operation.

- Sector ground mapping;
- Synthetic aperture (SA) spot mode;
- Track-while-scan (TWS) mode engaging three separate targets.

The radar achieves this capability by interleafing the radar modulation required for each mode on a pulse-by-pulse basis and effectively operating as several radars in one. This offers immense flexibility to the aircraft as a weapons platform.

3.8 Target Tracking

During the pulsed radar tracking mode when the radar is locked on, it follows and automatically maintains key data with respect to the target:

- Tracking in range;
- Angle tracking in azimuth and elevation.

Tracking is maintained and the radar is said to have 'target lock' when all these loops are closed.

3.8.1 Range Tracking

Tracking in range is usually accomplished using a technique called range gating which automatically tracks the target as its range increases or decreases. The concept of the range gate is shown in Figure 3.28.

The radar return in the region of the target return will comprise noise and the target return. The range gating technique uses two gates, an 'early gate' and a 'late gate'. The early gate

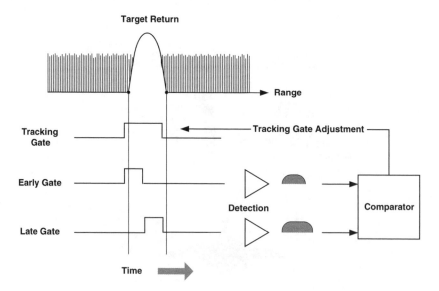

Figure 3.28 Range gate tracking.

is positioned near the leading edge of the target echo and detects and captures energy from the early part of the target return. Conversely, the late gate is positioned near the trailing edge of the target echo and detects and captures the energy from the trailing edge of the target return.

The detected signals from the early and late gate are compared and the result is used to position the tracking gate so that it is coincident with the target return. In the example shown, both the early and late gates are positioned early (to the left) of the target return and the tracking gate is also incorrectly positioned. Consequently, the early gate detects less energy than the late gate. Identifying this discrepancy will cause the energy from early and late gates to be equalised and the tracking gate to be moved to the right (down range) so that it correctly coincides with the target echo. While the radar maintains target lock this process will be continued, maintaining the tracking gate at the same range as the target echo.

3.8.2 Angle Tracking

During the radar tracking mode the radar tracks the angle to the target in azimuth and elevation. In other words, the line-of-sight (LOS) to the target and the radar boresight are kept as close as possible. The LOS needs to be established within a frame of reference and usually the radar is stabilised in roll and pitch using attitude data from the aircraft attitude sources: inertial reference system (IRS) or secondary attitude and heading reference system (SAHRS). The final axis in the orthogonal reference set is usually the aircraft centre-line/heading.

There are three main methods of angle tracking that are commonly used, these are:

- Sequential lobing;
- Conical scan (conscan);
- monopulse.

3.8.2.1 Sequential Lobing

One of the first tracking radar principles adopted was sequential lobing, which in its earliest form was used in a US Army angle tracking radar air defence radar. The principle of operation of sequential lobing is shown in Figures 3.29a and 3.29b.

To track a target in one axis, two lobes are required; each lobe squints off the radar boresight. The centre point of where the two lobes overlap represents the boresight of the antenna and this is the LOS that the radar antenna is trying to maintain. It can be seen that, when the signal return from the target is the same in both beams, the LOS to the target has been achieved. As the target moves, continual error signals will be sensed and the antenna servo system responds by nulling the error and maintaining LOS to the target. If four lobes, A and B and C and D, are positioned as shown in Figure 3.29a, then lobes A and B provide tracking in elevation and lobes C and D provide tracking in azimuth. The reflected signal from the target received in each of the four lobes is routed via a channel switching assembly sequentially switched into the receiver. In this way, each of the four lobe returns is measured and error signals are derived to drive the antenna elevation and azimuth drive servomotors.

Figure 3.30 illustrates how each of the four lobes is switched in turn into the receiver. In practice, the waveguide switching arrangement is cumbersome and prone to losses;

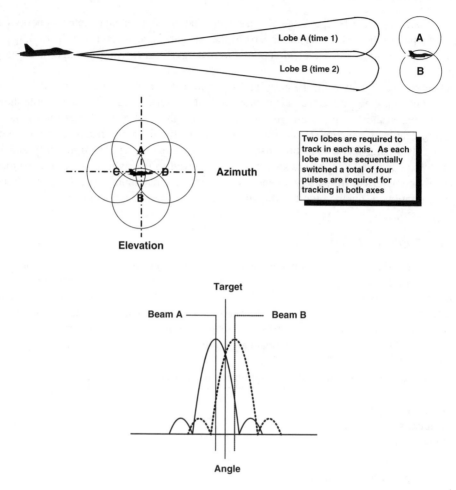

Figure 3.29 Sequential lobing – principle of operation.

therefore, radar performance is compromised. The other significant disadvantage suffered by this method relates to the time taken for the sequencing to occur. The radar PRF will determine the maximum time that the receiver will be switched to a particular lobe. Only when the radar has completed the range sweep for a particular PRF can the receiver sequence to the next lobe. Furthermore, the elevation and azimuth error can only be updated once per cycle, and this adversely affects update rate and tracking error.

Sequential lobing can be detected by a target and transmissions can be devised that will cause the radar to break lock. Transmitting on all four beams and receiving only on one may counter this. This technique is called lobe on receive only (LORO).

3.8.2.2 Conical Scan

Conical scan (conscan) is the logical development of the sequential lobing scheme already described. Conscan is another example of the earliest form of angle tracking – used because

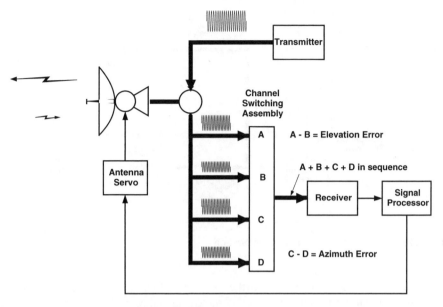

Figure 3.30 Sequential lobing – tracking configuration.

it was the easiest to use with the technology available at the time. The concept of conscan is depicted in Figure 3.31.

In conscan, one lobe is used that squints off-boresight. The lobe is then rotated such that the target return is enclosed within the imaginary cone that is swept out around the antenna boresight. In a parabolic antenna the rotating conscan beam is achieved by rotating the antenna feed at the desired rate. In more sophisticated arrangements the feed may be nutated, and this can achieve a better tracking performance than the straightforward rotating feed at the expense of a more complex feed mechanism. Typical conscan rates may be in the range up to 50 Hz.

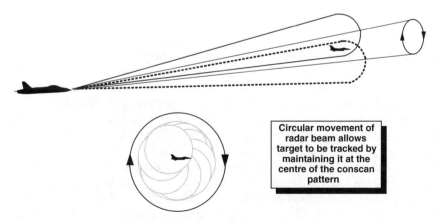

Figure 3.31 Conical scan – principle of operation.

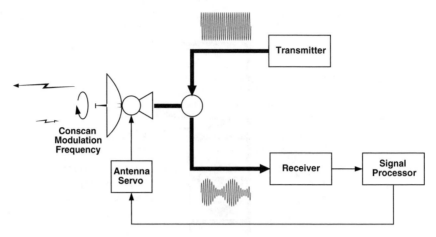

Figure 3.32   Conical scan – tracking configuration.

If the target is located off-boresight, the receiver will receive an amplitude-modulated signal at the conscan frequency. By detecting the signal and resolving the error components, drive signals are fed to the azimuth and elevation servo motors to move the antenna such that the target is back on boresight. The architecture of a conscan tracker is shown in Figure 3.32. As for the sequential lobing technique, there is only one receiver, but this is continuously fed with the conscan signal and the sequencing delays experienced with sequential lobing are avoided.

Conscan does offer these performance advantages over the sequential lobe technique but does itself suffer from a major disadvantage. Potential foes can identify the conscan frequency and can radiate a signal modulated with the conscan frequency that can cause the radar to break lock. Conscan is therefore susceptible to electronic countermeasures. This deficiency may be overcome if the target is illuminated with a non-scanning beam and conscan is used only for the receive channel. In this way, adversaries do not know they are being tracked by conscan means since the tracking operation is opaque to them. This technique is known as conical scan on receive only (COSRO).

3.8.2.3 Monopulse

Monopulse is the preferred tracking method, and most tracking modern radars use it out of choice. The term monopulse means that a tracking solution may be determined on the basis of a single pulse rather than the beam sequence (sequential lobing) or a complete conical scan. The tracking data rate is therefore much higher and therefore potentially more accurate. Another advantage is that the tracking is based upon the simultaneous reception of the target return in all four channels and any variation in the echo in time can be readily accommodated. This is not the case with the other techniques.

Monopulse uses four simultaneous beams as shown in Figure 3.33, in which the beams are stacked in elevation and side by side. All four beams squint away from the antenna boresight by a small amount. Comparison of the target returns in all four channels is undertaken and error signals are derived to drive the antenna azimuth and elevation drive servo motors as appropriate. Monopulse techniques may use either phase or amplitude comparison to

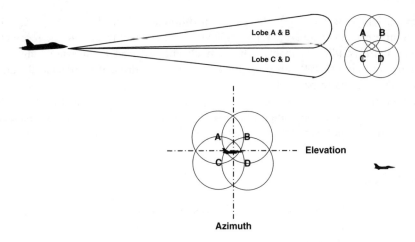

Figure 3.33 Monopulse – principle of operation.

perform the tracking task. Of the two, amplitude comparison is generally preferred. A former UK AI radar (AI23b) used in the Lightning aircraft employed amplitude comparison in the elevation channel and phase comparison in the azimuth channel.

All four channels transmit the same signal. The target return is received in each part of the monopulse array and fed into waveform junctions called hybrids which perform the sum and differencing function. A simplified portrayal of this arrangement is shown in Figure 3.34. Downstream of the summing and differencing process, three RF channels are formed: the

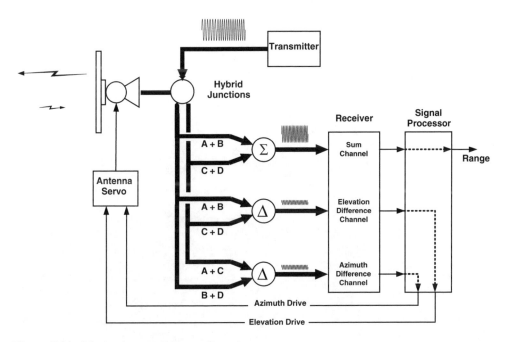

Figure 3.34 Monopulse – tracking configuration.

sum channel $(A + B + C + D)$; the elevation difference channel $(A + B) - (C + D)$; the azimuth difference channel $(A + C) - (B + D)$. Each channel is fed into the receiver which has three corresponding channels. The sum channel is used to measure range, and the elevation and azimuth difference channels are used to drive the antenna elevation and azimuth servo drives respectively.

In the early days of monopulse radar the provision of three identical receiver channels caused problems and some compromises were sought that multiplexed two channels. The radars of today do not experience this problem.

References

Kayton, M. and Freid, W.R. (1997) *Avionics Navigation Systems*, 2nd edn, Wiley-Interscience.
Lovell, B. (1992) *Echoes of War – The Story of H₂S Radar*, Adam Hilger.
Pilot's Handbook – Honeywell Radar RDR-4B.

4 Advanced Radar Systems

4.1 Pulse Compression

In Chapter 3 the determination of range resolution for a simple pulsed radar was shown as being dependent upon the pulse width, τ. In fact, the expression for minimum range resolution is given by the following:

$$\text{Minimum range } R_{\text{res}} = \frac{c \times \tau}{2}$$

where c is the speed of light $(3 \times 10^8 \text{ m/s})$ and τ is the pulse width (s).

There are practical limits as to how small the pulse width may be made. As was seen in Chapter 3, the theoretical receiver bandwidth required to pass all the components of a pulse of width τ is $2/\tau$, or $1/\tau$ if a matched filter is used. Therefore, narrow pulses need a wider receiver bandwidth which leads in turn to more noise and a greater risk of interference. However, perhaps more troublesome is the fact that, as the pulse width reduces, so peak power must increase to keep the average power constant. There are clearly definite physical limits as to how high the peak power can be. Therefore, to reduce the range resolution, a solution has to be sought that does not lie in the direction of ever-reducing pulse widths.

In fact, techniques exist that permit the range resolution to be determined to a much finer degree, although sophisticated modulation and signal processing has to be employed. The technique called pulse compression (sometimes colloquially known as 'chirp') is able to both improve range resolution and help in extracting target echoes from noise.

In pulse compression the RF carrier is not a fixed frequency modulated by the pulse envelope, rather the RF carrier is modulated with a particular characteristic. In the simplest form of pulse compression, the carrier is frequency modulated according to a linear law, in fact the carrier frequency increases in linear fashion for the duration of the transmitted pulse.

Military Avionics Systems   I. Moir and A. Seabridge
© 2006 John Wiley & Sons, Ltd

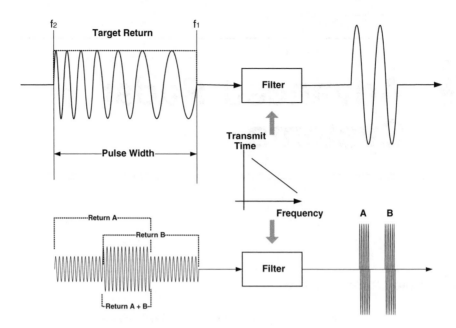

Figure 4.1 Principle of pulse compression.

In Figure 4.1 the frequency increases from f_1 at the pulse leading edge to f_2 at the trailing edge. The target return contains virtually the same modulation within the target echo. In the radar receiver the signal is passed through a filter that has the property of speeding up the higher frequencies at the trailing edge of the pulse so that they catch up with the lower-frequency components at the leading edge. The overall effect is to compress the signal to a width of $1/B$, where B is the bandwidth of the transmitted pulse, equal to $f_2 - f_1$. This narrow, processed signal in fact has the form $\sin x/x$ and is also increased in power by a factor equivalent to the pulse compression ratio. Therefore, pulse compression, as well as greatly improving range resolution, also greatly increases the signal power and hence target detection. Pulse compression ratios for a practical radar system can easily be achieved in the region of 100–300 and therefore the advantages can be considerable.

It has been mentioned that the compressed pulse has the form $\sin x/x$, the same format as for the radar antenna main beam and sidelobes. Therefore, pulse compression does have the disadvantage that it produces range sidelobes as well, and these may cause difficulties in some applications. A further potential problem can occur if the target echo contains a significant Doppler shift when range errors may be experienced.

The lower part of Figure 4.1 shows another advantage of pulse compression. In this example, two target returns are overlaid, and, using conventional signal processing, it would not be possible to discriminate between them. However, using pulse compression and the accompanying signal processing, both echoes are enhanced and may be separated.

Although linear FM pulse compression as described above is the most widely used, there are several other techniques that achieve a similar outcome. In recent years the viability and application of surface acoustic wave (SAW) technology has allowed the pulse compression process to be implemented in a cost-effective manner.

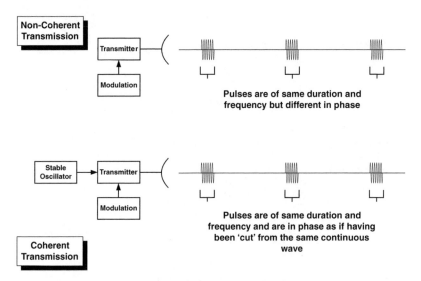

Figure 4.2 Comparison of non-coherent and coherent transmission.

Pulse compression is used as part of the signal processing associated with synthetic aperture radar (SAR) and some advanced forms are used in active electronically scanned arrays (AESAs) where compression ratios up to 1000 may be attained using sophisticated signal processing methods.

4.1.1 Coherent Transmission

The description of the pulsed radar in Chapter 3 related to the transmission of non-coherent carrier waveforms. That is, the successive pulses of energy that were transmitted were unrelated in phase from one pulse to the next. Pulse Doppler techniques using coherent radar transmissions enable much more data to be extracted. Coherent transmission can be likened to a transmitter that is transmitting continuously while being switched in and out of the antenna (Figure 4.2).

In Figure 4.2, non-coherent transmission is shown at the top of the diagram. The transmitter is controlled by the modulator, and a series of pulses of the same duration but differing in phase is transmitted towards the target. The reflections from the target will likewise be unrelated in phase.

By contrast, the coherent transmission example shown in the lower part of the picture shows a stable local oscillator or 'STALO' being modulated to transmit the same series of pulses. The STALO runs continuously and an associated power amplifier is switched on and off to produce pulses of the appropriate pulse width and pulse repetition frequency (PRF). Therefore, within these pulses, the carrier is in phase, almost as though sections of the carrier have been 'cut' from the same continuous wave. The reflected energy from the target largely preserves this phasing in the received target echo, and this property provides very useful features for the radar designer.

The properties and composition of a non-coherent and coherent pulse train for a carrier of frequency f_0 are quite different, as shown on Figure 4.3. Each pulse train comprises a series of pulses with pulse width τ and a pulse period of $1/f_r$, equivalent to a PRF of f_r.

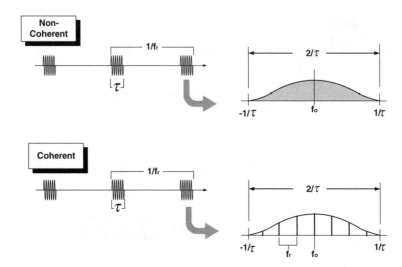

Figure 4.3 Characteristics of non-coherent and coherent pulse trains.

The non-coherent pulse train transforms into a continuous $\sin x/x$ spectral waveform with bandwidth $2/\tau$, represented as a width $\pm 1/\tau$ centred on the carrier frequency f_0.

The coherent pulse train spectral response is bounded by the same $\sin x/x$ envelope of width $\pm 1/\tau$ centred on the carrier frequency. However, in this case the energy is reflected by spectral lines each separated by f_r, the pulse repetition frequency. In fact these spectral lines are minispectra rather than spectral lines since the length of the pulse has an important effect, as will be seen.

Figure 4.4 illustrates the effect of varying the length of a series of coherent pulses. The longer the pulse train, the narrower are the resulting frequency spectra. In each case the frequency spectrum is represented by the $\sin x/x$ profile, but, as the length of the pulse train increases, this merges to form an apparent single spectral line. In the figure above, an infinitely long coherent pulse transforms into a single spectral line. Clearly, this is not practical in reality. A long coherent pulse train transforms into a narrow $\sin x/x$ response and the short coherent pulse train into a broader response. There is obviously a trade-off needed in the design to enable the most desirable performance to be achieved, bearing in mind practical constraints that apply. In a radar system where a target echo needs to be detected, the pulse train has to be of a sensible length. On the other hand, the longer the pulse train can be made within reason, then the narrower the frequency spectrum and therefore the greater the ability to reject unwanted signals.

Figure 4.5 depicts the difference between a single pulse and a pulse train comprising eight pulses. The single pulse of length τ transforms into a single $\sin x/x$ spectrum of width $2/\tau$ (ignoring sidelobes). The short pulse train transforms into a series of $\sin x/x$ spectra whose width is determined by the pulse width of the pulses spaced at intervals equal to the PRF, f_r. The individual spectra are bounded by a $\sin x/x$ envelope, this envelope being of width $2/\tau$, where τ is the length of the transmitted pulse train.

It may be seen that PRF is an important parameter in the pulse train spectrum. The lower the PRF, f_r, the closer together each of the spectra will be and after a certain point the sidelobes will begin to interfere with the adjacent spectra, distorting the signal content.

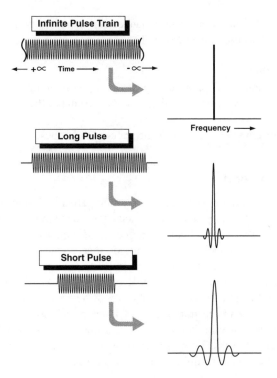

Figure 4.4 Effect of pulse train length.

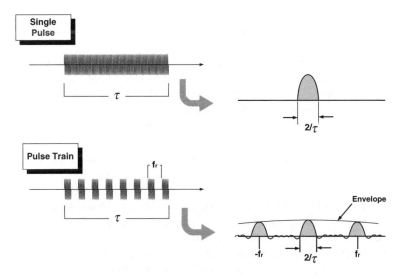

Figure 4.5 Comparison of a single pulse and a pulse train.

The use of pulsed transmissions in this manner can therefore be used in airborne radars provided a number of criteria are satisfied. These are as follows:

1. The radar is coherent.
2. The PRF is high enough to spread the spectral lines sufficiently far apart.
3. The duration of the pulse train is sufficiently short to make the spectral lines reasonably narrow.
4. Doppler filters are devised to filter out the spectral sidelobes.

4.1.2   Fourier Transform

The Fourier series, or Fourier transform as it is often called, is a mathematical relationship that identifies all those frequency components necessary to synthesise a particular waveform. The use of frequency-related analytical techniques is of immense importance in manipulating the signals that are frequency dependent, or that comprise important target-related frequency components. The mathematical theory is described in virtually every pure mathematics textbook, or for that matter most radar textbooks, and therefore will not be expounded in this publication. Rather, the subject will be addressed in sufficient detail such that readers can comprehend the importance of Fourier techniques as they apply to radar signal processing, particularly in Doppler radar and the associated applications.

The technique called the fast Fourier transform (FFT) is a useful method for creating a series of Doppler filters to discriminate a velocity return from ground clutter. The FFT process lends itself readily to implementation using digital computing techniques and therefore is commonly used in modern digital radars for Doppler filtering, Doppler beam sharpening (DBS) and synthetic aperture radar (SAR) modes. The FFT technique is also used in electronic warfare (EW) to analyse the characteristics of the opponent's radar systems in real time.

4.2   Pulsed Doppler Operation

The operation of a pulsed Doppler radar opens totally new areas that need to be considered. The derivation of target velocity using the Doppler shift has significant advantages and also permits low-level targets to be detected and tracked when flying in ground clutter regions. However, the radar sidelobes also collect a significant amount of ground clutter over a range of Doppler frequencies, and the characteristics of this clutter need to be fully understood for the best performance to be achieved.

Figure 4.6 shows the principal clutter returns for a pulse Doppler radar. These are as follows:

1. Mainlobe clutter (MLC) in the direction of the main beam. This is related to the velocity of the aircraft/radar, as has already been described in the Doppler navigator in Chapter 3. The velocity component of the mainlobe clutter reduces as the antenna boresight is moved to the left or right of the aircraft track. This happens as the component of forward velocity in the Doppler shift is reduced by the cosine of the angle on the antenna boresight with respect to aircraft track. Similarly, the size of the MLC Doppler shift is modified according to the cosine of the antenna look angle (depression angle).

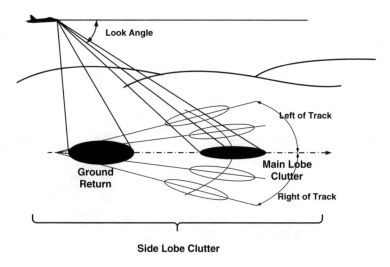

Figure 4.6 Principal returns for a pulse Doppler radar.

2. An altitude or ground return resulting from stray energy being reflected from the terrain directly underneath the aircraft. Since the terrain producing this return is directly below the aircraft, this return includes no Doppler component when the aircraft is flying straight and level over flat terrain. It therefore usually represents the zero Doppler shift position of the spectrum. However, there will be a Doppler bias if the aircraft is ascending or descending or the terrain below the aircraft is not level.

3. Returns are experienced across a whole area extending from ahead of the main beam clutter to well behind the aircraft ground return. This return is due to energy entering the system via the antenna sidelobes and is therefore known as sidelobe clutter. The sidelobe clutter region extends over a region that approximates to $\pm 2 \times V_r / \lambda$.

The typical Doppler spectrum resulting from an aircraft flying at velocity V_r, with pulsed Doppler antenna transmitting over terrain, is shown in Figure 4.7. On the diagram, Doppler frequency is increasing from left to right, starting with the extreme negative frequencies on the left, through the altitude line with zero Doppler shift to the extreme positive frequencies on the right. On the example shown there are the three main components of clutter already described.

The beginning of the sidelobe occurs at a point called the opening rate, below which targets flying much slower than the radar may become disengaged from the most negative aspect of the sidelobe clutter (SLC). At this point the Doppler shift from the receding terrain is negative and is approximately equal to the velocity of the radar $(= -2 \times V_r / \lambda)$. The other extent of the SLC is called the closing rate, which is a positive Doppler shift $(= +2 \times V_r / \lambda)$. The altitude line approximates to zero Doppler shift, while the main lobe clutter (MLC) has a positive shift but less than the closing rate (depending upon the antenna look angle).

The diagram shows a total of five targets from left to right (with the magnitude of the target echoes greatly exaggerated):

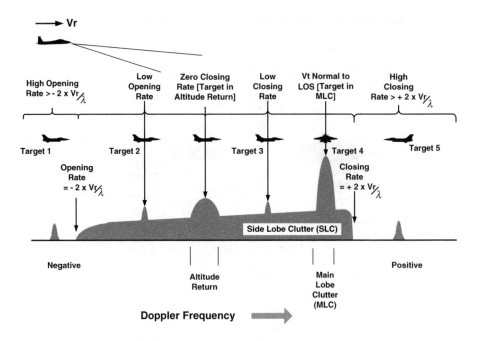

Figure 4.7 Typical Doppler spectrum.

1. Target 1 has a high opening rate (Doppler shift $> -2 \times V_r/\lambda$) and appears to the left of the negative SLC.
2. Target 2 has a low opening rate and appears through the negative SLC.
3. Target 3 has a low closing rate and appears through the positive SLC.
4. Target 4 is flying at the same velocity or tangentially to the radar and is masked by the MLC.
5. Target 5 is approaching at high speed with a high closing rate and appears in the clutter-free zone to the right of the SLC (Doppler shift $> +2 \times V_r/\lambda$).

It should be noted that target echoes appearing outside the SLC, altitude line and MLC regions will still have to be detected among the receiver noise, as for standard pulsed radar.

It can be seen how many variables affect where clutter and the target appear with respect to each other: radar velocity and radar and target relative velocities, antenna position, target geometry, etc. As many of these variables alter, as they rapidly will in a dynamic combat situation, the shape and relative positions of the target echoes and clutter regions will change quickly with respect to one another.

In certain situations, especially when operating at medium PRF, unwanted returns can be received in the sidelobes from very large ground targets. Industrial plants such as refineries and chemical processing facilities can produce significant returns that can be captured in this way, even if the sidelobe gain is 30 dB below the main beam. A solution to this problem is to use a guard horn and guard receiver channel, (Figure 4.8).

The guard horn has a broad low gain response and the gain is selected such that it is positioned below the antenna main beam response but just above the response of the first

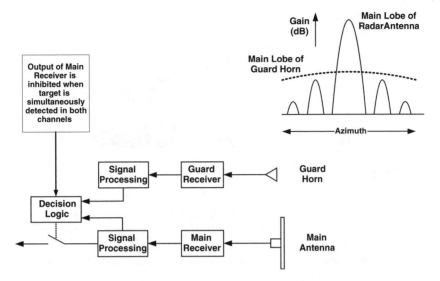

Figure 4.8 Use of a guard channel to reduce sidelobe clutter.

sidelobes. The receiver architecture features two parallel channels for the main antenna and the guard horn. Normally, the main returns from targets will be detected via the main beam and usual signal processing will occur. Target returns will be received in the guard channel but, owing to the low gain of the guard horn, will not be detected. When a target is detected simultaneously through the guard horn and the main antenna channels, decision logic will cause the main receiver output to be inhibited. Therefore, any targets that would have been detected from the sidelobes may be suppressed from the display. This technique may also be used as an electronics countermeasure tool to negate unwanted jamming entering via the antenna sidelobes.

4.2.1 Range Ambiguities

The effect of range ambiguities in a basic pulsed radar was discussed in Chapter 3 and shown in Figure 3.17. However, that diagram portrayed an air-to-air engagement where no ground clutter was present. The situation becomes more complex in pulsed Doppler when significant ground clutter has to be taken into account. The problem is simply stated in Figure 4.9.

This diagram shows a situation where a pulsed Doppler radar is looking down at three targets: two aircraft and a moving ground vehicle. The range return comprises three main elements:

- The altitude return on the left;
- The first target, clear of the MLC;
- The second and third targets obscured by the MLC;

If a high PRF is being used, then the range sweep may be less than the total range shown in the diagram.

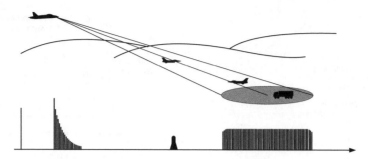

Figure 4.9   Representative flight profile.

It may be assumed that, owing to the use of the high PRF, the range is split into effectively three unambiguous sectors or zones. The overall effect is to superimpose the range zones on top of each other as far as the receiver is concerned. This situation is more complex than the example given in Chapter 3 as there is also the altitude return and MLC to consider. The overall effect is shown in Figure 4.10.

The effect of superimposing the range zones leads to the composite return at the lower right; this is far more difficult to unscramble than for the simple air-to-air non-clutter case. The one target that was detectable outside clutter has now been totally subsumed. This extremely simple example shows how altering the PRF – in this case increasing it, probably for good reason – has had the effect of losing the target in a combination of the altitude return and MLC.

4.2.2   Effect of the PRF on the Frequency Spectrum – Doppler Ambiguities

The effect of the PRF on the frequency spectrum also needs to be considered. In an earlier description the frequency spectrum of a short coherent pulse train was shown in Figure 3.4 to

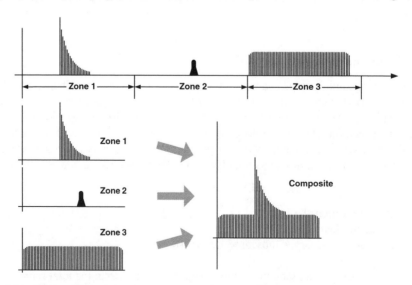

Figure 4.10   Effect of superimposing range zones.

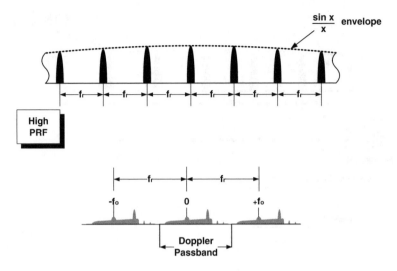

Figure 4.11 Effect of a high PRF on the frequency spectrum.

be a series of $\sin x/x$ frequency responses repeated at an interval determined by the PRF (f_2). This was an ideal case and represented the ideal pulse response. As we have seen in Figure 4.7, the true Doppler frequency response is in reality far more complex. Figure 4.11 shows the real situation for a high PRF.

The spectrum comprises replica sets of the true Doppler frequency response repeated at intervals determined by the PRF and modulated by a $\sin x/x$ envelope of width $2/\tau$, where τ is the width of the pulse train. In this case the full Doppler passband, containing all the components of the Doppler spectrum, is clearly seen, as the high PRF spaces out the Doppler spectra so that there is no mutual interference with the sidebands (which are also replicas of the Doppler frequency response).

If the PRF is reduced, then the situation portrayed in Figure 4.12 can occur. The $\sin x/x$ envelope is unaltered since the length of the pulse train is unchanged. However, the true Doppler return is now overlapping with upper and lower sidebands, giving the very confusing and ambiguous composite Doppler profile shown at the bottom of the figure.

It is clear that, when considering the operation, of pulsed Doppler in a look-down mode of operation, the system design characteristics need to be chosen with care. In particular, the selection of PRF is particularly crucial.

4.2.3 Range and Doppler Ambiguities

The operation of a pulse Doppler radar can be affected by both range and Doppler ambiguities. The range ambiguity is determined by the $1/x$ relationship already described in Chapter 3 (Figure 3.17) and is a fairly straightforward relationship. The determination of Doppler ambiguity is more involved since it depends on the PRF, the velocity of the radar and the carrier frequency (wavelength) being used (Figure 4.13). The gap between two adjacent Doppler spectra is determined by the value of the radar PRF such that the altitude

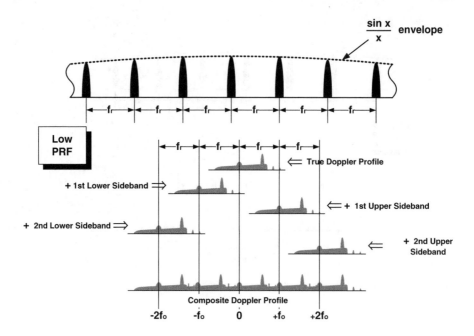

Figure 4.12   Effect of a low PRF on the frequency spectrum.

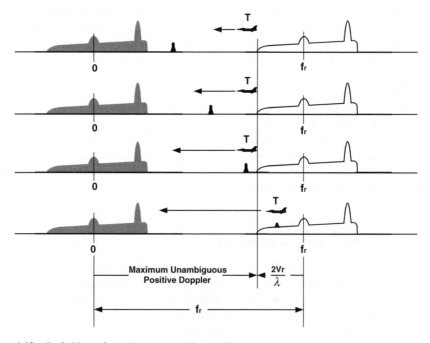

Figure 4.13   Definition of maximum unambiguous Doppler.

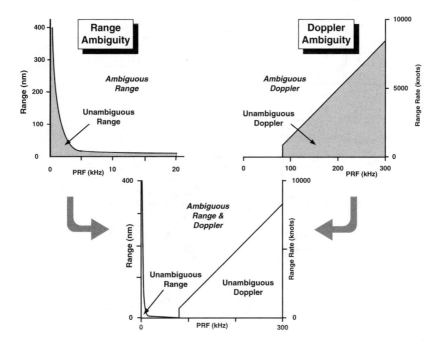

Figure 4.14 Areas prone to range and Doppler ambiguity.

line of the second spectrum will occur at frequency f_r after the altitude line for the first spectra (the altitude line also represents zero Doppler shift). However, the negative SLC extends back from the second altitude line. The maximum unambiguous positive Doppler is defined as $f_r - 2 \times V_r/\lambda$, as shown in the figure. The reason can be seen from the diagram. It shows four closing targets with increasing closing velocity from top to bottom. As the target velocity increases, so the target Doppler shift increases in frequency and moves further to the right until finally, in the last example, the target return has been subsumed by the negative SLC of the previous pulse.

The effect of ambiguous range and ambiguous Doppler is shown in Figure 4.14. In the top left of the diagram the effect of range ambiguity with respect to range and PRF is shown. The range area in which no ambiguity occurs is represented by the shaded portion shown at the bottom left part of the plot, adjoining the axis. Everything to the right of the curve represents areas where the range is ambiguous. This simple $1/x$ relationship is unaffected by changes in V_r or λ.

The areas affected by Doppler ambiguity are shown in the top right picture. The unambiguous Doppler areas are shown as the shaded portion at the bottom right. Every point on the diagram to the left of the diagram represents ambiguous Doppler.

At the bottom of Figure 4.14, both diagrams are combined to give a total picture of range and Doppler ambiguous and unambiguous areas. The shape of this diagram depends upon the carrier frequency and therefore wavelength and radar velocity. For the example shown, λ is 3 cm, which is equivalent to a carrier frequency of 10 Ghz, typical of a fighter radar; V_r is assumed to be 1000 knots. Therefore, the diagram is based upon realistic figures relating to a supersonic engagement.

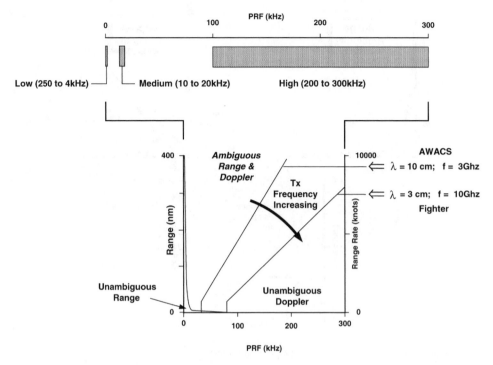

Figure 4.15   Factors affecting the unambiguous Doppler zone.

Both range and Doppler ambiguities may be resolved by changing the PRF at which the radar is operating, and the use of staggered or multiple PRFs is often used for this purpose.

Increasing λ increases the size of the unambiguous Doppler envelope, while decreasing λ has the reverse effect. Therefore, an AWACS radar operating at ~3 GHz/10 cm will have a much bigger unambiguous Doppler envelope than that shown in Figure 4.14. Decreasing radar velocity has the reverse effect, and therefore a fighter radar closing at only 500 knots will have a much smaller unambiguous Doppler envelope than the one shown in Figure 4.14.

Figure 4.15, shows the difference between the unambiguous Doppler zones for an AWACS and a fighter aircraft. The choice of PRF is crucial to obtaining the optimum performance of the radar. Normally, the three PRF bands shown at the top of the figure are considered:

- Low PRF ~250–4000 Hz;
- Medium PRF ~10–20 kHz;
- High PRF ~100–300 kHz.

These figures are indicative; precise figures may vary from radar to radar, depending upon the design drivers and the precise performance being sought.

The advantages and disadvantages of each of the PRF bands depend in large measure upon the type of radar mode being used and the nature of the target engagement. For an extensive review of the benefits and drawbacks of each PRF type, see Stimson (1998) and Skolnik (2001).

4.3 Pulsed Doppler Radar Implementation

The pulsed Doppler radar has a similar layout to the pulsed radar, but there are significant differences (Figure 4.16). The key differences are as follows:

1. A computer called the radar data processor (RDP) has been added and is central to the operation of the radar.
2. The exciter performs the function of stimulating the transmitter in order that coherent transmissions may be maintained.
3. The synchroniser function has been integrated into the exciter and the RDP. The exciter provides reference signals to the receiver local oscillators and the synchronous detection function. The RDP performs the control functions for the radar and interfaces with the radar controls and other on-board avionics systems and sensors.
4. The modulator task has been included in the transmitter.
5. A digital signal processor has been added that provides the processing power to undertake the necessary filtering and data manipulation tasks.
6. The indicator function has been moved to a multifunction display.

The transmitter and receiver paths are similar to the basic pulsed radar shown in Figure 3.15. The major functional differences are:

• The inclusion of the exciter, which performs the STALO function for the transmitter and provides the necessary local oscillator (LO) and reference signals for the receiver;

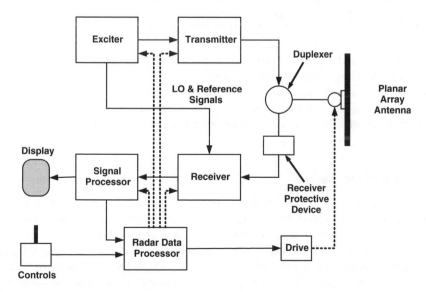

Figure 4.16 Pulsed Doppler radar.

- The use of the RDP to perform a control and synchronisation function with the other major units: exciter, transmitter, receiver and signal processor;
- The replacement of a video processor with a digital signal processor.

The key to understanding how the pulsed Doppler radar works lies in the interaction of the receiver, RDP and signal processor, and each are described more comprehensively below.

4.3.1 Receiver

The receiver block diagram is shown in Figure 4.17. The receiver takes the raw target reflected return received by the antenna and passed via the receiver protection device. Usually the signal is passed through a low-noise amplifier (LNA) which improves the signal-to-noise ratio before the signal enters the receiver by amplifying the target signal proportionately more than the noise.

The resulting signal is mixed with the LO1 reference signal, f_{LO1}, from the exciter, producing the first intermediate frequency, f_{IF1}, which is amplified by the first IF amplifier. The amplified IF1 signal is mixed with the LO2 reference signal, f_{LO2}, from the exciter, producing the second intermediate frequency, f_{IF2}, which enters the second IF amplifier. The resulting signal is fed into the synchroniser video detector which also receives reference signals from the exciter.

The synchronised video detector performs the function of detecting the signal and resolving it into in-phase (I) and quadrature (Q) components, thereby preserving the phase relationship of the incoming signal (Figure 4.18). The synchronous detector compares the incoming IF signal with two reference signals to determine the magnitude of the I and Q components. The Q component is determined by comparing the incoming IF signal with the reference signal from the exciter. This enables the magnitude of the Q (y axis) component to be measured and passed into the Q channel A to D converter. Similarly, the incoming IF signal is compared with the exciter reference signal, phase shifted (delayed) by 90°, which enables the magnitude of the I (x axis) component to be measured and passed into the I channel A to D converter. Both A to D converters sample the Q/I channel video at an interval approximately equal to the pulse width.

The magnitude of the received signal is determined by the vector sum of the Q and I components. The phase angle φ is determined by $\tan^{-1}(Q/I)$. The digitised Q and I video components are passed from the receiver to the signal processor.

If the radar is required to perform a monopulse angle tracking function, at least two and possibly three identical receiver channels are required depending upon the tracking scheme adopted.

4.3.2 Signal Processor

The signal processor architecture is shown in Figure 4.19. The signal processor sorts the digitised Q and I information by time of arrival and hence by range. This information is stored in range increment locations called range bins. The signal processor is then able to sort out the majority of the unwanted ground clutter on the basis of the Doppler frequency content. The processor forms a series or bank of narrowband Doppler filters for each range bin; this enables the integration of the energy from successive pulses of the same Doppler frequency. An example of clutter rejection is shown in Figure 4.20.

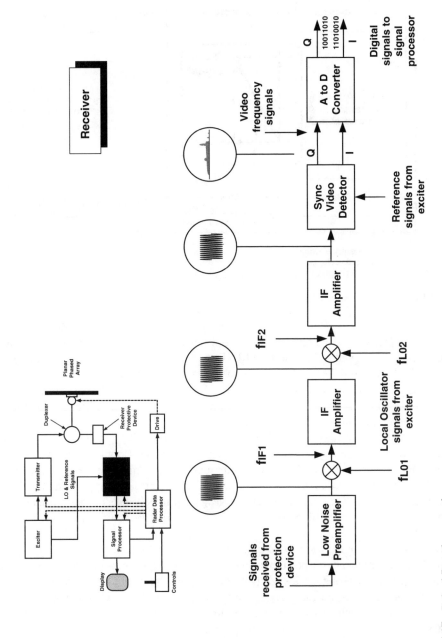

Figure 4.17 Pulsed Doppler – receiver.

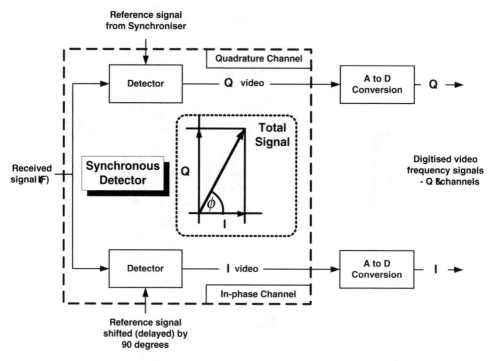

Figure 4.18 Synchronous video detector.

In this example, successive return signals from range R are fed into the clutter rejection system. Pulse n is delayed by a period T (1/PRF) and subtracted from the succeeding pulse $(n + 1)$ at the summation device. In the case of the central MLC and other unwanted clutter at repetitions of f_r, the effect is to cancel out the clutter. However, in the case of frequency components around $f_r/2$, the effect will be a summation. This has the effect of rejecting or cancelling clutter and reinforcing the desired signal.

The use of fast Fourier transform (FFT) techniques using the digital computation capabilities of the signal processor greatly facilitates the clutter rejection and Doppler filtering functions.

In the same way that the basic pulsed radar may track a target in range and antenna LOS, the Doppler radar is in addition able to track the velocity of a target using a velocity gate technique. This is very similar to the range gate principle except that, instead of an early and late gate, a pair of filters is used to track the target velocity instead, as shown in Figure 4.21.

The velocity gate comprises a low-frequency and a high-frequency filter, and the exact tracking Doppler frequency is positioned where the voltage from both filters is equal. In the example given, the target has a slightly higher Doppler frequency than the velocity gate, with a resultant error ε. Accordingly, an error voltage, ΔV, is generated and used to reposition the filters such that the error is minimised and the velocity gate represents the correct value.

The signal processor identifies the level of noise and clutter and sets a threshold such that the genuine target echoes may be detected above the background. The resultant detected target positions are fed to a scan converter and stored. The scan converter is able to identify

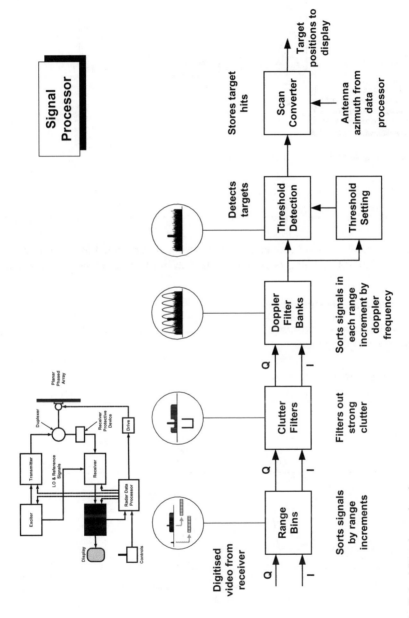

Figure 4.19 Pulsed Doppler – signal processor.

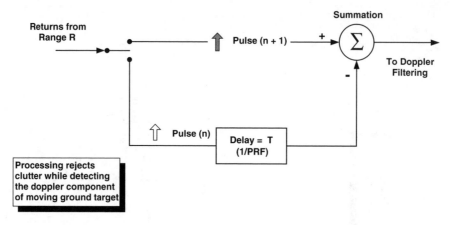

Figure 4.20 Simple clutter cancellation scheme.

the targets in memory and present them on a multifunction display in the appropriate position.

The advantage of the scan converter is that targets are presented as if on a bright raster (TV) screen without the fading target echoes that would feature on an analogue display.

4.3.3 Radar Data Processor

The radar data processor is shown in Figure 4.22 and is the heart of the radar. The RDP serves the following functions:

1. It receives commands from the radar control panel and avionics systems to exercise control over the radar modes and submodes, as determined by the combat situation.

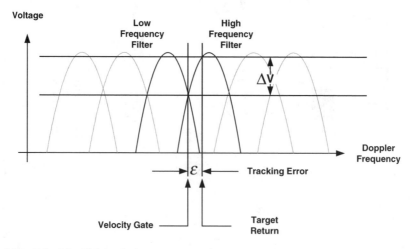

Figure 4.21 Principle of the velocity gate.

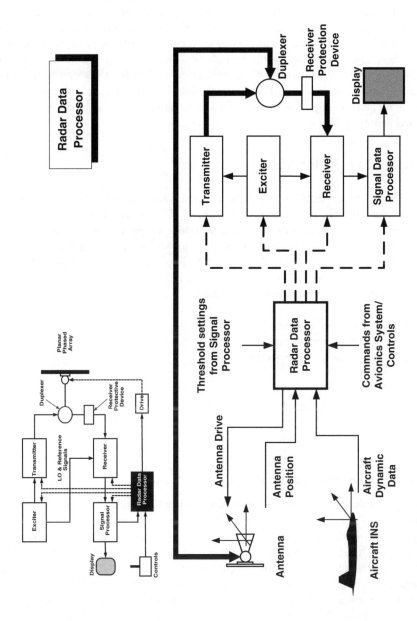

Figure 4.22 Pulsed Doppler – radar data processor.

2. It accepts aircraft attitude and stabilisation data in pitch roll and yaw axes to maintain the correct antenna position; it also receives velocity and acceleration data from the aircraft Inertial Navigation System (INS).
3. The RDP provides antenna drive commands to align the antenna axis with the horizontal axis, vertical axis and lateral axis (usually heading) as an orthogonal axis set. The RDP receives antenna positional data to close the antenna stabilisation loops such that the radar antenna boresight is stabilised and aligned to aircraft/geodetic axes to facilitate weapon-aiming functions accounting for the attitude, velocity and acceleration of the aircraft.
4. The RDP provides control outputs to the transmitter and exciter to determine the nature of the radar transmissions: carrier frequency, modulation, PRF, etc. It also provides data to the receiver and signal processor to enable the target reflections to be processed, filtered, detected and displayed.

4.4   Advanced Antennas

In Chapter 3 the characteristics of the radar antenna were described. The description of the antenna implemented the beam shaping, and the 'gathering' function was not explored in depth. In this section it is intended to move beyond a simple comparison of the parabolic reflector and planar array antenna to examine how technology assists in the development of new antennas that can offer significant benefits beyond merely increasing the antenna beamwidth and minimising the sidelobes.

The following antenna types will be described:

- Further examination of the planar array;
- The electronically steered array (ESA);
- The active electronically scanned array (AESA).

For each of these antenna types the principles of operation will be described; examples of deployment given and an overall comparison made.

All of these antenna types use some form of phasing within the antenna and may be generically termed 'phased arrays'.

4.4.1   Principle of the Phased Array

The principle of the electronically steered phased array is shown in Figure 4.23. A phased array comprises a number of radiating elements – effectively miniature antennas – each of which is able to radiate independently. The relative phase of each element decides the direction in which the array radiates. Figure 4.23 represents a radiating array in which the elements are all radiating at the same frequency but each equally displaced in phase from the others. In detail A, each element of radiated energy will be in phase with the other elements once a distance A has been travelled. Therefore, at angle A off-boresight all the radiating elements will be in phase and a phase front or radiated waveform will be directed perpendicular to the phase front. In detail B the distance has changed to distance B and the angle to B. In this case phase front B will be formed.

By altering the phase of the respective radiating elements within the antenna feed, the antenna beam may be shaped and electronically steered in any desired direction in elevation

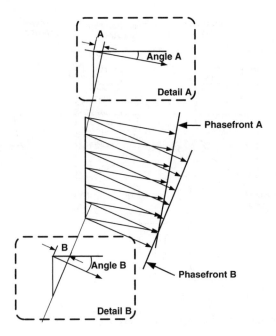

Figure 4.23 Principle of the phased array.

or azimuth. This is achieved by placing phase-shifting elements in the transmit line of each radiating element, and electronic control of these devices can be very rapidly executed and the ESA can therefore achieve scan rates far exceeding that which a mechanically scanned array can achieve.

4.4.2 Planar Arrays

The planar array was briefly described in Chapter 3, and most radars produced over the past 30 years or so have adopted this format. The planar array represents the simplest implementation whereby physical phasing of the individual RF distribution system shapes the antenna radiating pattern, analogous to the parabolic antenna previously described. The radiating surface of the planar array comprises several hundred radiating slots that are clearly visible on the front face of the antenna. These slots radiate the transmitted radar energy and receive the energy returning as the target echo. These slots behave as dipoles and at the relatively high frequencies commonly used are easier to fabricate than a dipole. Whereas the parabolic reflector used the physical distance between the antenna focal point and the parabolic reflector to shape the beam, the planar array uses a more subtle technique. This can be best understood by referring to Figure 4.24. A planar array is mechanically formed (via phasing) and mechanically steered by means of conventional antenna scanning servo techniques.

On the left are shown the radiating slots on the front face that have already been described. Behind the slots and within the array itself is a carefully arranged series of feeders that each feed sections of the slot array. The diagram at the top right shows how these feeders connect

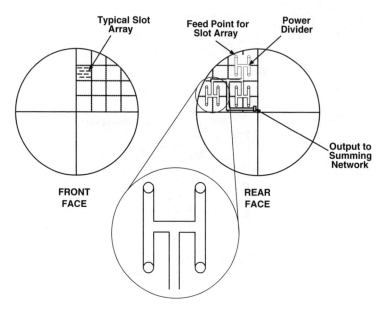

Figure 4.24   Physical characteristics of a planar array.

into the H-shaped power dividers, each of which feeds a group of four radiating slot elements. The H-shaped power dividers in turn connect to feeders so that eventually the whole planar array slot matrix is connected to the radar summing network.

The system described above distributes the radiated energy across the antenna face and gathers the incident energy of the reflected target echo. The power dividers maintain the necessary phasing to ensure that the beam is shaped or focused, and thus provides a narrow beam with low sidelobes as required by the radar.

The planar array does not use electronic switching means to scan the beam as is the case in the ESA. Rather, the beam is formed and shaped using phasing techniques embedded within the design of the antenna. To scan targets in the outside world, the planar array must be mechanically scanned and therefore cannot meet the high slew rate scanning capabilities of the ESA. Nevertheless, the planar array is likely to be less expensive than the ESA, not requiring the expensive phase-shifting elements, and is still used for many airborne radar applications.

4.4.3   Electronically Scanned Array

The principle of the phased array and the electronically scanned array (ESA) has already been outlined in the description of the phased array, and a top-level diagram showing an ESA radar is presented in Figure 4.25.

The key point about the ESA is that the remainder of the radar is relatively conventional in terms of the transmitter and receiver. It is the individual phase shifters and radiating elements that provide the enhanced capability, especially in terms of scan rate. The arrangement shown above is called a passive ESA as there are no active or transmitting elements within

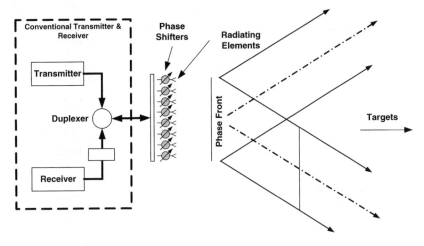

Figure 4.25   Electronically scanned array.

the antenna; the phase shifters are purely passive devices. This passive configuration will be compared with that of an active ESA in a later section.

Some of the advantages conferred by the beam agility are listed below:

1. Tracking can be initiated as soon as the target is detected.
2. The antenna can scan 100° in a matter of milliseconds, whereas a mechanically scanned array would take 1 s or more.
3. Targets may be illuminated even when outside the search volume.
4. The time on target or dwell time may be optimised according to target type and requirements.
5. Specialised detection and modulation techniques may be used to assist in extracting target signals from noise.
6. Terrain-following techniques may be improved owing to the flexibility in adapting scan patterns and beam shaping.
7. It permits electronic countermeasures (ECM) techniques to be employed anywhere within the field of regard of the antenna.

The ESA removes many of the components that contribute to failures in a conventional mechanically scanned antenna. Rotating waveguide joints, gimbals, drive motors, etc., are all removed. Consequently, the reliability of the ESA is improved. Failures of the phase shifters may be easily accommodated as the antenna can stand up to perhaps 5% of these failing before the radar performance is adversely affected.

The B-1B Lancer employs a passive ESA radar that uses ~1500 phase control devices. The antenna is movable and is capable of being locked in a detent such that the antenna points forwards, sideways or vertically downwards. The AN/APQ-164 employs a totally dual-redundant architecture – with the exception of the antenna – to improve mission availability. The B-2 Spirit stealth bomber also uses a passive ESA, in this case the AN/APQ-181.

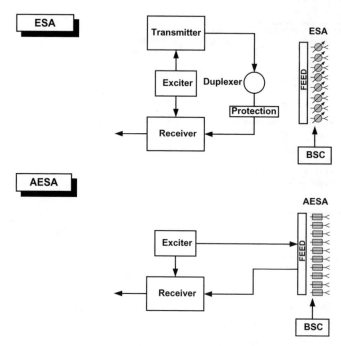

Figure 4.26 Comparison of passive and active ESA.

4.4.4 Active Electronically Steered Array (AESA)

The passive and active ESA top-level architectures are compared in Figure 4.26. As has already been seen, the passive ESA is fairly conventional, with the exception of the antenna which includes the many passive phase-shift elements. The active ESA, referred to as an active electronically steered array (AESA), is quite different.

The AESA includes multiple individual active transmit/receive (T/R) elements within the antenna. Depending upon the precise implementation, there may be anywhere between 1000 and 2000 of these individual T/R elements which, together with the RF feed, comprise the AESA antenna. As for the passive ESA, these elements are highly redundant and the radar can continue to operate with a sizeable percentage of the devices inoperative. This graceful redundancy feature means that the radar antenna is extremely reliable; it has been claimed that an AESA antenna will outlast the host aircraft.

The fact that the transmitter elements reside in the antenna itself means there is no stand-alone transmitter – there is an exciter but that is all. As before, there is clearly a need for a receiver as well as an RDP and signal processor. The active T/R elements are controlled in the same way as the phase shifters on the passive ESA, either by using a beam-steering computer (BSC) or by embedding the beam-steering function in the RDP.

The T/R elements are very small but encompass significant functionality. The architecture of a typical T/R module is shown in Figure 4.27. An early generation T/R module is also pictured in the lop left-hand part of the diagram. The T/R module is quite small, measuring in the region of 0.7×2.0 in and ~0.2 in deep, and is very reliable, having an extremely high mean time between failure (MTBF). Latest implementations fabricate the T/R module as part of a 'tile', where a tile comprises a module of perhaps four T/R devices.

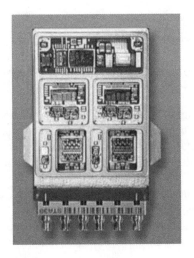

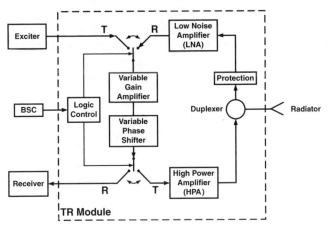

Figure 4.27 TR module architecture.

It can be seen from the figure that the T/R module is a miniature RF subsystem in its own right. The beam-steering computer or RDP controls the operation of the upper and lower T/R switches as follows:

1. *Transmit.* During transmit, the exciter is switched via the upper switch in the T position and passes through the variable gain amplifier and variable phase shifter. The transmit signal is routed through the lower switch, also in the T position, and passed to the high-power amplifier (HPA). The transmitted power is passed through the duplexer to the radiator element.
2. *Receive.* Received energy from the target is passed via the duplexer and a protection device to the low-noise amplifier (LNA). It is then routed through the upper switch, now in the R position, and through the variable gain amplifier and variable phase shifter to the lower switch which is also now in the R position. This routes the received signal to the receiver where the demodulation, detection and signal processing can be undertaken.

The T/R module architecture described above is implemented today in a hybrid chip integrating a series of monolithic microwave integrated circuits (MMICs) to perform the RF functions.

The ability to control many individual T/R modules by software means confers the AESA with immense flexibility of which only a few examples are listed below:

1. Each radiating element may be controlled in terms of amplitude and phase, and this provides superior beam-shaping capabilities for advanced radar modes such as terrain-following, synthetic aperture radar (SAR) and inverse SAR (ISAR) modes.
2. Multiple independently steered beams may be configured using partitioned parts of the multidevice array.
3. If suitable care is taken in the design of the T/R module, independent steerable beams operating on different frequencies may be accommodated.
4. The signal losses experienced by the individual T/R cell approach used in the AESA also bring considerable advantages in noise reduction, and this is reflected in improved radar performance.

Figure 4.28 shows the comparative losses for a passive ESA on the left versus an active ESA on the right. In each case the losses per device in dB are notified on the left with the device

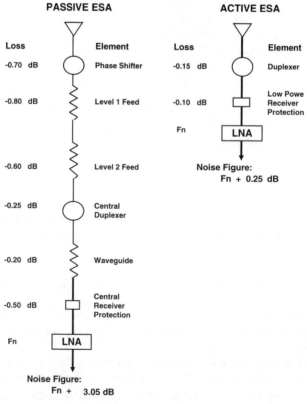

Figure 4.28 ESA and AESA front-end loss comparison.

Table 4.1 US fighter aircraft being fitted or retrofitted with AESA radars

Platform	Radar	Number of TR modules	Status
F-22 Raptor	AN/APG-77	1500	Entering service
F-18E/F Upgrade	AN/APG-79	1100	Entering service
F-16E/F (Block 60)	AN/APG-80	1000	Entering service
F-15C	AN/APG-63(V)2	1500	In service
F-35	AN/APG-81	1200	In development

named on the right of the RF 'chain'. The total noise figures for each case are totalled at the bottom. The results are:

$$\text{Passive ESA:} \quad \text{LNA noise figure } F_n + 3.05 \text{ dB}$$
$$\text{Active ESA:} \quad \text{LNA noise figure } F_n + 0.25 \text{ dB}$$

This is a dramatic improvement in the noise figure; it is especially significant achieving such an improvement so early in the RF front end. This results in a remarkable range improvement for the AESA radar.

A number of US fighter aircraft are being fitted or retrofitted with AESA radars. These are shown in Table 4.1.

The increase in AESA radar performance versus conventional radars has been alluded to earlier in the section. Figure 4.29, portrays the comparative ranges against a target where the cross-sectional area, σ, is normalised at $1\,\text{m}^2$, the size normally ascribed to a small airborne target such as a cruise missile target. The aircraft to the left of the dotted line are fitted with

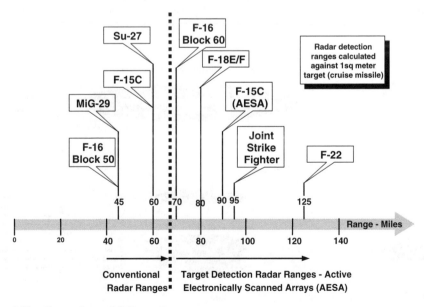

Figure 4.29 Comparison of fighter radar ranges.

conventional radars, while those to the right are AESA equipped. All the AESA-equipped aircraft show a significant range advantage.

Of particular interest are those data points that show a comparison of two models of the same aircraft: one fitted with conventional radar and the other with an AESA radar. The F-16 Block 60 (now the F-16E/F) shows an improvement from 45 to 70 nm (+55%), while the F-15C range has increased from 60 to 90 nm (+50%). Apart from the obvious improvement in range, it has been stated by a highly authentic source that AESA radar confers 10–30 times more in radar operational capability compared with a conventional radar (Report of the Defense Science Board Task Force, 2001).

4.5 Synthetic Aperture Radar

One of the important requirements of a modern airborne radar is that of providing high-quality radar mapping such that topographical features and potential targets may be identified and classified. The quality of the radar map depends upon its ability to resolve very closely spaced objects on the ground – in the most demanding case down to within a few feet.

The problem surrounding the need to resolve to such high resolution may be understood by reviewing the ground footprint of a radar beam radiated directly abeam the aircraft as shown in Figure 4.30. The oval footprint may be resolved into smaller cells or pixels. Along the radar boresight (perpendicular to the aircraft track) the range swath may be resolved into finer range resolution cells. Across the radar beam (along the aircraft track) the increments are azimuth increments as far as the radar is concerned.

The range resolution of a pulsed radar was described in Chapter 3. It will be recalled that, for a pulse of width τ, the range resolution increases as the pulse width becomes smaller. For a 1 µs pulse, the range may be resolved to 500 ft ($\sim$150 m); for a 0.1 µs pulse this would be reduced to 50 ft ($\sim$15 m); for a 0.01 µs pulse the resolution would reduce to 5 ft ($\sim$1.5 m) and so on. The major limitation to this approach to increasing range resolution relates to the frequency spectrum associated with a very short pulse. In Chapter 3 the receiver bandwidth required to pass all the frequency components is determined by the 3 dB bandwidth which in

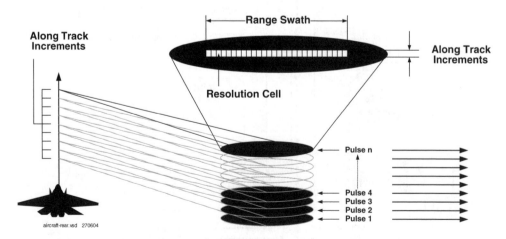

Figure 4.30 Principle of synthetic aperture radar.

turn is related to the pulse width, where $B = 1/\tau$. For the 0.01 μs pulse the bandwidth is 100 MHz. The impact of bandwidth depends upon the value of the carrier frequency; for an AI radar operating at 10 GHz, a 100 MHz bandwidth is not excessive, but, as frequency decreases, it may become more of a problem. Pulse compression techniques as described earlier in the chapter can achieve significant improvements in range resolution by a factor in the hundreds. This does not improve the bandwidth, however, as the bandwidth must still be sufficient to accommodate the compressed pulse.

Azimuth resolution is more problematical. The along-track resolution is determined by the azimuth bandwidth using the following formula:

$$\theta_{3\,\text{dB}} = \lambda/L \quad (\text{rad})$$

wherev λ is the wavelength and L is the antenna length. The range resolution cell is $R \times \theta_{3\,\text{dB}}$, where R is the range at which the resolution cell is being considered.

For a for sideways looking airborne radar (SLAR) with a 5 m (16.5 ft) long antenna operating at 3 GHz ($\lambda = 10$ cm) and at a range of 50 nautical miles (90.9 km) the azimuth range resolution cell is ~60 ft (18.2 m), still at least an order of magnitude away from the desired performance.

The answer to this problem is to use a technique called synthetic aperture radar (SAR). In this solution the aircraft forward movement is used to create a large synthetic or artificial aperture, as can be seen in Figure 4.30. The aircraft shown on the left of the diagram is transmitting a series of pulses at constant intervals, which equates in turn to constant along-track increments. A series of patches of ground will be illuminated abeam the aircraft and the return from pulses $1, 2, 3, 4, \ldots, n$ may each be detected and collected in a series of range bins. The principle of this technique, easily implemented in a digital computer, is shown in Figure 4.31.

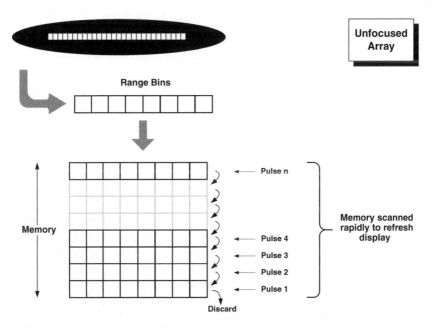

Figure 4.31 Collection and storage of successive pulse returns.

After a series of pulse returns has been stored, the oldest are eventually discarded; the number of pulse returns retained before discarding is a function of the algorithm being used. This memory bank of the returns of the previous n pulses may be rapidly scanned to present a strip-line picture of the area of the terrain being mapped. For reasons that will be described shortly, this fairly crude SAR processing technique is referred to as an unfocused array.

The key point about the forward movement of the aircraft is that it allows the signal processing to synthesise an aperture much larger than the real aperture (5 m or 16.5 ft) in the previous example. In a typical SAR application, an array length equivalent to 50 m or more may be synthesised, and in this case the azimuth resolution may be determined by the following approximate formula:

$$\text{SAR azimuth resolution} \approx \frac{\lambda \times R}{2 \times L}$$

where λ is the wavelength (m), R is the range (m) and L is the synthetic array aperture (m).

Using the figures in the former illustration, and using the synthetic array aperture figure, a SAR azimuth resolution of 3 ft or about 0.9 m may be achieved at 50 nm (90.9 km). This is an order of resolution that allows roads and, in some cases, road vehicles to be discriminated. Also, for range resolution – receiver bandwidth considerations apart – it was shown earlier that similar resolutions may be achieved by using pulse compression techniques. In this case, using a pulse width of 1 μs and pulse compression ratios of 100 or more, 3 ft range resolution may also be achieved. A combination of pulse compression to improve range resolution (across track) and synthetic aperture techniques to improve azimuth resolution (along track) means that very high resolutions in both directions may be achieved at long range.

As an aside, working numbers to determine the minimum resolution requirements for ground features are:

$$\text{Roads and map details}: \quad 30\text{–}50\,\text{ft}(10\text{–}15\,\text{m})$$

$$\text{Shapes/objects}: \quad \frac{1}{5}-\frac{1}{20} \text{ of the major dimension}$$

Therefore, an SAR radar providing 3 ft resolution at 50 nm would be able to distinguish freeways and trucks with ease, although compact cars may escape resolution.

The point was made earlier that the array configuration described in Figures 4.30 and 4.28 represented an unfocused array; the term 'unfocused array' will now be explained, with reference to Figure 4.32. The figure shows an unfocused array at the top of the diagram. If a linear array of length L is radiating abeam the aircraft, all the successive pulses will be radiated at right angles to the array and the pulse paths will effectively be in parallel. Another way of viewing this, considering an optical simile, is that the array will be focused at infinity. Therefore, point P will not be at the focus of the array; this effect is accentuated the closer P is to the array, or the shorter the range. These effects become more pronounced the larger the array and limit the effectiveness of an unfocused array. As a rule of thumb, the most effective azimuth resolution yielded is approximately equal to 0.4 times the array length.

The focused array is depicted in the lower part of the diagram – this has a slightly curved configuration to emulate a parabolic reflector. In this situation all the successive pulses are effectively focused at point P and the unfocusing errors will be negated. To achieve this effect, two adjustments need to be made. Allowance has to be made for the fact that pulses at

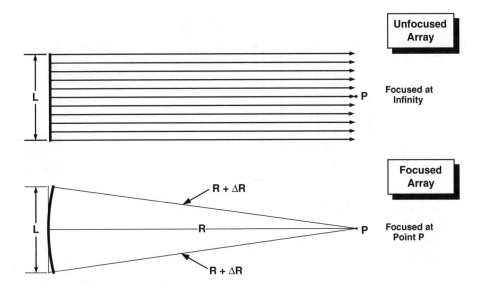

Figure 4.32 Unfocused versus Focused SAR Array.

the extremity of the array have to travel a small but finite distance, ΔR, further to reach point P than pulses originating from the centre of the array. Also, pulses originating from the leading edge of the array have to be biased by a minute angular increment to the rear (clockwise), while those from the trailing edge have to be biased by a similar angular increment forwards (counterclockwise). If these adjustments can be made during the digital processing operation by taking account of the motion of the aircraft, the SAR array can be focused at point P.

The computations undertaken to formulate the focused array are outlined in Figure 4.33. The top matrix comprises range bins that are populated during each successive range sweep. The contents are incremented until eventually the oldest return is discarded. The data in each column are read every time the array advances by the minimum azimuth resolution distance. Each array is focused by applying the necessary corrections and summed individually for each bin. The magnitude of each range bin is fed into the display memory. Once the last memory location has been filled, the display memory is decremented and the oldest data discarded. The display memory contents are fed to a display or recording media device.

In effect, this process is similar to the unfocused array data manipulation, except the focused array is continually updated. The process of summing the columns in the range bin array is called azimuth compression and the effect is to synthesise a new array every time the radar advances by the incremental along-track distance.

As for all these transformations, a number of trade-offs need to be made. However, after taking into account a range of factors, the ultimate azimuth resolution achieved by using a focused array is given by the following formula:

$$\text{Minimum azimuth resolution distance} = \frac{\text{length of the real antenna}}{2}$$

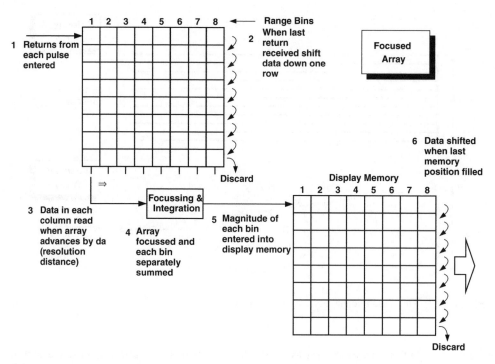

Figure 4.33 Computation associated with formulating a focused array.

Therefore, in real practical terms, even with a focused array, the minimum azimuth resolution distance is limited; in the example previously quoted with a 5 m (16.5 ft) real array, the azimuth resolution is limited to 2.5 m or ~8 ft.

The key to azimuth resolution is the Doppler processing and signal integration process. Figure 4.34 shows the Doppler shift history as a radar approaches, passes abeam and then recedes from the target.

As the aircraft flies by the target, the Doppler shift starts at $+2 \times V_r/\lambda$, reduces through zero as the aircraft passes abeam and then decreases to $-2 \times V_r/\lambda$. As the target passes abeam, the rate of change in frequency is maximum, and immediately adjacent to zero frequency the lines are virtually straight. Also, for a number of evenly spaced points positioned near to each other the frequency difference between them, Δf, will be constant. Therefore, each of the returns will have a different frequency, determined by its azimuth position, and by using Doppler filtering techniques that azimuth position may be measured. The way in which this processing is achieved is shown in Figure 4.35.

The incoming video returns are modified by applying the necessary phase corrections as described above to focus the array. Then each point has a constant Doppler frequency which can be discriminated from the others and which relates to its azimuth position. Each time the radar travels a distance equal to the array length to be synthesised, the phase-corrected signals that have been gathered in the range bin row are fed to the array of Doppler filters in the columns. The integration time for the filters is the same time it takes the radar to transit the array length, and the number of Doppler filters in each column thus depends upon the length of the array. The greater the array length, the longer is the integration time and the

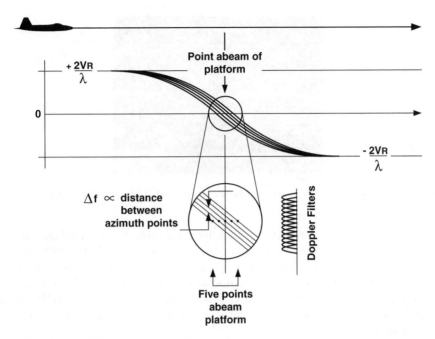

Figure 4.34 Doppler history of radar passing a target.

greater the number of Doppler filters formed for a given frequency coverage. Therefore, as the array length increases, the filters become narrower and azimuth resolution becomes finer. As in this region abeam the target the points on the ground are evenly spaced, and the frequencies are evenly spaced, a fast Fourier transform (FFT) may be used to form the filters which minimises the amount of computing required.

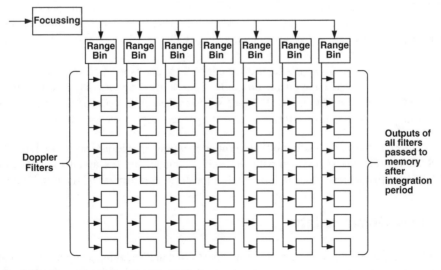

Figure 4.35 Example of Doppler processing.

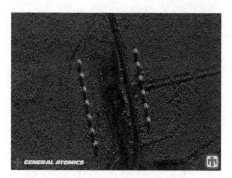

Figure 4.36   SAR picture of tanks with 1 ft resolution (Sandia National Laboratories)

The use of synthetic aperture imaging is an important asset in today's airborne platforms. Figure 4.36 is an example of an SAR picture that shows two rows of tanks with a resolution equivalent to 1 ft. The picture shows one of the curious features of SAR images: each tank has a large shadow in the six o'clock area, showing that the targets were illuminated from the twelve o'clock position. Therefore, although sophisticated data processing enables the target to be viewed in high resolution at long range, it cannot negate the fact that the target was illuminated at a low grazing angle.

The SAR principles have been described for a fixed antenna scanning terrain abeam of the aircraft. There are other modes that are commonly used:

1. *Spotlight mode.* In the spotlight mode the look angle of the real antenna is altered so that it always illuminates the target. This has a number of advantages. Since the real antenna is always trained on the target, the length of the synthetic array is not limited by the beamwidth of the real antenna. Also, the fact that the target is viewed from different aspects helps to reduce the graininess of the response.
2. *Doppler beam sharpening.* Doppler beam sharpening (DBS) is a subset of SAR operation, lacking several signal processing refinements, and as a result it lacks the performance of the more sophisticated SAR mode. Nevertheless it provides a significant improvement in performance over the real-beam mapping that is the baseline mapping mode in many radars.
3. *Inverse SAR.* Inverse SAR (ISAR) is a variant of SAR that is used against moving targets that have a rotational component; normally, SAR is used against fixed ground targets. ISAR effectively uses the minor Doppler shifts caused as a result of target movement. ISAR may be used against aircraft or moving ground targets such as ships.

The difference between the main modes of ground mapping, including real beam, DBS, SAR and ISAR, are difficult to envisage without a direct comparison. For an air-to-ground radar with the basic parameters listed in Table 4.2, a top-level comparison is given in Table 4.3.

4.6   Low Observability

The importance of a radar system in detecting, identifying and engaging a target is undeniable. However, for years the aircraft designer has been attempting to minimise the ability of a radar

Table 4.2   Baseline air-to-ground mapping radar specification

Frequency	10 GHz
Antenna	Flate plate; 3° × 3° beamwidth; 30 dB gain; −30 dB sidelobes
Transmitter	Travelling wave tube (TWT); 10 kW peak power; 5% duty cycle
Receiver	5 dB noise figure; dual conversion; RF preamplifier; STC; AGC
A/D conversion	I/Q 8 bits; 120 MHz maximum rate
Signal Processor	100 MFLOPS/s; 8 Mbytes RAM
Radar data processor	2.5 MIPS; 64 kBytes RAM
Communications	1553 data bus
Display	512 × 512 × 8 bits; monochrome
Exciter	Up to 250:1 pulse compression; variable CHIRP rate; 0.01 μs minimum to 10 μs maximum

to detect the aircraft. This art is known as low observability or, more colloquially, stealth, and has been utilised for many years, although perhaps with more prominence with the introduction into service of the US Air Force stealth aircraft: F-117 Nighthawk, B-2 Spirit and F-22 Raptor (Figure 4.37).

Table 4.3   Comparison of ground-mapping modes

Parameter	RBGM	DBS	SAR	ISAR
Azimuth				
Centre	Heading stabilised	Selectable ±60°	Selectable 60° ±60°	Selectable ±60°
Swath	±60° (max.)	45°	2.5 nm × 2.5 nm	N/A
Scan rate	60 deg/s	Varies with azimuth 60–5 deg/s	Spotlight mode	Spotlight mode
Resolution	Real beam	20:1 beam sharpening 512 azimuth bins	25 ft cross-range resolution 512 azimuth bins	Target motion dependent 128 doppler bins, 0.25–2.0 Hz resolution
Range				
Scale	Selectable 5–160 nm	Selectable 5–40 nm	Range/azimuth centre is designated	Range/azimuth centre is designated
Resolution	R_{max}/256 : 950 ft at 40 nm	$\Delta R = R \times \Delta a_z \times$ 325 ft for 30 nm	25 ft, 512 range bins	5 ft, 512 range bins
PRF	PRF = 2 kHz, P-to-P frequency agility	Variable PRF: 2.5 kHz to 500 Hz	Variable PRF: 1.5–3 kHz, PRF = 1700 Hz at 30°, 25 nm, 300 knots	PRF = 800 Hz at 100 nm
Signal processing	STC, AGC, NCH integration	32-point FFT	512-point FFT	128-point CFT

F.117 A Nighthawk
(US Air Force photo
by Staff Sgb
D Allmen 11)

B-2 Spirit (US Air Force
photo by Master Sergeant
Val Gempis)

F.22 (US Air Force photo)

Figure 4.37   US Air Force stealth aircraft.

The history of the use of stealth techniques stretches back before these aircraft were developed (Figure 4.38). This shows that the development of stealthy techniques stretches back to the mid-1950s. The evolution of stealthy aircraft can be characterised by three distinct generations:

1. *First generation.* The original strategic reconnaissance aircraft developed by the Lockheed 'Skunk Works' – the U-2 and the SR-71 Blackbird.

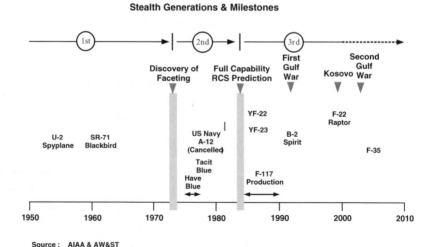

Figure 4.38   History of the development of stealth aircraft.

2. *Second generation.* In the early 1970s the concept of 'faceting' was invoked, first on the two Have Blue demonstrator aircraft and then applied in production on the F-117. This technique involved the use of angular faceted areas designed to redirect or deflect the radar energy away from the emitter.
3. *Third generation.* This generation was enabled when computational techniques had developed such that the entire aircraft could be predicted and evaluated using numerical means. These aircraft are soft blended shapes quite unlike the faceted approach. The Tacit Blue demonstrator proved these techniques which were subsequently adopted on the US Air Force advanced tactical fighter (ATF) fly-off aircraft: the Lockheed Martin YF-22A and the Northrop YF-23A. Subsequently, these techniques have been applied to the Northrop B-2 Spirit stealth bomber which is in service and the Lockheed Martin F-22A Raptor which is just entering service. The F-35 [previously the joint strike fighter (JSF)] also applies similar techniques, looking very similar in appearance to the F-22.

Figure 4.38 also shows the milestones of the campaigns in which the stealth aircraft have been successfully deployed: the F-117 in the first and second Gulf Wars and Kosovo and the B-2 during Kosovo and the second Gulf War. Despite flying thousands of missions, the only casualty was an F-117 shot down near Belgrade during the Kosovo campaign.

Recalling the basic radar range equation discussed in Chapter 3, the radar range is proportional to $\sigma^{1/4}$ where σ is the radar cross-section (m^2). Reducing the cross-sectional area therefore affects radar range, although only according to the fourth root. However, by carefully designing an aircraft, the value of σ may be reduced by many decades (and the reflected signal by many dB), and therefore reduction in the radar cross-section is an attractive and effective approach to reducing the detection range of an aircraft by radar.

4.6.1 Factors Affecting the Radar Cross-section

The radar cross-section may be thought of as comprising three elements:

- Geometric cross-section;
- Directivity;
- Reflectivity.

There is a limit to what can be done regarding the first component since the size of an aircraft will be dictated by the role, weapons payload and range. However, a combination of low-directivity and low-reflectivity techniques has been successfully developed, as the degree of stealth offered by some of the aircraft testifies.

The major areas on conventional fighter aircraft that contribute to a high radar cross-section are shown in Figure 4.39. The radar, cockpit, engine intakes, drop-tanks, engine exhaust and rudder/elevator combination can all produce large returns, primarily because they can act as radar reflectors. External carried ordnance and the wing leading edge can produce large reflections on occasions. The aircraft fuselage and small blended inlets like a gun muzzle produce relatively low radar reflections, as do minor air inlets or blended inlets on top of the aircraft.

For a stealthy shape to be achieved, these factors need to be taken into account at the design stage. The design features needed to ensure a low-observability aircraft are shown in Figure 4.40. These include:

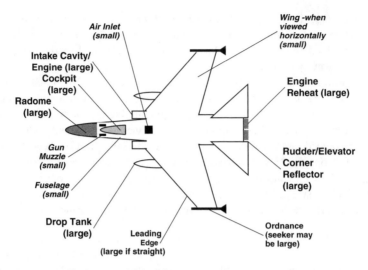

Figure 4.39   Areas contributing to a high radar cross-section.

1. Blended wings and fuselage, and the use of blended chines at the side of the fuselage. These features were originally incorporated on the SR-71 and also utilised on later aircraft, particularly on the Northrop YF-23A and B-2, minimising all necessary bumps and protuberances including external antenna.
2. Swept leading edges, not necessarily linear, to reduce 'end-fire' array effects or to ensure that constant angles are maintained.
3. A low-profile 'blended cockpit' to avoid the cockpit and pilot acting as a corner reflector.

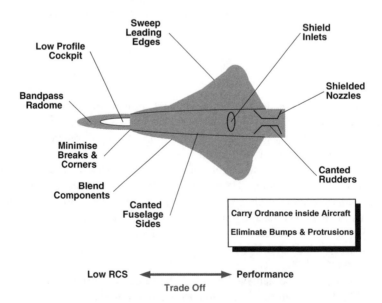

Figure 4.40   Low-observability design objectives.

4. Ordnance. On the most stealthy aircraft the weapons carriage is internal.
5. Bandpass radome and other radar RCS reduction techniques which will be described later.
6. Canted rudders such as those used on the YF-22A, YF-23A, F-22A and F-35A/B/C assist in not providing a corner reflector effect. The Northrop YF-23A was an extreme example of this feature, and its stablemate, the B-2, being a flying wing, has no rudders at all.
7. Shielded engine nozzles. Note the B-2 overfuselage engine intakes and exhausts.

Unfortunately, many if not all of these features mitigate against aircraft performance, so in reality the RCS and performance have to be the subject of trade-offs. Many of the stealthy aircraft are unstable and require high-integrity sophisticated flight control or fly-by-wire (FBW) systems. In most other respects the aircraft systems are fairly conventional and have often borrowed and adapted major subsystems from non-stealthy combat aircraft.

The benefits conferred by the design principles outlined above are augmented by the use of radar-absorbent material (RAM) to improve the low-observability features. An example of the combination of both techniques may be gained by examining the intake design on the F-117 and F-35 fighters.

The F-117 intake uses a combination of single reflections, multiple reflections and RAM to lower the engine intake RCS (see Figure 4.41 which shows the intake grill and a diagram

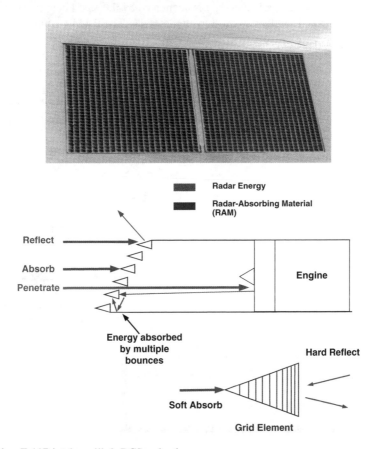

Figure 4.41 F-117 intake grill & RCS reduction.

portraying the intake construction). The intake grill is actually coated with RAM and is triangular shaped. Energy incident upon the intake grill may be reflected from one of the triangular grid elements, being dissipated and reflected away from the threat radar. This is indeed how faceting is used across the entire external surface of the F-117 with considerable success. Some incident energy will impinge directly upon the grid element and will be absorbed. A proportion of the energy will pass through the grill and enter the intake. After being reflected within the intake, it will undergo multiple bounces against the RAM-coated rear of the grid element and will mostly be dissipated within the engine intake. Such minor amounts of energy that do escape the intake grill will be of very low power and randomly scattered.

The F-35 intake is reported to be more sophisticated, being a serpentine duct rather than a direct, more conventional intake. Although at first glance the relatively open intake would appear to suffer from a high RCS, sophisticated techniques are used that lower the cross-section over a range of frequencies. The intake is designed to counter radar threats at three wavelengths loosely termed long ($\sim$30 cm), medium ($\sim$10–20 cm) and short ($\sim$3 cm), equating to 1 GHz (long-range surveillance radar), 1.5–3 GHz (AWACS radar) and 10 GHz (fighter radar) respectively.

At long wavelengths the duct behaves as indicated in Figure 4.42. The wavelength is too large to propagate effectively down the inlet and a minimal amount of energy reaches the engine/blocker. Most of the incident energy is attenuated by the RAM coating around the inlet lip and the remainder reflected away from the threat radar. The RAM coating around the inlet lip is optimised to have a maximum effect at $\sim$1 GHz.

For medium wavelengths the wave is able to propagate effectively down the inlet and proceeds down the serpentine duct unimpeded. As it reaches the engine, it impinges upon the RAM-coated blocker which is tuned to absorb the maximum amount of energy at this frequency and to attenuate subsequent reflections or bounces. Most of the energy is dissipated at this point; the very small amount of residual energy that does remain is reflected out of the duct (Figure 4.43).

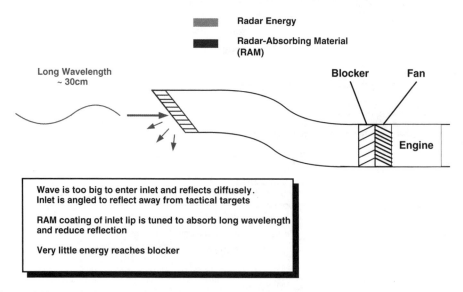

Figure 4.42   F-35 engine inlet duct response to long wavelengths.

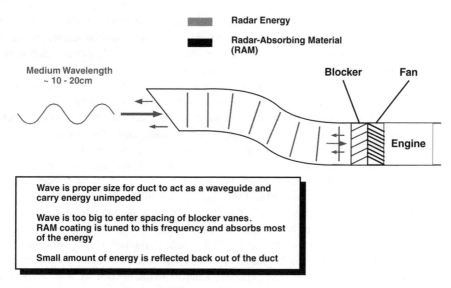

Figure 4.43 F-35 engine inlet duct response to medium wavelengths.

Short-wavelength energy incident upon the duct travels readily down the duct, reflecting off the RAM coating on the inside of the duct walls. This RAM is tuned to achieve maximum attenuation at 10 GHz and therefore heavily attenuates the energy. The energy that does reach the engine becomes trapped and is mostly dissipated in the blocker/engine labyrinth. A small portion of the energy is reflected out of the front of the duct, as shown in Figure 4.44.

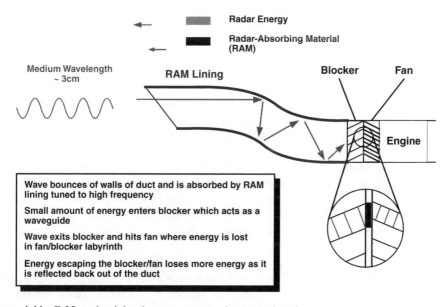

Figure 4.44 F-35 engine inlet duct response to short wavelengths.

From the foregoing it can be seen that the shape and size of the inlet, the nature of the serpentine duct, the use of the blocker fan and the judicious use of specialised RAM coating material greatly reduce the intake RCS, thereby enabling an apparently normal engine inlet to have a low RCS response to different types of radar.

The maintenance and preservation of the radar-absorbing coatings on a stealth aircraft pose a major servicing penalty. Every time an aircraft panel is removed for maintenance it has to be resealed – effectively recaulked – in order to preserve the low-observability signature of the aircraft. This takes time enough, but in many cases the sealant used has a long curing time so it can take some time before the aircraft may be used on a mission. Also, certainly during the F-117 production run and possibly with the B-2 bomber, different aircraft were coated with different coatings or sealants as production proceeded and improved treatments became available. This created a configuration control issue with several different variants of coating standard across the fleet. It is understood that modification programmes are in place to ameliorate this problem.

The stealthy aircraft creates a problem for the radar and radio designers as a vast proliferation of antennas have to be fitted on to the aircraft in order for it to fulfil its mission, each with its own particular system performance goals. In Chapter 2, the wide array of antennas performing the primary radar, CNI, and EW functions on a JIAWG/F-22 RF architecture was outlined, and the trend towards shared apertures for JAST/F-35 applications was described. The radar, being at the front of the aircraft, faces particular problems that need to be addressed to maintain the overall RCS signature of the aircraft.

4.6.2   Reducing the RCS

Some of the particular problems that the radar confronts in minimising the RCS are as follows:

1. *Antenna mode reflections*. The antenna mode reflections mimic the antenna main beam and sidelobes. Therefore, merely positioning the antenna such that the main beam is not pointing towards the threat radar is insufficient.
2. *Random scattering*. This is caused if the antenna characteristics are not uniform across the antenna. The solution is to maintain close production tolerances and hence uniformity across the array.
3. *Radar antenna edge diffraction*. Mismatches of impedances at the perimeter of the antenna can cause reflections called edge diffraction. In effect the outer perimeter of the antenna acts as a loop and reflections tend to be abeam of the antenna rather than fore and aft.
4. *Coupling of structural modes into the antenna*. In this case the antenna can effectively 'mirror' the incident energy from the threat radar, thereby compromising low RCS. In an ESA, active or passive, this can be readily overcome by rotating the array such that it points slightly down and the reflected energy is directed away from the threat radar. This technique, although effective, has the disadvantage of reducing antenna effective area by the cosine of the depression angle.

The subtle and different effects of the antenna reflection effects of antenna mode, random scattering and edge diffraction are shown in Figure 4.45. Antenna mode reflection is minimised by ensuring correct RF matching of the feed elements to avoid reflections.

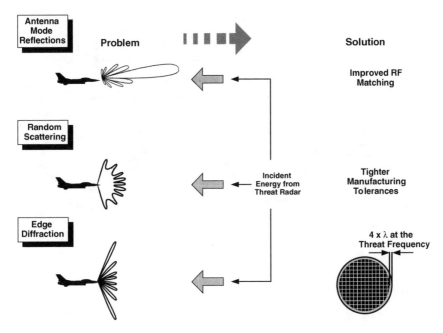

Figure 4.45 Factors affecting the antenna RCS.

Random scattering is minimised by maintaining close production manufacturing tolerances to maintain uniformity across the array. Edge diffraction effects may be reduced by placing RAM material around the perimeter of the array to minimise the loop reflection effect. This should be a width identically equal to $4 \times \lambda$ at the threat radar operating frequency, and so a significant penalty may apply as the antenna size and therefore gain is reduced accordingly.

4.6.3 Comparative RCS Values

A crude comparison of the RCS value for a range of objects is shown in Figure 4.46. The comparison is shown with two scales: the upper scale depicts the RCS in square metres while the lower scale depicts dB/m^2, that is, dB power reduction of a return related to a reference square meter of target. The 1 m^2 target is the reference point for a small airborne target such as a cruise missile.

The stealth aircraft B-2, F-22, F-117 and F-35 appear on the left-hand side of the diagram in the -30 to -40 dB region. Conventional aircraft range from the F-18E/F at 0 dB to bombers and transport aircraft at $+30$ dB. Ships are a huge target at 10^4 m^2 or more. Normal everyday objects range from insects at -30 dB, through birds to humans at -10 dB. On this diagram it could be deduced that an F-22 has an RCS of 1×10^{-7} (or one ten-millionth) of that of a B-52 bomber.

Although these comparisons serve some purpose in indicating some of the relative magnitudes involved, they do not present the whole picture. In fact, no aircraft may be made invisible, and stealth aircraft may be detectable at lower than conventional radar frequencies. They may, however, be very difficult to track or engage even if detected.

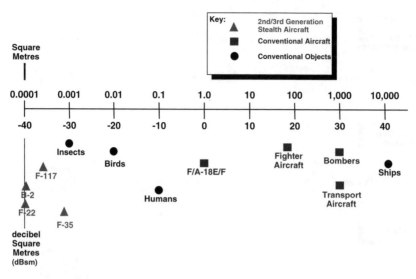

Figure 4.46 Comparative RCS values.

4.6.4 Low Probability of Intercept Operation

While the reduction of the antenna RCS is an important consideration, thought also has to be given to the energy radiated when the radar is operating. An important feature is low probability of intercept (LPI) which minimises the possibility that the radar is detectable by a foe using a radar warning receiver (RWR). The easiest way of avoiding detection is to use some type of emission control (EMCON) restricting the radar emissions. However, if this strategy were to be carried to extremes, then there would be little point in fitting the radar. Since the radar needs to be used to detect, identify and engage targets, the question of LPI has to be seriously addressed.

There are a number of ways by which probability of intercept may be minimised:

1. Reduce peak power by increasing integration time. The RWR is normally a broad-beam device, usually with four sets of antennas giving 360° panoramic cover in azimuth (a typical RWR antenna with 90° angle cone coverage has a gain of only ∼7 dB). The RWR also has to receive a number of successive pulses to determine parameters such as frequency, pulse width, PRF, etc., and be able to quantify the characteristics of the radar and identify the likely platform. RWR operators have the advantage that the radar energy only has to travel one way to them. However, they are using an antenna with low gain and have to have a relatively long sample time to make intelligent deductions about the pulses of energy they are receiving. The emitting radar is not so constrained.
2. Balance bandwidth against lower peak power. To be able to discriminate between signals in frequency, the RWR has to have a relatively narrow bandwidth. For the emitting radar, bandwidth is a function of radar design, and therefore, by using a broad-bandwidth receiver and spread spectrum modulation techniques, detection may be avoided.
3. Maximise antenna gain. Although this aids the RWR during transmission, it helps the radar during reception. Moreover, the antenna sidelobes should be reduced as far as

possible – for a LPI radar design the sidelobes should be reduced to 55 dB below the main beam.
4. Minimise system losses and noise. For the reception of the target echo, range will be maximised when front-end losses are kept to a minimum and the receiver noise figure is reduced. The advantage of using an AESA over a passive ESA design was demonstrated in Figure 4.25, where the use of the individual T/R modules provided a 3 dB noise advantage.
5. Use different forms of modulation. The modulation types described in this book are fairly well understood and widely used. However, there a plethora of more advanced modulation techniques that may be used either to extract low-level target signals from noise or use transmission patterns that a RWR may not be able to detect, let alone characterise.

It has to be assumed that the modern AESA radars being deployed on a range of US fighters use these and many other techniques to maximise range while avoiding detection. It has been stated publicly that the F-22 AN/APG-77 radar can operate with varying modulation and frequency on a pulse-by-pulse basis. It can also operate using multiple beams in multiple radar modes using the concept shown in Chapter 3 (Figure 3.27). It is claimed that the AN/APG-77 is able to detect, identify and engage a target without the foe ever being aware that an enemy is present. For further information on stealth RF design, see Lynch (2004).

References

Lynch Jr, D. (2004) *Introduction to RF Stealth*, SciTech Publishing Inc.
Report of the Defense Science Board Task Force (April 2001) Future DoD airborne high-frequency radar needs/resources.
Skolnik, M.E. (2001) *Introduction to Radar*, 3rd edn, SciTech Publishing Inc.
Stimson, G.W. (1998) *Introduction to Radar Systems*, 2nd edn, SciTech Publishing Inc.

5   Electrooptics

5.1   Introduction

The use of electrooptic sensors in military avionics systems has steadily evolved over the past three decades. Infrared (IR) missiles were originally produced during the 1950s with missiles such as Sidewinder in the United States and Firestreak and Red Top in the United Kingdom. Television (TV) guidance was used on guided missiles such as TV-Martel developed jointly by the United Kingdom and France during the 1960s and AGM-62 Walleye in the United States. Lasers were used for target illumination during the latter stages of the Vietnam War. Forward looking IR (FLIR) imaging systems were developed and deployed during the 1970s, and third-generation systems are now taking the field. Infrared track and scan (IRTS) systems followed. Now, integrated systems are in operation that combine a number of sensor types to offer a complete suite of capabilities.

This chapter describes the following electrooptic technologies that are to be found on a range of modern military, law and order and drug enforcement platforms:

- Television (TV) – day, low-light and colour (section 5.2);
- Night-vision goggles (NVG) (section 5.3);
- IR imaging including forward looking infrared (FLIR) (section 5.4);
- IR tracking systems including IR-guided missiles and infrared track and scan (IRTS) (section 5.5);
- Lasers – target illumination, range-finding and smart bomb guidance (section 5.6);
- Integrated systems (section 5.7) (usually carried in external pods or multiaxis swivelling turrets; on stealth aircraft, carried internally to preserve aircraft low-observability characteristics).

The characteristics and range of all the electromagnetic sensors used in modern military avionics systems was described in Chapter 2 (Figure 2.2), spanning 10 decades of the electromagnetic spectrum. Electrooptic sensors and systems cover the top two decades from 10^4 GHz to 10^6 GHz in frequency (Figure 5.1). In fact the categorisation of devices in Hz at

Military Avionics Systems   I. Moir and A. Seabridge
© 2006 John Wiley & Sons, Ltd

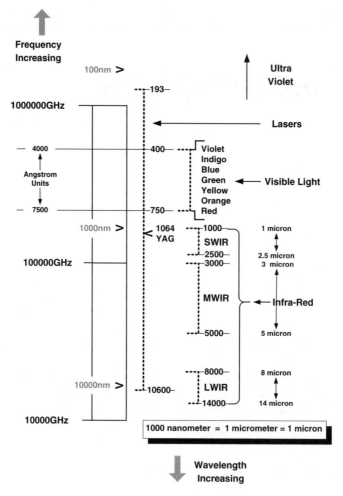

Figure 5.1 Electromagnetic Spectrum for electrooptic systems.

this point in the spectrum becomes meaningless because of the huge numbers involved, so that wavelength is more usually used. Therefore, wavelength is referred to in microns, where $1 \, \mu m = 1 \times 10^{-6}$ m or one-millionth of a metre. In the visible light portion of the spectrum that the human eye uses, angstrom units are sometimes used, especially in the scientific community, where $1 \, \mathring{A} = 10^{-2}$ m or $10^4 \, \mu m$.

The three specific bands of interest in Figure 5.1 are:

1. The visible light spectrum from 750 to 400 nanometres covering red/orange/yellow/green/blue/indigo/violet.
2. The IR spectrum from 1 to 14 µm which itself is subdivided into three regions:

   - Shortwave IR: 1–2.5 µm;
   - Medium-wave IR: 3–5 µm;
   - Long-wave IR: 8–14 µm.

3. The spectrum within which airborne lasers operate from 193 nanometres near the ultraviolet part of the spectrum to 10 600 nanometres approaching the far IR spectrum. Of particular interest is the frequency at which yttrium aluminium garnet (YAG) operates, namely 1064 nanometres, this being one of the most common lasers used.

As the area of the spectrum in which electrooptic sensors operate is close to and includes visible light, so they experience many of the same shortcomings to a greater or lesser extent: obscuration due to water vapour or other gases, scattering due to haze and smoke, etc. The near-IR part of the spectrum such as the shortwave IR region suffers most, although sensors operating in the long-wave IR band are less affected.

Therefore, a major problem confronting the use of sensors in the IR region in particular is the severe attenuation that occurs in certain parts of the spectrum, allowing only certain windows to be used. This is similar to the radar attenuation described in Figure 3.4 of Chapter 3. The IR transmission characteristics in the atmosphere are presented in Figure 5.2. This shows the percentage transmission over the IR band from 1 to 14 μm at sea level. As shown in the figure, at sea level there are a number of areas where the attenuation is significant, particularly in a region between 6 and about 7.6 μm where there is no transmission at all – mainly owing to water vapour and CO_2. At altitude the situation is much improved, although there are one or two unfavourable attenuation 'notches', again owing to H_2O or CO_2. Therefore, atmospheric attenuation is most likely to affect systems being used at low level, while air-to-air missiles being used at medium or high altitude will be less affected.

In fact, IR systems tend to use the IR windows or bands already described to avoid the worst of the problem using the SWIR (1–2.5 μm), MWIR (3–5 μm) and LWIR (8–14 μm) regions, and all the sensors utilised in avionics systems operate in these bands.

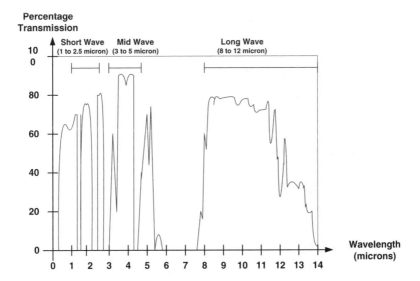

Figure 5.2 Transmission characteristics of infrared (sea level).

5.2   Television

The visible light sensors used in electrooptics are available to assist and augment the platform operator's 'Mk 1 eyeball'. The main categories used are:

- Direct vision optics, effectively a direct optics system offering image magnification in the same wave as a conventional telescope or binoculars;
- Television (TV) imaging in monochrome and, more recently, colour;
- Low-light TV (LLTV) for viewing in low-light conditions.

TV camera sensors used in military applications existed for many years prior to the 1980s when the sensing technology was imaging tubes. Charge coupled devices (CCDs) have now largely replaced the imaging tube, and these are described below.

The CCD imaging device comprises a number of resolution cells or pixels that are sensitive to incident light. Each pixel comprises elements of polysilicon placed upon a layer of silicon dioxide. Below the silicon dioxide is a silicon substrate layer. As the incident light arrives at the polysilicon, a 'well' of charge is built up according to the level of the incident light. The pixels are arranged in X rows and Y columns to provide the two-dimensional CCD array as shown in Figure 5.3. In each pixel the charge according to the amount of incident light is captured by a potential barrier that surrounds the well and maintains the charge in place. The pixels are separated by an interelement gap and respond to light below a wavelength of 1100 nm (visible light lies between 400 and 750 nm). Therefore, visible spectrum energy incident upon the array will provide a pattern of charge within each pixel of the CCD array that corresponds to the image in view. The CCD array is an optical plane that

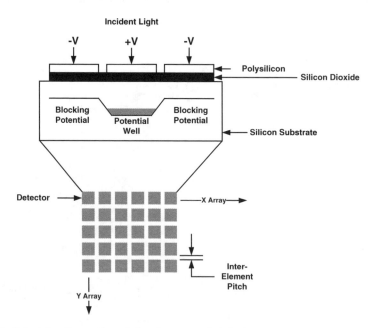

Figure 5.3   Principle of operation of the charge coupled device.

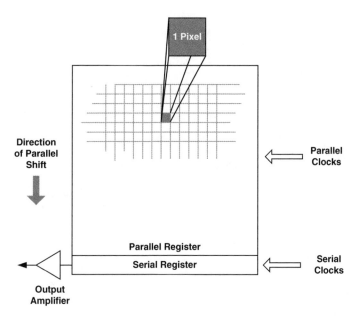

Figure 5.4 Full frame CCD.

preserves the image in the same way as a frame of film, and developments of this technology are now to be found in commercial digital cameras – replacing wet-film technology.

Once the array of retained charge has been established, the image needs to be scanned. There are a number of ways of accomplishing this, two of which are described below:

- Full frame devices;
- Frame transfer devices.

A full frame CCD device is shown in Figure 5.4. The charge associated with a row of pixels is shifted sequentially to the serial register, whereupon the row is read out as a stream of data. This is repeated on a row-by-row basis until the complete array of pixels has been read off the imaging device or chip. As the parallel register is used for both imaging and read-out, a mechanical shutter or illuminated strobe must be used to preserve the image. The simplicity of this method means that operation is simple and provides images of the highest resolution and density.

The frame transfer architecture is shown in Figure 5.5 and is similar to that of the full frame transfer except that a separate identical parallel register, called a storage array, is provided that is not light sensitive. The captured image is quickly read into the storage array from whence the image may be read out as before while the next image is being formed. The advantage of this approach is that a shutterless or strobeless imaging process is possible, resulting in higher frame rates. Performance is sacrificed as imaging is still occurring while the image is being read to the storage array, resulting in smearing of the image. Twice the area of silicon is required to fabricate an imaging device of comparable imaging coverage, so the frame transfer array provides lower performance at a higher cost.

The precise optical configuration of a CCD depends entirely upon the intended operational use. CCD devices may be used in a tactical platform such as a battlefield helicopter or as a

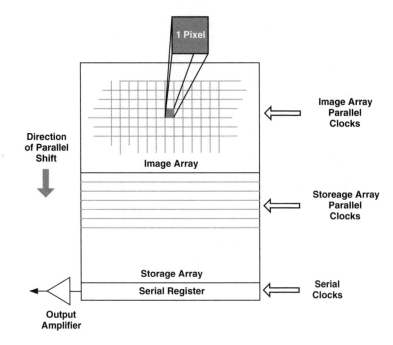

Figure 5.5 Frame transfer CCD.

high-resolution imaging system for a theatre reconnaissance aircraft or UAV. The imaging requirements and field of view (FOV) will be different in each case. Looking at the capabilities of two CCD imaging systems illustrates this fact. The examples chosen are the systems on board the AH-64 C/D Longbow Apache and RQ-4 Global Hawk:

1. *AH-64 C/D Apache*. The target acquisition designation sight (TADS) or AN/ASQ-170 is part of the overall turreted EO sensor system on the Apache battlefield helicopter. The TV element is part of the lower turret assembly which is described more fully in Chapter 9. The display is presented to the pilot/gunner by means of a TV raster display, originally on a CRT, but more recently this has been upgraded to a colour active matrix liquid crystal display (AMLCD). Military display technology is described in Chapter 11. The FOV options available for the TV are:

Underscan	0.45°
Narrow	0.9°
Wide	4.0°

The TADS turret is capable of traversing ± 120° in azimuth and + 30 to –60° in elevation with respect to the aircraft axes. Aircraft manoeuvring may reduce the turret FOV in certain circumstances.

2. *RQ-4 Global Hawk*. The CCD imaging system is a theatre reconnaissance high-resolution imaging system designed to provide high-quality images for intelligence purposes. The field of view is specified in milliradians (mrad), where a radian is equivalent to 57.3°,

therefore 1 mrad is equivalent to 0.0573° (1 mrad is the angle subtended by an object of 1 m length at a range of 1000 m). The specified FOV for the CCD imaging device, part of the integrated sensor system (ISS) is 5.1 × 5.2 mrad (0.3° × 0.3°). The platform is capable of providing the following coverage in a 24 h period:

Mode	Coverage/quality
Wide area survey	138 400 km²/24 h to NIIRS 6 quality
Spotlight	19 002 km × 2 km spot images/24 h at NIIRS 6.5 quality

The national imagery interpretability rating scale (NIIRS) is an imaging classification for radar, IR and visual imaging systems. The scale is from 0 to 9, where 0 represents unusable and 9 represents the highest quality. The scale equates to qualitative criteria to which the image interpreter can easily relate. For example: NIIRS 6 quality allows the spare tyre on a medium-sized truck to be identified; NIIRS 7 quality allows individual railway sleepers (rail ties) to be identified.

Low-light TV is accomplished by the use of image intensifier tubes or devices that amplify the image incident from the scene in view. Today, the use of night-vision goggles (NVGs) provides operators with a look-up, look-out capability that is far more flexible than the use of a dedicated low-light TV (LLTV) sensor, and this technology is described in the next section.

5.3   Night-vision Goggles

Night-vision devices gather ambient light and intensify the image by using a photocathode device to produce electrons from the photons in the incident light. The electrons are multiplied and then used to bombard a phosphor screen that changes the electrons back to visible light. The sensors are normally configured as a monocular or binocular 'goggle' which is attached to the operator's helmet, and through which he views the external low-light scene. However, as will be described in Chapter 11, care has to be exercised to ensure that the NVGs are compatible with the cockpit displays and night lighting.

Figure 5.6 shows the principle of operation of the image intensifier. The unenhanced image being viewed passes through a conventional objective lens to focus the incoming photons that represent the image upon the photocathode. The gallium arsenide (GaAs) photocathode produces a stream of electrons that are accelerated with the assistance of an applied voltage and fired towards the microchannel plate. Electrons pass through the microchannel plate – a small glass disc that has many microscopic holes (or channels) – and the effect is to multiply the electrons by a process called cascaded secondary emission. This multiplies the electrons by a factor of thousands while registering each of the channels accurately with the original image.

The electrons then impact upon a phosphor screen that reconstitutes the electrons into photons, thereby reproducing the original image but much brighter than before. The image created on the phosphor screen passes through an ocular lens that allows the operator to focus and view the image. As the image is recreated using a green phosphor, the intensified image possesses a characteristic green-hued outlook of the object being viewed.

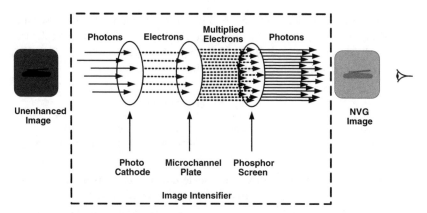

Figure 5.6 Principle of image intensification.

Night-vision devices (NVDs) were originally used over 40 years ago and employed an IR source to illuminate the target which was then viewed using an image intensifier. Apart from the low reliability of the early imaging systems, the use of an IR illumination source clearly provided the enemy with information about the user's whereabouts, and later systems obviated the need for IR illumination. This technology is sometimes termed generation 0. This led to the development of passive imaging devices which today are classified by four generations, each generation providing significantly improved performance over the preceding generation. A summary of these generations and the technology advancements associated with each is given below:

1. *Generation 1.* This generation used the ambient illumination provided by the moon and stars. The disadvantage was that the devices did not work well on cloudy or moonless nights. Essentially the same technology was used as for generation 0 so that poor reliability was also an issue.
2. *Generation 2.* Major improvements in image intensifier technology enhanced the performance of the image intensifier and improved the reliability. The biggest improvement was the ability to see images in very low ambient light conditions such as may be found on a moonless night. This technology introduced the microchannel plate which not only magnified the electrons but created an image with less distortion because of the 'channelling' effect.
3. *Generation 3.* This has minor technology additions by way of manufacturing the photocathode using gallium arsenide (GaAs), a material that is very efficient at converting photons into electrons and makes the image sharper and brighter. Another improvement was coating the microchannel plate with an ion barrier to improve reliability. Generation 3 technology is typical of that in use with the US military today.
4. *Generation 4.* Generation 4 is referred to as 'filmless or gateless' technology. The ion film was removed to allow more electrons to pass through the microchannel plate, thereby improving the tube response. Gating of the power supply allows the tube to respond more quickly to rapid changes in lighting conditions.

The use of NVGs is usually achieved by clamping them on to the operator's helmet as shown in Figure 5.7.

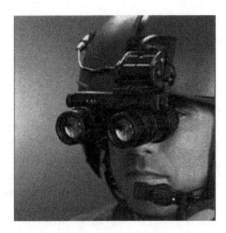

Figure 5.7 Helmet with mounted night-vision goggles. (Infrared 1)

Figure 5.8 shows a typical NVG image of a AV-10 in flight. Therefore, as well as providing images to survey and attack an enemy, NVGs can be of considerable use in allowing friendly forces to operate covertly, such as air-to-air refuelling at night without lights.

More recently, the use of helmet-mounted displays (HMDs) incorporating night-vision devices has been adopted in the military combat aircraft community, and these are discussed in detail in Chapter 11.

5.4 IR Imaging

IR imaging by day or by night has become one of the most important sensing technologies over the past 30 years. In that time, technology has advanced appreciably up to the point where the quality of IR imaging and visible light imaging is virtually indistinguishable.

As has already been described, the visible light spectrum from 400 to 750 nanometres, SWIR from 1000 to 2500 nanometres (1–2.5 μm), MWIR from 3000 to 5000 nanometres (3–5 μm) and LWIR from 8000 to 14 000 (8–14 μm) are very close to one another in the

Figure 5.8 Typical NVG image (AV-10 in flight). (Infrared 1)

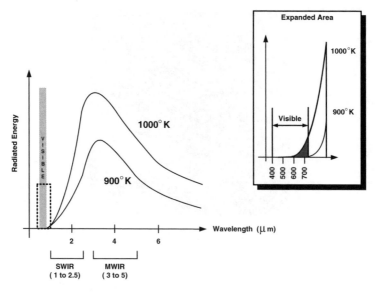

Figure 5.9 Black body radiation.

electromagnetic spectrum. It is therefore hardly suprising that many of the characteristics are very similar.

Sensing in IR wavelengths is basically about sensing the radiation of heat. All objects radiate heat primarily depending upon their temperature but also to some extent upon the material and the nature of the surface. In classical physics the emission of thermal energy is referenced to that from a black body which is the ideal thermal emitter. Figure 5.9 shows typical plots of radiated energy versus wavelength for a black body whose temperature is 900 K (627°C) and 1000 K (727°C) respectively. The higher the temperature, the higher are the levels of radiated energy. It can also be seen that the peak value of the radiated energy moves to the left (decreases in wavelength) the hotter the object becomes. This characteristic is called Wien's law and will be examined in more detail shortly.

The problem with this model is that not all objects are black and are perfect radiators. This can be accounted for by applying an emissivity coefficient that corrects for an imperfect radiator. Table 5.1 tabulates the emissivity coefficient for some common materials.

It can be seen that most building materials have fairly high emissivity coefficients. On the other hand, metals have a relatively low value when polished, but this increases appreciably when the surface oxidises. Aluminium is slightly different as it has a higher value when anodised which decreases if the surface is oxidised.

The other effect that Figure 5.9 illustrates is the fact that, if an object gets sufficiently hot, it emits visible light. It can be seen that an object at 1000 K is beginning to radiate energy in the red portion of the visible light spectrum. If the object were to be heated further, this area would encroach to the left. Eventually, if the object were heated to a sufficiently high temperature, then it would emit energy right across the visible light spectrum in which case it would appear white. This tallies with what everyone knows: if you heat an object it will first begin to appear red ('red hot'), and if the object is heated to a high enough temperature it will eventually appear white ('white hot').

Table 5.1 Emissivity coefficients for some common materials

Surface material	Emissivity coefficient
Black body (matt)	1.00
Brick	0.90
Concrete	0.85
Glass	0.92
Plaster	0.98
Paint	0.96
Water	0.95
Wood (oak)	0.90
Plastics (average)	0.91
Aluminium (oxidised)	0.11
Aluminium (anodised)	0.77
Copper (polished)	0.04
Copper (oxidised)	0.87
Stainless steel (polished)	0.15
Stainless steel (weathered)	0.85

The effect of Wien's law is presented in Figure 5.10 which shows power density versus wavelength for different body temperatures. As can be seen, the wavelength of the power density peak decreases as the temperature of an object increases. Summarising the data:

Temperature (K)	Wavelength of power density peak (nanometres)
6000	483
5000	580
4000	724
3000	966

The reverse side of this law is that the peak value falls off very rapidly as the temperature of the object decreases. The peak power density wavelength for an object at 373 K (or 100°C) is 7774 nanometres (7.77 μm); for an object at 290 K (~17°C or room temperature) it is 10 000 nanometres (10 μm).

Wien's law provides a formula for the peak wavelength as follows:

$$\lambda_{\text{peak}} = \frac{2900}{T}$$

where λ_{peak} is the peak wavelength (μm) and T is the temperature of the object (K).

This suggests that, theoretically, to obtain a maximum response from the IR sensor in the region where people and vehicles radiate, the wavelengths calculated above should be chosen, i.e. ~8000–10 000 nanometres (8–10 μm) at the lower end of the LWIR band. All things being equal, this should be the case. However, other factors such as availability and maturity of sensors to operate in the band and the effect of scattering due to haze or smoke may also have an impact. Also, as the wavelength increases, the size of the optics should also

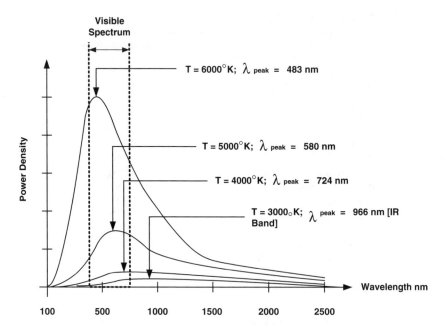

Figure 5.10 Wien's law.

increase (similarly to radar since both IR and radar emissions are electromagnetic waves), and therefore angular resolution will reduce with increasing wavelength unless the optics are scaled proportionately. As ever, space and volume are at a premium in a military avionics installation, and in some cases the increase in volume to accommodate larger optics is unlikely to be acceptable. Medium-wave operation is generally preferred both in high-temperature humid (tropical) and in arid (desert) conditions owing to the 3–5 μm window. The US Army generally has preference for LWIR operation which is better with haze and smoke (being at the radar end of the IR spectrum, this band has characteristics closer to that of radar). Some sophisticated systems provide dual-band operation – MWIR and LWIR – to enjoy the best of both worlds.

5.4.1 IR Imaging Device

A generic IR imaging device is shown in Figure 5.11. The target is shown emitting radiation on the left side of the diagram, with the radiation spectral energy determined by a combination of absolute temperature and emissivity. The radiated power has to compete with a number of extraneous sources: background radiation, sky radiance, the sun, reflections from clouds and other unwanted signal sources that generate clutter against which the target has to be detected. As well as the clutter, the energy radiated by the target is subject to atmospheric attenuation which can be particularly acute at low level and at certain frequency bands.

The incoming energy is focused by an appropriate set of optics, and in most cases some scanning arrangement is necessary to scan the target on to the detector array. Some arrays called 'staring arrays' do not need the optical scanning, and these will be described later in the chapter. Once the detector has formulated the IR image, the result is read out in a similar

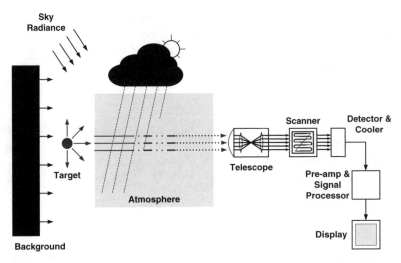

Figure 5.11 Generic IR imaging system.

way to the CCD sensor and the resulting data are amplified, processed and displayed. Most sensors need cooling in order to operate – usually to around 77 K – and special cooling systems are needed to perform the cooling task. A range of sensor materials can be used, all of which have their own particular advantages and bands of operation.

Three typical detector configurations are shown in Figure 5.12, and each type is used for different types of IR operation. These are:

1. *Linear array.* The linear array is used to form an image strip, and a scene image may be generated by successively adding the strips together. The 1024 × 8 array illustrated is one used by BAE SYSTEMS.

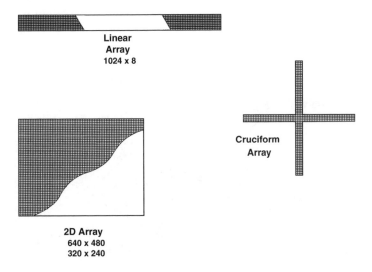

Figure 5.12 Typical IR detector configurations.

2. *Two-dimensional array.* The two-dimensional array forms an *X–Y* matrix that lends itself readily to generating a rectangular image in a similar way to the CCD array described earlier. The 640×480 and 320×240 two-dimensional arrays portrayed are typical of state-of-the-art third-generation systems in service today.
3. *Cruciform array.* This array is used to accomplish an IR tracking function and will be described later in the chapter.

The scanning configuration adopted augments the detector configuration used to provide an IR image that may be examined for strategic reconnaissance, intelligence gathering, battle damage assessment or for a platform operator to prosecute an engagement. Three basic IR scanning techniques for imaging will be described. These are the rotating scanner, planar array and focal plane array (FPA).

5.4.2 Rotating Scanner

One of the first techniques to be employed was the rotating optics method which was also known as linescan. In Figure 5.13 the platform is flying from left to right as the scanner rotates about an axis parallel with the aircraft heading. In this case the scanner is rotating in a clockwise direction looking forwards, and successive strips of ground are imaged as the imaging mirror sweeps from right to left. Therefore, as the platform flies forwards, the series of image strips may be recorded and an area image may be constructed of the ground that has been scanned.

This technique was one of the first to be used for area IR imaging and was evolved using very small arrays as the technology was not available to produce large linear arrays. The image suffers from distortion towards the horizontal limits of the scan as the sightline moves appreciably away from the vertical. Furthermore, for clear images, the relationship between aircraft velocity, *V*, and height above the terrain, *h*, or *V/h*, has to be closely controlled or successive imaging strips will not be contiguous or correctly focused. Another disadvantage of this method when it was first introduced into service was that there were no high-density digital storage devices available. The images were therefore stored on film which had to be developed after the sortie before analysis could begin. Early IR linescan systems such as

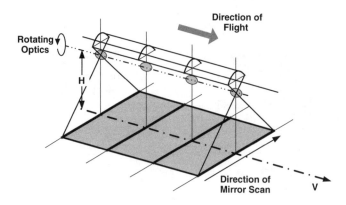

Figure 5.13 Rotating scanner (IR linescan).

those carried by the UK F-4K Phantom carried the system in a large pod beneath the aircraft centre-line. This technique has been likened to a whisk broom where the brush strokes are sequential right to left movements.

The Royal Air Force Tornado GR4a and Jaguar GR3 reconnaissance variants use an embedded IR linescan VIGIL system produced by Thales/Vinten. This system has the following attributes:

Detector	Single cadmium mercury telluride (CMT) (CdHgTe) detector operating from 8 to 14 μm
Scan rate	600 lines/s
Angular resolution	<0.67 mrad
Pixels	8192/line or ~4.9 Mpixels/s
Weight	23 lbs

5.4.3 Planar Image

The planar image technique is shown in Figure 5.14. By comparison with the rotating scanner system just described, this is called a 'push broom' system since it is analogous to a broom being pushed forwards. This system uses line detector arrays as outlined above. As the aircraft moves forwards, the optics allow the strip detector to image the area of interest as a series of strips that can then be formed into a continuous area image. This type of scanning arrangement lends itself to high-altitude imaging systems on platforms such as the Global Hawk. The main operational capabilities of the Global Hawk EO system are outlined below (NIIRS is the national image interpretability rating scale):

Detector	Indium antimonide (InSb) detector operating from 3.5 to 7 μm (MWIR)
Field of view	Wide area scan: 5.5 × 7.3 μrad
Performance	Wide area scan: NIIRS 5
	Spotlight: NIIRS 5.5

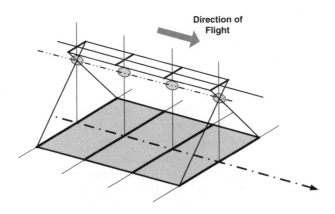

Figure 5.14 Planar image.

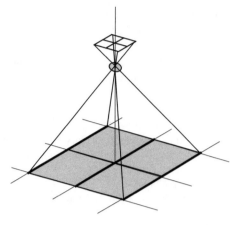

Figure 5.15   Focal plane array.

The performance of the IR imaging is almost as good as the CCD visual imaging system for the Global Hawk described earlier in the chapter, where the corresponding figures were NIIRS 6 and 6.5.

5.4.4   Focal Plane Array or 'Staring Array'

The focal plane array (FPA), often referred to as a 'staring array', is portrayed in Figure 5.15. The FPA provides an image on to a focal plane that coincides with the sensing array, most usually a two-dimensional array whose dimensions scale easily to a standard rectangular display format: NTSC, PAL and, more recently, VGA and XVGA and above, greatly simplifying the optics. Although the figure depicts the focal plane array with a vertical axis, in tactical systems the array face is usually facing directly towards the target. In most cases the forward looking IR (FLIR) sensor is looking forwards, the term being relative as it is usually mounted upon a gimballed assembly that has extreme angular agility and slew rates in order to be able to track targets while the platform is manoeuvring. As will be seen later in the chapter, several EO sensor systems are commonly physically integrated into the co-boresighted sensor set to aid sensor fusion and allow target data to be readily handed off from one sensor type to another.

In an array the entire surface is not given over to IR energy sensing. There is a certain overhead involved with interconnecting the array which prevents this from being the case. In a practical array the overhead is represented by a term called the fill factor which describes the useful portion of the array as a percentage. On modern state-of-the-art arrays, the fill factor is usually around 90 %.

The array is effectively read in a sequence of frames in the same way as any other real-time imaging device. Therefore, the time between successive read-outs of the array image is the time available for the array to 'capture' the image, and this is referred to as the integration time and permits successive images of the target to be generated.

The key element in the performance of any IR imaging device lies in the performance of detectors and the read-out of the imaged data in a timely fashion. There are many sensor types and technology issues to be considered, and some of the detector technology issues are outlined briefly below.

Table 5.2 Overview of IR FPA detector technologies

Technology	Wavelength (μm)	Typical array (FPA)	Cooling (K)	Application
Lead silicide (PbSi)	1–5	Not generally used for military applications		
Indium antimonide (InSb)	3–5	640 × 512 640 × 480 512 × 512 320 × 240	78	Tactical UAV; ATFLIR AN/ASQ-228
Cadmium mercury telluride (CMT) (CaHgTe)	8–12	640 × 480	77	Apache M-TADS Litening II pod
Lead tin telluride (LTT) (PbSnTe)	8–12		77	
Quantum well infrared photodetector (QWIP)(GaAs; AlGaAs)	9–10	320 × 240	70–73	Experimental for aerospace

Al, aluminium; As, arsenide; Ga, gallium; Hg, mercury; In, indium; Pb, lead; Sb, antinomy; Si, silicon; Sn, tin; Te, tellurium.

5.4.5 IR Detector Technology

The technology of the IR imaging detectors is rapidly moving in terms of materials and array size. Table 5.2 gives a brief overview of some of the key technologies for the FPA implementation in aerospace applications. Many of the materials developed for medical and industrial use may not be suitable for aerospace applications. This is a rapidly evolving area of technology and the details of new technologies are not always available in the public domain.

For reasons indicated earlier, most applications today are based in the MWIR and LWIR bands, although the band chosen will be dependent upon detailed specification requirements. There is a desire to move towards dual-band operation where the optimum wavelength may be chosen for the imaging task in hand. There is also an aspiration to introduce multispectral imaging technology to aerospace applications because of the increase in operational capability that would bring. At the moment, contemporary technology may find it difficult to discriminate targets hidden beneath camouflage nets or foliage. Multispectral sensing will provide battlefield commanders with sensors that would be able to overcome this deficiency. The typical desired capabilities of a modern sensor are summarised below:

Pixel pitch	~20–40 μm
Frame rate	50 Hz (PAL); 60 Hz (NTSC) with a desire to go to 100 Hz and above
Maximum integration Time	99% of frame time
Data rate	10 MHz upwards
Array size	640 × 480 (VGA resolution), heading towards 1000 × 1000 (1 Mpixels) or above in next generations: F-35 and space applications

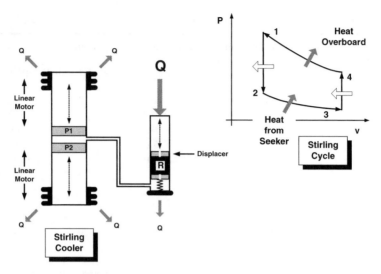

Figure 5.16 Stirling cycle cooler.

It can be seen that all the sensor detector types require cooling, and there are two ways of doing this. Originally, cooling was achieved using a Dewar flask together with a liquid nitrogen cryogenic coolant. More recently, miniature refrigerator devices have been developed that work on a Stirling cycle principle. The Stirling machine and the associated cycle is shown in Figure 5.16. The Stirling machine comprises a compressor cylinder with two moving pistons on the right; these pistons can be moved by means of linear electrical motors. This cylinder has finned heat exchangers to assist in dumping heat overboard. In the second cylinder on the right, the heat load is mounted on a 'cold finger', at the top of the cylinder. This cooling cylinder contains a regenerator device which is free to move up and down but which is normally biased to the top of the cylinder by means of a spring. The regenerator device has the ability temporarily to store heat, accepting heat from the cycle and donating it back to the cycle during different phases. The Stirling machine operates in four discrete changes in pressure/volume (*P/V*) during one cycle, and the cycle has the overall effect of extracting heat from the cold finger abutting the sensor and rejecting it from the machine by means of heat exchangers on the compressor and regenerative cylinders. The linear motors are powered by an aircraft or pod power supplies which draw relatively small amounts of power.

The principle of operation of the Stirling cycle is described below. At the start of the cycle the pistons P1 and P2 are at the top and bottom of the compression cylinder respectively and the regenerator is at the top of the cooling cylinder:

1. *Phase 1.* The linear motors compress the gas, and the heat so generated is dissipated in the heat exchangers. The black arrow on the subdiagram at top right shows that heat is rejected from the cooler.
2. *Phase 2.* The pistons remain in their compressed position so that the volume of the shared gas is constant. The gas above the regenerator expands while moving the regenerator down, compressing the spring and releasing heat into the regenerator (white arrow).
3. *Phase 3.* The pistons are returned to their original positions at the top and bottom of the compression cylinder by the linear motors, increasing the volume of the

shared gas. Heat is rejected from the heat load/seeker assembly into the cycle (black arrow).

4. *Phase 4.* The regenerator releases heat into the shared gas (white arrow) at constant volume and therefore pressure increases. The spring biases the regenerator to the top of the cooling cylinder.

The cycle is repeated continuously and a heat load is withdrawn from the seeker assembly, causing it to cool down rapidly. The characteristics of a typical Stirling cooler are:

- Input power $\sim$30–50 W;
- Heat load (seeker) $\sim$0.5–1.5 W;
- Seeker operating temperature $\sim$77 K;
- Cool down time 5–10 min;
- No seals, no lubricants, no maintenance, sealed for life;
- MTBF $\sim$5000–10 000 h.

IR detector packaging for second-generation arrays is now possible in a number of forms that are illustrated in Figure 5.17. These are direct hybrid, indirect hybrid, monolithic and Z technology (in all cases except the monolithic method, the electrical connection for the detector chip is made by means of indium 'bumps' which provide a soft metal interconnect for each pixel):

1. *Direct hybrid.* In this configuration the chip is connected to an array of preamplifiers and row and column multiplexers to facilitate the read-out.

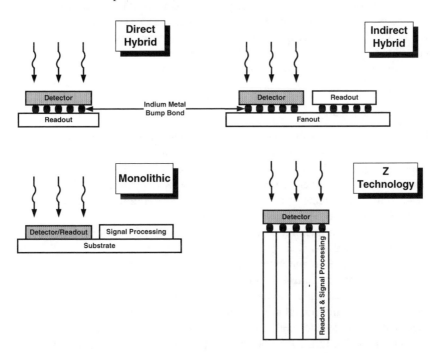

Figure 5.17 IR detector packaging schemes.

2. *Indirect hybrid.* This is similar to the direct method except that the detector and read-out electronics are interconnected by fan-out which connects the two chips electrically by means of a metal bus on a fan-out substrate. This has advantages in testing the detector array and allows the size of the preamplifiers to be increased to improve dynamic performance.

3. *Monolithic.* In this method both detector and signal processor are combined in the same chip which in turn is mounted on the same substrate as the signal processing. In fact, the two do not have to be packaged on to the same substrate but can be segregated in terms of substrate and operating temperature, thereby possibly reducing the cooling load by cooling the detector alone.

4. *Z technology.* This provides additional signal processing space on a pixel-by-pixel basis in the Z direction (as opposed to the $x–y$ array direction). This is used when the detected output of every pixel is to be individually processed, as is the case in multispectral and hyperspectral applications (Chan *et al.*, 2002; Bannen and Milner, 2004; Carrano *et al.*, 2004).

5.5   IR Tracking

The use of IR seeker heads to track and engage targets has been in use in military systems for many years. The Raytheon AIM-9 Sidewinder missile was one of the first of many such systems to be deployed. It is still in service today with many air forces around the world, and the latest version AIM9-X is about to enter service with the US Armed Forces. Petrie and Buruksolik (2001) give an interesting perspective on the history of the Sidewinder. The introduction of simple man-launched surface-to-air missile (SAM) such as Stinger has been a feature of the use of IR technology. The threat of such weapons is still very much with the aviation community today when used by renegade or terrorist organisations to attack unarmed military or civil transport aircraft. IR search, track and scan (IRSTS) systems are used as a primary sensor system on many fourth-generation fighter aircraft.

5.5.1   IR Seeker Heads

To illustrate some of the capabilities and limitations of IR tracking devices, the use of IR seekers in an air-to-air missile context will first be examined.

Reticle tracking is achieved by rotating a small disc or reticle with clear and opaque segments in front of the seeker detector cell. In early IR tracking heads the detector would have been a simple arrangement; later versions used more complex detector arrays, and the very latest missiles use an FPA array with a better FOV.

A simple example of a tracking reticle is shown in Figure 5.18. Either the disc rotates or the IR image is rotated by means of a rotating mirror. Whichever method is used, the objective is to scan the IR image with relative rotary movement of the reticle and modulate the IR return. Figure 5.18 shows a simple reticle that is translucent and allows 50 % transmission on one half while alternately chopping the image on the other between clear and opaque sectors. This modulation technique yields a series of pulses of IR energy that is detected by the detector cell. By carefully choosing the characteristics of the reticle and therefore the resulting modulation, an error signal may be derived which allows the seeker head to track the target by suitable servo drive systems. The reticle scan rate of early

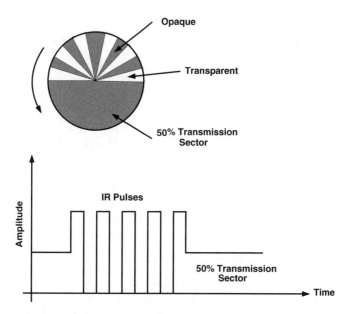

Figure 5.18   Simple reticle tracker.

seeker heads was ∼50–70 Hz, not dissimilar from the radar conscan tracking described in Chapter 3.

The choice of the reticle type determines the kind of modulation that is employed to track the target – most seeker heads use either amplitude modulation (AM) or frequency modulation (FM). A commonly used technique employs a wagon wheel stationary reticle with nutating optics scanning in a rosette pattern, rather like the petals of a flower. The type of reticle, type of modulation and frequency of rotation/nutation for a given application are usually not advertised, as to do so would reveal key characteristics of the head and its performance. Unlike radar tracking techniques such as conscan, where the characteristics of the radiated power reveal the angular scan rate, IR scanning is passive and therefore more difficult to counter by deception means.

Earlier, a cruciform detector was shown in Figure 5.12. The operation of a cruciform or cross-configured seeker is shown in Figure 5.19. This system uses a stationary element with nutating optics which scans the image in a circular fashion over the arms of the cruciform. If the target is located on the boresight of the seeker, as shown on the left, the time between pulses received from elements 4 and 1 will equal that between pulses received from elements 3 and 4. If the target drifts off boresight in a 2 o'clock direction as shown on the right, the pulses will be an unequal distance apart. Successful tracking is achieved by using pulse period measuring techniques with the appropriate servomechanisms to maintain the target on-boresight.

Early seeker heads possessed a limited capability, only able to engage the target from the rear aspect where the IR tracker had clear sight of the engine jet-pipe and exhaust plume. Part of the limited performance was that early missile detectors were uncooled lead sulphide (PbS) elements, so the sensitivity was very low. The instantaneous field of view was ∼4° with a seeker head FOV of ∼25°. Sightline tracking rates were also low (∼11 deg/s), so the engagement of manoeuvring targets was not possible.

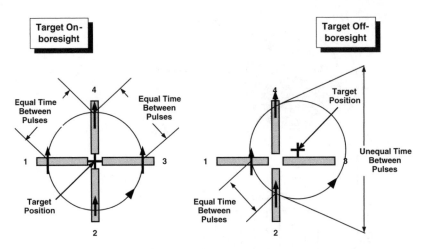

Figure 5.19 Tracking using a cruciform detector array.

With developments introduced in the 1960s, cooled arrays were introduced using a 'one-shot' liquid nitrogen bottle located in the missile launcher. The coolant bottle was renewed before each sortie and contained enough coolant to allow the missile head to operate for ~2.5 h. Modern systems are capable of full-aspect engagements; that is, they are sufficiently sensitive to acquire and track the target aircraft from any position.

In close encounter engagements the sightline spinrate of the target may be high as it crosses rapidly in front of the aircraft/seeker head. This can occur so quickly that the seeker head is unable to acquire the target. The solution to this problem is to use one of the aircraft sensors to track the prospective target and slave the missile to that sightline. The Royal Air Force used this technique to slave the Sidewinder missile sightline to the AWG 11/12 radar sightline in the F-4M Phantom in a mode called Sidewinder expanded acquisition mode (SEAM). The technique is still used today, except that the missile seeker sightline is slaved in a more sophisticated manner to the system or pilot cues. Most fighter aircraft entering service in the last 10 years are apt to have a means of slaving the seeker boresight to steering cues given by a helmet-mounted sight (HMS) projecting directly on to the pilot's visor/sight. See the discussion on this topic in Chapter 11.

5.5.2 Image Tracking

The high fidelity of IR imaging systems as already described opens up the possibility of image tracking and also image recognition, although the algorithms involved with the latter function can be quite complex. The resolution available with imaging systems is now close to or approaching that available with visible range sensors, and therefore specific objects may be easily tracked once identified and designated by the operator. As will be seen, this is an important feature in engaging a target, as sensor fusion using a combination of sensors, trading off relative strengths and weaknesses, is in many cases an important feature in the successful prosecution of a target.

TV and IR imaging provides good resolution in an angular sense but not in range. Radar and lasers offer good range resolution but poor angular resolution. Using the right combination of sensors provides the best of both.

Typical tracking algorithms include the following:

1. Centroid tracking, where the sensor tracks the centre of the target as it perceives it. This is particularly useful for small targets.
2. Correlation techniques that use pattern matching techniques. This is useful to engage medium to large targets but can be difficult if the target profile alters drastically with aspect, for example an aircraft.
3. Boundary location or edge tracking can be used where the target can be split into segments, the arrangement of the segments providing recognition features.

The use of a human operator is most useful in ensuring that the correct target is identified and tracked. Target recognition is also vital under most rules of engagement where it is essential to have the ability to fire without positive identification in order that no 'blue-on-blue' or friendly fire incidents occur. Again, correlation of imagery with other sources/ sensors can be of great assistance. However, in high-density dynamic target situations the human operator will soon reach saturation and automatic target tracking will be essential.

5.5.3 IR Search and Track Systems

IR search and track systems (IRSTS) have been used for air-to-air engagements for some time. The US Navy F-14 Tomcat has such a system, and the Soviet-designed aircraft MIG 29, SU27 and SU35 all used first-generation systems. The function of IRSTS is to perform a function analogous to the airborne radar TWS mode where a large volume of sky is searched and targets encountered within the large search volume are characterised and tracked. The major difference is that, whereas the radar TWS mode is active, IRSTS is purely passive.

The key requirements of an IRSTS are:

- Large search volume;
- Autonomous and designated tracking of distant targets;
- Highly accurate multiple-target tracking;
- Passive range estimation or kinematic ranging where sightline spin rates are high;
- Full integration with other on-board systems;
- FLIR imaging;
- High-definition TV imaging.

A state-of-the-art implementation of IRSTS is the passive infrared airborne tracking equipment (PIRATE) developed by the EUROFIRST consortium which will be fitted to the Eurofighter Typhoon. Figure 5.20 shows the PIRATE unit and the installation on Typhoon of the left side of the fuselage. The equipment uses dual-band sensing operating in the 3–5 and 8–11 μm bands. The MWIR sensor offers greater sensitivity against hot targets such as jet engine efflux, while the LWIR sensor is suited to lower temperatures associated with frontal engagements. The unit uses linear 760×10 arrays with scan motors driving optics such that large volumes of sky may be rapidly scanned. The field of regard (FOR) is stated to be almost hemispherical in coverage. The detection range is believed to be $\sim$40 nm.

Figure 5.20 PIRATE seeker Courtesy Trailers phonics and installation on Eurofighter Typhoon (Eurofighter GmbH).

The operational modes of PIRATE are:

1. Air-to-air:

   - Multiple-target tracking (MTT) over a hemispherical FOR – the ability to track in excess of 200 individual targets, with a tracking accuracy better than 0.25 mrad;
   - Single-target track (STT) mode for individual targets for missile cueing and launch;
   - Single-target track and identification (STTI) for target identification prior to launch, providing a high-resolution image and a back-up to identification friend or foe (IFF).

2. Air-to-ground:

   - Ability to cue ground targets from C^3 data;
   - Landing aid in poor weather;
   - Navigation aid in FLIR mode, allowing low-level penetration.

The sensor data may be displayed at 50 Hz rates on the head-down display (HDD), head-up display (HUD) or helmet-mounted display (HMD), as appropriate.

5.6 Lasers

Lasers – the term stands for Light Amplification by Stimulated Emission of Radiation – have been used in military systems for almost four decades. The US Air Force used laser-guided bombs (LGBs) during the later stages of the Vietnam War, and European avionics systems such as Jaguar and Tornado were adopting laser systems during the late 1960s and early 1970s. These systems are now commonly used as range finders, target designators and missile/bomb guidance. Laser systems may be fitted internally within the aircraft such as in aircraft like the Tornado GR4 and F-18. They may be housed in pods for external carriage on weapons/stores stations on fighter aircraft, or they may be housed in highly mobile swivelling turrets for helicopter and fixed-wing use.

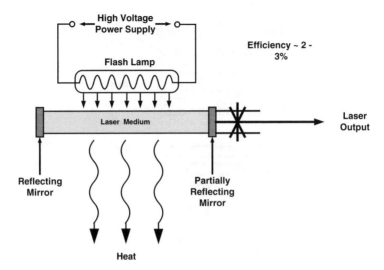

Figure 5.21   Principles of operation of a laser.

The major advantage of using lasers is the fact that they can provide range information that passive systems such as visible light and IR radiation cannot. Lasers are therefore particularly useful when used in conjunction with these other technologies to provide sensor fusion, blending and merging the advantages and disadvantages of the different capabilities. Some of these integrated systems are described later in the chapter.

5.6.1   Principles of Operation

The principles of operation of a laser depend upon exciting the energy levels of electrons within specifically 'doped' materials. Electrons within the material are stimulated to higher energy levels by an external source; when the electrons revert to a lower energy level, energy of specific wavelength is emitted, depending upon the material and the energy supplied.

A laser works on this principle but has other unique properties. Specifically, the energy that is emitted is coherent; i.e. the radiated energy is all in phase rather than being randomly related as may be the case during light emission. Figure 5.21 shows a diagrammatic representation of a laser device.

The laser medium may be liquid or solid; most of the lasers used in military systems are based upon glass-like compounds. At one end of the medium is a reflecting mirror, at the other a partly reflecting mirror. The laser medium is stimulated by an input of energy from a flashlamp or other source of energy, which raises the energy levels of the electrons.

Figure 5.22 shows the various stages that occur for a laser to 'strike':

1. Stage 1. This is the initial quiescent condition with the electrons all at a natural low-energy state.
2. Stage 2. The flash tube is illuminated, stimulating the electrons and exciting them to a higher energy state – this phenomenon is known as population inversion and is an unstable state for the electrons.

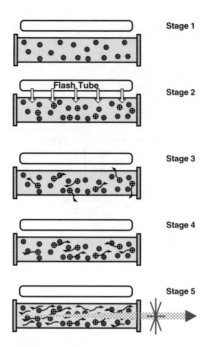

Figure 5.22   Stages leading to a laser strike.

3. Stage 3. The electrons remain in this state for a short time before decaying down to their original, lower and stable energy state. This decay occurs in two ways:

   • Spontaneous decay in which electrons fall down to the lower state while randomly emitting photons;
   • Stimulated decay in which photons released from spontaneously decaying electrons strike other electrons, causing them to revert to the decayed state – in these cases photons are emitted in the direction of the incident photon and with the same phase and wavelength.

4. Stage 4. Where the direction of these photons is parallel to the optical axis of the laser, the photons will bounce back to and fro between the totally and partially reflecting mirrors. This causes an avalanche effect in which the photons are amplified.
5. Stage 5. Eventually, sufficient energy is built up within the tube for the laser to strike, causing a high-energy burst of coherent light to exit the tube.

The wavelength of the emitted light is dependent upon the nature of the material being used in the laser medium since the energy released is specific to the energy levels within the atoms of that material. Also, since the amount of energy released is in the same discrete bundles, the emitted light is very stable in wavelength and frequency as well as being coherent (Figure 5.23).

Typical pulsed solid-state lasers used in aerospace use the following compounds:

1. *Ruby* is chromium-doped sapphire, while sapphire itself is a compound of aluminium and oxygen atoms. The formula for ruby is $Al_2O_3Cr^{+++}$, where Cr^{+++} indicates the

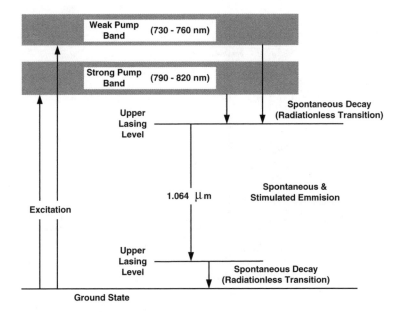

Figure 5.23 Energy levels of a YAG laser.

triply-ionised state of the chromium atom. The ruby laser mode radiates primarily at 694.3 nm. The characteristics of this material make it suitable only for pulsed operation.

2. *Neodymium:YAG* lasers, where YAG stands for yttrium aluminium garnet ($Y_3Al_5O_{12}$). This is a popular choice for airborne systems as it may operate in both pulsed and CW modes. The YAG laser radiates at 1060 nm.

3. *Neodymium:glass* lasers may sometimes be used, but, glass being a poor conductor of heat, they are not suitable for continuous operation. Nd:glass operates on a wavelength of 1060 nm, the same as for YAG.

Figure 5.23 shows the energy transition levels for a YAG laser. During the excitation state, electrons are raised to two energy bands. These are the weak pump band (730–760 nm) and the strong pump band (790–820 nm). Electrons in both bands spontaneously decay with a radiation-less transition to the upper lasing level. A combination of both spontaneous and stimulated emissions occurs as the electrons decay to the lower lasing level. During this phase the device radiates energy at a wavelength of 1060 nm or 1.06 µm. Thereafter all electrons spontaneously decay to the ground state.

The stimulation source for lasers is usually a xenon or krypton flash tube. Xenon lamps are the best option for ruby lasers, and krypton lamps are a better match for Nd:YAG and Nd:glass lasers but are more expensive and so are seldom used. The problem with using a flash lamp as the excitation source is that it is very inefficient. Lasers that are lamp pumped are very inefficient (~2–3% efficient). The rest of the energy can only be dissipated as heat, which causes real problems for the aerospace designer. The reason for this can be seen from Figure 5.24.

The reason for these very low efficiencies is that the flash lamp spectrum is wide compared with the narrow band in which the desired spectrum lies. Therefore, the lamp energy is

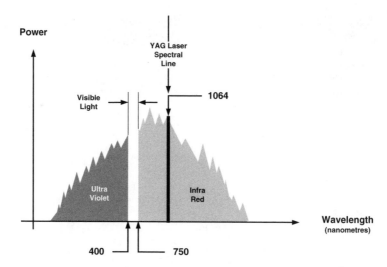

Figure 5.24 Flash lamp spectral characteristics.

poorly matched to the band of interest. Modifications can be carried out to shift the lamp spectrum more into the red region, but this still presents problems.

The solution to this problem is to use laser diodes rather than flash lamps to excite the laser medium. Laser diodes lend themselves to be more easily tuned to the frequency of interest. Figure 5.25 depicts a configuration in which laser diodes are used instead of a flash lamp, and this results in higher efficiencies. The higher efficiencies result in a lower unwanted heat load that a design has to dissipate, and therefore reliability may improve at the same time as performance. Diode-pumped lasers are now used, and this has allowed greater usable power output, permitting a designating aircraft to fly much higher while illuminating the target, thereby allowing greater stand-off ranges.

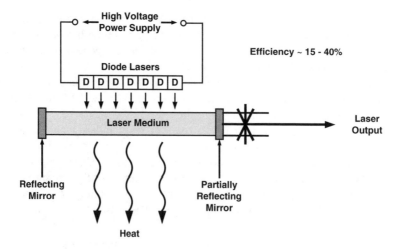

Figure 5.25 Diode-pumped laser.

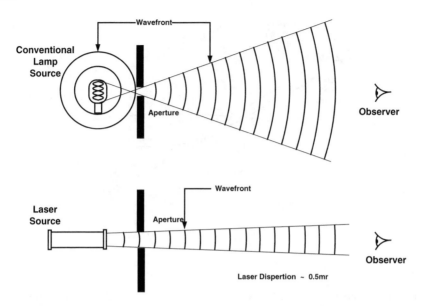

Figure 5.26 Properties of laser emissions.

As well as the properties of having a very stable, discrete wavelength and coherent transmission, laser emissions possess another important property, that is, the property of low dispersion. Figure 5.26 shows a comparison between a light-emitting source and a laser emission. The conventional lamp light emits light in all directions, rather like ripples in a pool. Even after passing through an aperture, the light diverges into a relatively wide wavefront. The laser source has a much narrower beam after passing through the aperture and therefore has low divergence. As a result the laser beam still has relatively high beam intensities far away from the emitter. Therefore, the laser is able to transmit coherent energy, at a fixed stable frequency and with much lower beam divergence than conventional high-intensity light sources.

5.6.2 Laser Sensor Applications

The beamwidth of a typical laser is ~0.25 mrad, and this means it is very useful for illuminating targets of interest that laser-tuned seeker heads can follow. For example, a laser designator illuminating a target at 10 000 ft will have a spot of 2.5 ft in diameter. The first deployment of laser-guided bombs (LGBs) used this technique. As the laser designator illuminates the target, energy is scattered in all directions in a process called specular reflection. A proportion of this energy will be reflected in the director of an observer who may wish to identify or engage the target.

The laser can operate as a range finder when operating in the pulsed mode. Pulsed operation has the capability of delivering high power densities at the target, for example, a laser delivering a 100 mJ pulse in a 20 ns pulse has a peak instantaneous power of 5 MW. In this sense the laser operation is analogous to radar pulsed modes of operation and mean power, duty cycle and peak power are equally as important as they are in radar design. Even allowing for atmospheric attenuation, a laser can deliver reasonable power densities at a target, albeit for very short periods. The narrow pulse width allows accurate range

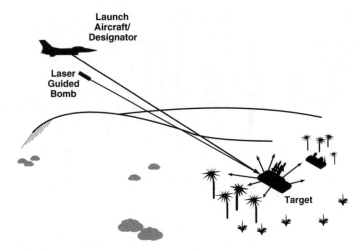

Figure 5.27   Laser guidance of an LGB.

measurements to be made. The 20 ns pulse mentioned above allows range resolutions to within ~10 ft.

Therefore, the laser offers a number of options to enhance the aircraft avionics system weapon-aiming performance in the air-to-ground mode:

- Laser designation in CW mode to guide a missile;
- Laser reception of a target position marked by a third party;
- Laser ranging to within a few feet.

Examples of these engagements are shown in Figures 5.27 and 5.28. Figure 5.27 illustrates a system where the aircraft has a self-designation capability. The aircraft can designate the

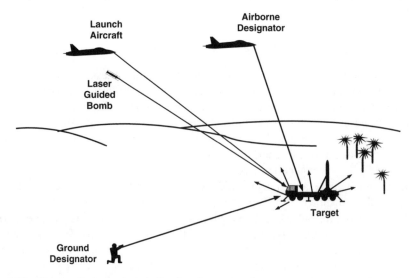

Figure 5.28   Third-party laser designation by air or ground means.

target and launch the LGB to engage the target. The designation/launch aircraft illuminates the target until the LGB destroys it. This type of engagement is used when the attacking force has air supremacy, the launch aircraft is free to fly without fear from counterattack by surface-to-air missiles (SAMs) and the target is one that is easy to identify – such as a bridge. Third-generation laser systems are capable of engaging from a height of 40 000–50 000 ft and perhaps 30 nm from the target.

In other situations, third-party designation may be easier and more effective. If the target is one that is difficult to detect and identify from the air, it may be preferable to use some ground forces such as the Special Forces to illuminate the target of interest. The laser designator signal structure allows for codes to be set by both ground or air designator and launch aircraft so the LGB is used against the correct target. In other cases the designator aircraft may possess a higher-quality avionics system than the launch aircraft and may act as a force multiplier serving a number of launch aircraft. This technique was used during Desert Storm when F-111 aircraft designated targets for F-16s and the RAF Buccaneers designated targets for the Tornado GR1s.

The LGBs are not always special bombs but in some cases free-fall bombs fitted with a laser guidance kit. This adds a laser seeker head and some guidance equipment to the bomb, allowing it to be guided to the target within specified limits, known as a 'footprint'. Therefore, provided the LGB is launched within a specified speed and altitude launch envelope with respect to the target, it should be able to hit the target if all the systems work correctly. The operation of an LGB guidance system is illustrated in Figure 5.29.

The reflected laser energy passes through optics that are arranged to produce a defocused spot on a detector plane. Detector elements sense the energy from the spot and feed four channels of amplifiers, one associated with each detector. In an arrangement very similar to the radar monopulse tracking described in Chapter 3, various sum and difference channels are formed using the outputs of amplifiers A to D. These are multiplexed such that elevation and azimuth error signals are produced and then fed to the guidance system which nulls the error.

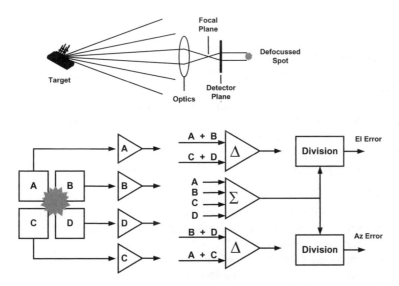

Figure 5.29 LGB Guidance.

Figure 5.30   F-18 internally fitted laser units.

The deployment of LGBs was graphically illustrated by media coverage of Desert Storm, although there was by no means a 100% success rate and a relatively small proportion of bombs dropped during that campaign were laser guided. By the time of the second Gulf War, a much higher proportion of laser-guided weaponry was used. However, as in all systems, there are drawbacks. Airborne tactical laser systems, by the very nature of their operating band very close to visible light, suffer degradation from the same sources: haze, smoke, precipitation and, in the desert, sandstorms. In other operating theatres, laser-guided systems may suffer more from weather limitations compared with operating in relatively clear conditions in the desert.

As was stated earlier, lasers may be fitted internally in some aircraft and in pods on others. Figure 5.30 shows the three units that comprise the laser system for the F-18.

5.6.3 US Air Force Airborne Laser (ABL)

The use of the lasers described so far relate to the tactical use of lasers to mark targets, determine target range and designate the target so that it may be engaged by a variety of air-launched weaponry. These lasers are not 'death rays' and, although they exhibit reasonably high energy levels, they are not sufficiently powerful to destroy the target by energy alone. They can, however, cause serious damage to the human eye, and laser safety issues are discussed later in the chapter.

The US Air Force airborne laser (ABL), designated as YAL-1, is a high-energy laser weapon system designed to destroy tactical theatre ballistic missiles. It is being developed by the Air Force in conjunction with a team comprising Boeing, Northrop Grumman and Lockheed Martin. The laser system is carried on board a converted Boeing 747F freighter and is curently undergoing test and evaluation. If successful, it is intended to procure several platforms with an initial operational capability of three aircraft by 2006 with a total capability of seven aircraft by 2010.

The ABL system actually carries a total of three laser systems:

1. A low-power, multiple set of laser target-illuminating beams comprising the target iluminating laser (TILL) to determine the range of the target and provide information on the atmosphere through which the primary beam will travel. The TILL provides the aiming data for the primary beam.
2. A beacon iluminating laser (BILL), producing power in kilowatts, reflects energy from the target to provide data about the rapidly changing nature of the atmosphere along the sightline to the target. This information is used to bias a set of deformation control mirrors

in the primary laser beam control system such that corrections are applied to the COIL laser beam as it engages the target.

3. The chemical oxygen iodine laser (COIL) is the primary beam generating the killer beam to destroy the target. This beam power is in the megawatt region and operates on a wavelength of 1315 µm. When a missile launch is detected, either by satellite or by AWACS, the target information is passed via data links to the ABL aircraft. The COIL beam is directed at the target by means of a large 1.5 m telescope mirror system at the nose of the aircraft which focuses the primary beam on the missile, destroying it shortly after launch.

5.6.4 Laser Safety

Pulsed solid-state lasers of the types commonly used in avionics applications have eye safety implications, as do many laser types. The peak powers involved are so high and the beams so narrow that direct viewing of the beam or reflections is an eye hazard even at great distances. Nd:YAG and Nd:glass lasers are particularly dangerous because their output wavelength (1064 nm) is transmitted through the eye and focused on the retina, yet it is not visible to the human eye. The wavelengths at which the human eye is most susceptible are between 400 and 1400 nm where the eye passes the energy and for the greater part the retina absorbs it. Above and below this band, the eye tissue absorbs rather than passes the energy. Lasers can be made to operate in a safe mode using a technique called Raman shift; in this way, Nd:YAG lasers can operate on a 'shifted' wavelength of 1540 nm, outside the hazardous band. In fact, lasers operating on this wavelength may be tolerated with power $\sim 10^5$ times that of the 1064 nm wavelength.

Military forces using lasers are bound by the same safety code as everyone else and therefore have to take precautions, especially when realistic training is required. The solution is that many military lasers are designated as being 'eye safe' or utilise dual-band operation. This allows personnel to train realistically in peacetime while using the main system in times of conflict.

To ensure the safe operation of lasers, four primary categories have been specified:

1. Class I. These lasers cannot emit laser radiation at known hazard levels.
2. Class IA. This is a special designation that applies to lasers 'not intended for viewing', such as supermarket scanners. The upper power limit is 4.0 mW.
3. Class II. These are low-power visible lasers that emit light above class I levels but at a radiant power level not greater than 1 mW. The idea is that human aversion to bright light will provide an instinctive reaction.
4. Class IIIA. These are intermediate-power lasers (CW – 1–5 mW) that are hazardous only for direct beam viewing. Laser pointers are in this category.
5. Class IIIB. These are for moderate-power lasers.
6. Class IV. These are high-power lasers (CW – 500 mW; pulsed – 10 J/cm^2) or diffuse reflection conditions that are hazardous under any conditions (either directly or diversely scattered) and are a potential fire and skin hazard. Significant controls are required of class IV facilities.

Many military lasers operating in their primary (rather than eye-safe) mode are class IV devices and must be handled accordingly.

5.7   Integrated Systems

As the electrooptic technologies used in avionics systems have matured, the benefits of integrating or fusing the different sensors have become apparent. Mounting the sensor on a rotating gimbal assembly allows the sensor(s) to track targets with high slew rates and with several degrees of freedom. Co-boresighting the sensors on this common gimbal assembly provides the ability to recognise targets and hand off target data from one sensor to another, thereby improving the overall capability of the integrated sensor set. Typical sensors that might be arranged in this fashion include:

- FLIR imaging using FPAs with several FOV options and, in some cases, dual-band sensors;
- CCD-TV with two or more FOV options using monochrome or colour displays and often a camera to record results;
- Laser target markers to illuminate targets, laser range finders to determine range and laser spot markers for target hand-off.

Such sensor clusters have to be carefully aligned or 'harmonised' to ensure maximum weapon-aiming accuracy. Also, given the high levels of resolution, the sensor cluster has to be stabilised to avoid 'jitter' and provide stable imagery. In some cases this stabilisation will be within $\sim$15–30 µrad.

These EO integrated systems may take any of the following forms:

- Installation in a pod to be mounted on a fighter aircraft weapon station;
- Installation in a turret for use on a helicopter or fixed-wing airborne surveillance vehicle;
- In stealthy aircraft: internal carriage to maintain low observability.

5.7.1   Electrooptic Sensor Fusion

The wavelength and frequency of the electromagnetic radiation comprising EO systems are contained within a relatively narrow portion of the spectrum:

- Visible light 0.400 µm (V) to 0.750 µm (R);
- IR bands 1.0 µm (lower SWIR) to 14.0 µm (upper LWIR);
- Airborne laser 1.064 µm and 1.54 µm (eye safe).

Therefore the total sensor set covers the relatively narrow band from 0.4 to 14.0 µm or a dynamic range of around 35:1.

Inspite of this relatively low dynamic range or coverage compared with radar and CNI, the properties of transmission of some of the sensing technologies are, however, quite different for the different bands/wavelengths being used. For example, a laser that would be extremely hazardous to the human eye when operated at 1.064 µm is eye safe when operated at 1.54 µm. Perhaps even more important, from the point of view of acquiring and engaging the target, the different fields of view (FOV) are quite different between laser and visible/IR sensors. The laser beam has very low divergence, typically of the order of 0.25 mrad

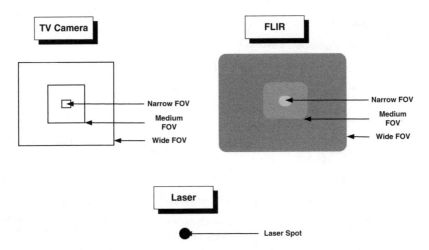

Figure 5.31 Typical EO sensor fields of view – Harmonisation.

$(2.5 \times 10^{-4}$ rad or $0.014°)$, whereas a navigation FLIR may have an FOV around $20° \times 15°$ ($\sim$1400 $\times$ 1050 times more). For target engagement activities, narrower FOV modes such as $4° \times 4°$ (MFOV) or $1° \times 1°$ (NFOV) may be used. The alignment or co-boresighting of the EO sensors must be carefully controlled (see Figure 5.31 which illustrates the principle of harmonisation but obviously does not show the respective fields of view to scale).

There are a number of systems issues that are very important and that must be taken into account if successful EO sensor fusion and weapons launch are to be accomplished. These factors include:

1. A relatively wide FOV for navigation and target acquisition ($\sim$20° $\times$ 20°) – in the case of the Apache TADS/PNVS, 40° $\times$ 30°.
2. A relatively narrow FOV for target identification and lock on ($\sim$4° $\times$ 4°) MFOV or (1° $\times$ 1°) NFOV.
3. A high target line of sightline slew rates, especially at short range, possibly >90 deg/s.
4. Relatively small angles subtended by the target and the need to stabilise the sensor package boresight within very tight limits (especially for long-range targets). A small bridge may represent a subtended angle of $\sim$0.024° at 40 nm, and a tank about the same angle at 10 nm. The problem may be likened to using binoculars when holding them in an unsteady manner. The enlarged image may be visible, but jitter renders the magnified image virtually useless unless the glasses can be steadied. Typical head stabilisation accuracies on third-generation sensor packages are of the order 15–30 μrad.
5. For a variety of reasons it is necessary to provide accurate inertial as well as sightline stabilisation for GPS/inertially guided weapons such as JDAM – often referred to as the J-series weapons. Accordingly, mainly advanced EO packages have a dedicated strapdown inertial navigation unit (INU) directly fitted on to the head assembly to improve pointing and positional accuracy and reduce data latency.

These technical issues are illustrated in Figure 5.32. These demanding technical requirements have to be met within a pod mounted under wing or under fuselage in a hostile

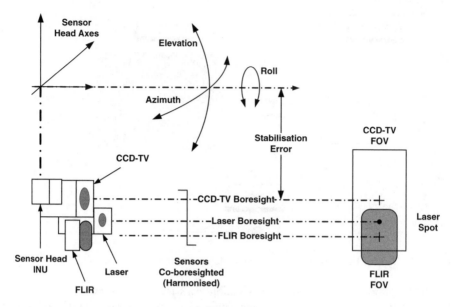

Figure 5.32   Boresighting, stabilisation and EO package sightline axes.

environment while the launch platform could be flying a supersonic speed. As will be seen, the space and weight available to satisfy these requirements are not unlimited, and modern EO sensor attack or surveillance packages represent very sophisticated solutions to very difficult engineering problems.

Figure 5.33 shows a typical integrated EO sensor for carriage within a 41 cm/16 in diameter pod, in this case the Northrop Grumman Litening II AT pod. This sensor pod contains the following:

- Strapdown INS;
- Wide FOV CCD camera/laser spot detector;
- Narrow FOV CCD camera;
- Laser designator/range finder;
- FLIR.

The entire sensor package is mounted on a gimbal assembly that is free to move in roll, elevation and azimuth.

5.7.2   Pod Installations

Podded installations are usually carried on fighter aircraft, although in certain circumstances they could be fitted to other aircraft. For example, targeting pods were trialled on the B-52 during the recent Iraq War, and plans are being reviewed to fit them to B-52 and B-1 bombers. As opposed to the turreted implementations described later in the chapter, EO pods lend themselves to be carried on certain weapons stations of fighter aircraft such that the aircraft can be reroled to perform a specific mission merely by fitting the pod.

(Courtesy Northrop Goumman)

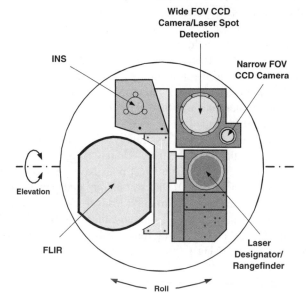

Figure 5.33 Typical pod-mounted sensor package (Litening II AT).

Apart from tailoring the mission-specific software to the requirements of the mission, no aircraft modifications are required. All the hardware and software modifications to adapt the pod to the aircraft baseline avionics system are in place such that all the subsystems and display symbology will be compatible to enable the mission to be performed. Therefore, EO pods give the battlefield commander additional flexibility in discharging his overall battle plan.

The first pods to be developed were the low-altitude navigation and targeting infrared for night (LANTIRN) system introduced into the US Air Force in the late 1980s. In truth, this system comprised two pods:

- The AN/AAQ-13 navigation pod containing a terrain-following radar (TFR) and an FLIR system to aid low-level navigation at night;

- The AN/AAQ-14 targeting pod comprising a targeting FLIR and laser designator/range finder.

The introduction of these pods in the later stages of the Cold War boosted the capability of the US Air Force accurately to attack ground targets in all weather conditions. These first-generation pods were fitted to a variety of aircraft including F-14, F-16C/D and F-15E in the United States and subsequently to a number of allied Air Forces. The almost simultaneous demise of the Soviet Union and the successful deployment and execution of Desert Storm in Iraq made military planners realise that most attacks using laser-guided munitions would need to be performed from 30 000 ft or above if reasonable aircraft survivability rates were to be achieved. The focus for the use of EO pods therefore shifted from low-level attack to attack from medium level.

To accomplish this new mission, performance improvements were needed to the targeting pod. Low-level ingress and egress to/from the target were not required as it was assumed that successful weapons launch could be made from over 30 000 ft above and beyond short-range and most medium surface-to-air missile (SAM) threats. The emphasis was on long-range target detection and identification and the increased deployment of weapons from long range with INS/GPS guidance. These drivers led to some of the performance improvements mentioned earlier, and with the 'third-generation' pods now entering service these requirements are largely satisfied.

Table 5.3 is a top-level comparison and summary of the most common targeting pods developed by contractors in the United States, United Kingdom, France and Israel and deployed from the late 1980s onwards. The latest pods have been designed with COTS in mind and are modular in construction so that technology and performance improvements may be readily inserted. The modular construction leads to easier maintenance, with faulty modules replaced on aircraft in a matter of minutes; high levels of built-in test (BIT) are provided readily to check out the system with high levels of confidence. Most of these pods have mean time between failure (MTBF) rates of a few hundred hours, roughly equivalent to one failure per year at peacetime flying rates.

Examples of EO targeting pods are shown in Figure 5.34.

5.7.3 Turret Installations

Whereas podded installations are useful for loading on to the weapon stations of a fast jet, turreted installations are more suited to permanent installations. For clear line of sight to the targets, they need to be located at the front of the aircraft, and for these reasons they are particularly well suited for installation on helicopters. Some turreted installations have been on fixed-wing aircraft such as the B-52H, the US Navy Orion P3-C and the Nimrod MR2 and have also been considered for the S-2 Viking. The last three aircraft have anti-submarine warfare (ASW) roles.

The first application of EO turrets on helicopters was with systems like the AH-64 Apache. Many different helicopters are now fitted with EO turrets, including special forces, coastguard and law and drug enforcement. In general, these systems are used in low-level and short-range engagements rather than the high-level long-range operation of podded fighter aircraft. The atmospheric conditions at low level, combined with other conditions such as smoke and haze, mean that LWIR systems often fare better than shorter-wavelength systems. However, MWIR with a shorter wavelength offers greater resolution. In some recent applications, dual-band MWIR and LWIR sensors are accommodated.

Table 5.3 Summary and comparison of EO targeting pods

Manufacturer	Product	Dimensions	Capabilities	Carriage aircraft
Lockheed Martin	AN/AAQ-13 navigation LANTIRN pod AN/AAQ-14 targeting LANTIRN pod	*L*: 199 cm/78 in *D*: 31 cm/12 in *W*: 211 kg/470 lb *L*: 251 cm/98 in *D*: 38cm/15 in *W*: 236 kg/524 lb	Terrain-following radar (TFR) Fixed FLIR: FOV 21° × 28° (640 × 512) FLIR: NFOV 4° × 4° (640 × 512); NFOV 1° × 1° Laser designator/ range finder Many upgrades during service, including third-generation FLIR; 40 000 ft laser; laser spot tracker; CCD-TV sensor; digital data recorder; geocoordinate generation for J-series weapons	F-16C/D F-15E F-14 + 12 international Air Forces Over 700 pods in service
BAE SYSTEMS	Thermal imaging and laser designator (TIALD)	*L*: 290 cm/114 in *D*: 35 cm/14 in *W*: 210 kg/462 lb	WFOV 10° × 10° MFOV 3.6° × 3.6° Electronic zoom ×2; × 4 Modification added a CCD-TV	Harrier GR 7 Jaguar GR1/ GR3 Tornado GR1/GR4
Northrop Grumman/Rafael Variants: Litening II – 1999 Litening II ER – 2001 Litening II AT – 2003	AN-AAQ-28(V) Litening II	*L*: 230 cm/87 in *D*: 40.6 cm/16 in *W*: 200 kg/440 lb	CCD-TV – MFOV and NFOV 640 × 480 FLIR operating on MWIR: WFOV 18.4° × 24.1° (Nav); MFOV 3.5° × 3.5°; NFOV 1° × 1° Laser spot tracker/range finder Laser designator > 50 000 ft/40 miles Litening III has dual mode (including eye safe)	US ANG F-16 block25/30/32 USMC; Spanish Navy; Italian Navy AV-8B; Spanish F/A-18; Israeli F-15I; Israeli-16C/D/I; German Navy and Air Force Tornado Total of 14 Air Forces

Table 5.3 (*Continued*)

Manufacturer	Product	Dimensions	Capabilities	Carriage aircraft
Lockheed Martin	Sniper XR (extended range) targeting pod Export version known as PANTERA	L: 239 cm/87 in D: 30 cm/12 in W: 200 kg/440 lb	CCD-TV: WFOV: 4° × 4°; NFOV: 1° × 1° 640 × 480 FPA FLIR operating on MWIR Laser – diode pumped; laser >40 000 ft; laser range finder/spot tracker; dual-mode laser (including eye safe); geo coordinate generation for J-series weapons	USAF F-16 block 50 ANG F-16 block 30 F-15E A-10
Raytheon	AN/ASQ-228 ATFLIR, recently named Terminator	L: 183 cm/72 in D: 33 cm/13 in W: 191 kg/ 420 lb	640 × 480 FPA FLIR operating in MWIR: WFOV 6° × 6°; MFOV 2.8° × 2.8°; NFOV 0.7° × 0.7°	F/A-18A +, C/D,E/F Replacement for AN/AAS-38 Nite-Hawk
Thales	Damocles	L: 250 cm/98 in D: not quoted W: 265 kg/ 580 lb	Third-generation MWIR FLIR sensor: WFOV 24° × 18° (Nav); MFOV 4° × 3°; NFOV 1° × 0.75° Laser range finder: 1.5 μm (eye-safe) Laser designator/ range finder/spot tracker: 1.06 μm	Super Entendard Mirage 2000 replaces Atlis, Expected to be fitted to Rafale

The increasing use of unmanned air vehicles (UAVs) in reconnaissance and combat roles has given significant impetus to the production of smaller, lighter systems suitable to be used as a UAV payload.

Figure 5.35 shows typical turrets with imagery examples.

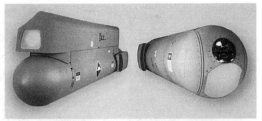

LANTIRN Pods (Lockheed Martin)

TIALD Pod (BAE SYSTEMS © 2005)

LITENING Pod (Courtesy Northrop Grumman)

SNIPER XR Pod (Lockheed Martin)

SNIPER XR Pod as file (Lockheed Martin)

Figure 5.34 Examples of EO targeting pods.

AH-64 C/D Apache TADS/PNVS
(Lockheed Martin)

Fur Turnet (L-3 WESCAM)

TADS/PNVS Arrowhead (Lockheed Martin)

MX-20 US Navy P-3
(Courtesy of UP-45 Squadron)

Tank (L-3 WESCAM)

Figure 5.35 Typical EO turrets and imagery examples.

5.7.4 *Internal Installations*

Stealthy aircraft such as the F-117 and F-35 incorporate EO sensor suites to assist in engaging ground targets. The F-35 in particular will incorporate an interesting internally carried system called the electrooptic sensing system (EOSS). This comprises two major functional elements:

Table 5.4 Summary of typical EO turreted systems.

Manufacturer	Product	Role	Capabilities	Carriage aircraft
Lockheed Martin	AN/AAQ-11 target acquisition and designator sight (TADS) AN/ASQ-170 pilot's night-vision sight (PNVS)	Navigation/ attack	Direct vision optics: WFOV 18° × 18°; NFOV 3.5° × 3.5° TV camera: WFOV 4° × 4°; NFOV 0.9° × 0.9°; underscan 0.45° × 0.45° FLIR: WFOV 50° × 50°; MFOV 10.2° × 10.2°; NFOV 3.1° × 3.1°; underscan 1.6° × 1.6° Laser range finder/designator Arrowhead modification program commenced in 2000: M-TADS and M-PNVS utilising RAH-66 Comanche technology. Features: LWIR FLIR using 640 × 480 FPA; stabilisation improvements; eye-safe laser; FLIR and IR sensor fusion; colour CCD-TV camera	AH-64C/D Longbow Apache Over 1000 systems delivered
Raytheon	AN/AAQ-27 Follow on to the AN/AAQ-16B/C/D	Surveillance and navigation/ attack	-16B variant: LWIR FLIR, dual FOV -16C variant: LWIR FLIR, dual FOV -16D variant: LWIR FLIR, three FOV with laser range finder/designator -27: MWIR FLIR using 640 × 480 FPA. Dual and three [AN/AAQ – 27 (3 FOV)] versions available	CH-53E; CH-47; MH-60L/M; SH-60B; podded version on F-18 as AN/AAR-50 MV-22 Osprey RAN Super Seasprite

Table 5.4　(*Continued*)

Manufacturer	Product	Role	Capabilities	Carriage aircraft
Wescam	AN/ASQ-4 using Wescam MX-20 turret	Surveillance	Features: high-quality gyrostabilisaton and picture quality; high-magnification step zoom capability on FLIR and TV sensors; geolocation for pinpointing ground target location; MWIR FLIR with 640 × 480 FPA: WFOV 12.0° × 9.3°; NFOV 2.9° × 2.3°	P-3C Orion S-2 Viking Nimrod MR-2

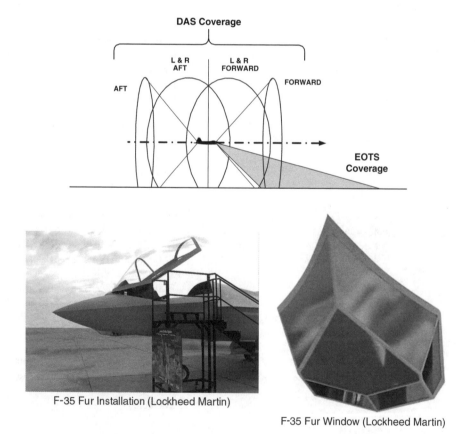

F-35 Fur Installation (Lockheed Martin)

F-35 Fur Window (Lockheed Martin)

Figure 5.36　F-35 EO sensor vertical coverage and EOTS installation.

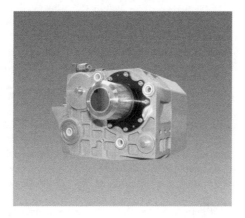

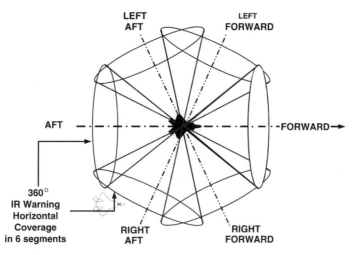

Figure 5.37   F-35 horizontal coverage using DAS sensors.

1. The electro-optic targeting system (EOTS) being developed by Lockheed Martin and BAE SYSTEMS. This is an internally carried EO targeting system that shares many common modules with the SNIPER XR pod already mentioned. The EOTS looks downwards and forwards with respect to the aircraft centre-line, as shown in Figure 5.36. The EOTS installation and EO sensor window are shown.

2. The distributed aperture system (DAS) being developed by Northrop Grumman together with BAE SYSTEMS comprises six EO sensors located around the aircraft to provide the pilot with 360° situational awareness information that is detected by passive means. The concept of horizontal coverage of the DAS is depicted in Figure 5.37. The six DAS sensors provide a complete lateral coverage and are based upon technology developed for the BAE SYSTEMS Sigma package (shown in the inset). Key attributes are dual-band MWIR (3–5 μm) and LWIR (8–10 μm) using a 640 × 512 FPA. Each sensor measures ∼7 × 5 × 4 in, weighs ∼9 lb and consumes less than 20 W. Sensor devices with megapixel capability (1000 × 1000) are under development and will be incorporated.

References

Atkin, K. (ed.) (2002–2003) *Jane's Electro-Optic Systems*, 8th edn.

Bannen, D. and Milner, D. (2004) Information across the spectrum. *SPIE Optical Engineering Magazine*, March.

Capper, P. and Elliott, C.T. (eds) (2001) *Infrared Detectors and Emitters: Materials and Devices*, Kluwer Academic Publishers.

Carrano, J., Perconti, P. and Bannard, K. (2004) Tuning in to detection. *SPIE Optical Engineering Magazine*, April.

Chan, Goldberg, Der and Nasrabadi (2002) Dual band imagery improves detection of military target. *SPIE Optical Engineering Magazine*, April.

Kopp, C. (1994) The Sidewinder story, the evolution of the AIM-9 Sidewinder. *Australian Aviation*, April.

Petrie, G. and Buruksolik, G. (2001) Recent developments in airborne infra-red sensors. *Geo Informatics*, February.

6 Electronic Warfare

6.1 Introduction

Warfare has always been conducted by adversaries who have been at great pains to understand their enemy's strengths and weaknesses in order to minimise the risk to their own forces and territory. The detection and interception of messages and the efforts to deceive the enemy have long been the task of the 'secret service'. The military aircraft in its infancy in World War I was used to detect troop movements and observe enemy movements, while on the ground the use of radio interception confirmed the aerial observations. As methods of communication developed, so too did methods of interception become more effective. Radar has developed from a mere detection mechanism to a means of surveillance and guidance.

Modern warfare is conducted in a rich electromagnetic environment with radio communications and radar signals from many sources. Figure 6.1 shows an example military situation with combined land, sea and air forces operating against an enemy territory which is, in turn, being defended by similar forces. The key players in this example include the following:

1. *Military planning* maintains communications with all forces either from the battlefield or from staff headquarters. Communications needs to be swift and secure at all times to include information from tactical units, from cooperating forces and from analysis databases. This communications network is vital to build an understanding of the tactical situation and to ensure that orders are received and placed, in the secure knowledge that the status and disposition of own forces is not disclosed to the enemy.
2. *Air defences* will be using radar to detect incoming airborne threats and will be making full use of available intelligence received by land line or data link. They will also be issuing orders by radio to fighter, missile and artillery defence systems.
3. *Air superiority aircraft* will be on quick reaction alert on dispersal or loitering on combat air patrol (CAP). They will be in constant radio communication and using their radar discretely to identify targets.

Military Avionics Systems   I. Moir and A. Seabridge
© 2006 John Wiley & Sons, Ltd

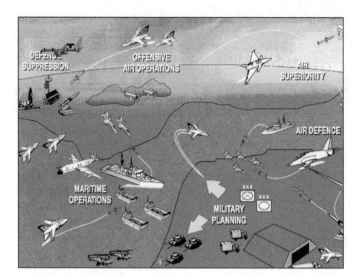

Figure 6.1   Typical battlespace scenario.

4. *Defence suppression* may be using radar for terrain following or for seeking targets.
5. *Maritime operations* in the form of rotary- or fixed-wing units will be conducting open
   ocean or sonar searches to locate and identify surface units or submarines. This will
   involve the use of radar and passive or active sonar, with intelligence sent to headquarters
   (HQ) by data link. Communication with other units will be controlled by use of high
   frequency (HF), very high frequency (VHF), ultrahigh frequency (UHF), shortwave
   marine band or data link. A radar altimeter enables the aircraft autopilot to maintain an
   accurate height over the sea surface regardless of changing atmospheric conditions.
6. *Offensive operations* will be using radar for detecting targets, and launched radar guided
   missiles will also be emitting.
7. *Naval forces* will be conducting their own operations in close cooperation with their own
   naval and marine forces. This will include the use of surveillance radar, self-defence radar
   and communications. Their communications include marine band shortwave for communication with merchant vessels or fisheries vessels, as well as very low frequency
   (VLF) for communicating with submarines.
8. *Land forces* will be similarly employed with their own units and deploying a wide range
   of radar and communications system.

As if this situation is not complex enough, modern warfare attracts the attention of the media
with their attendant TV and sound satellite links and mobile telephone traffic.

   The radio-frequency spectrum covered by the emitters used by these forces is broad, as
illustrated in Figure 6.2. No single item of equipment can cover this range for transmission or
reception. Hence, most communications and radar systems are designed for use in specific
bands. These bands are usually designated by international convention, as detailed in
Chapter 7.

   The main role of electronic warfare is to search these radio-frequency bands in order to
gather information that can be used by intelligence analysts or by front-line operators. The

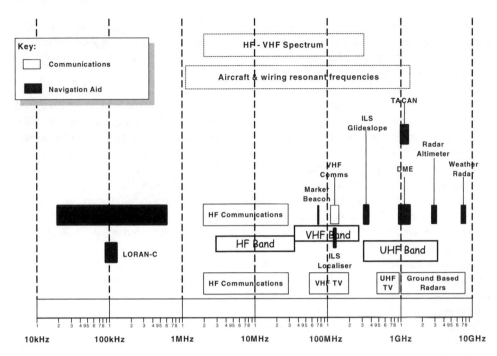

Figure 6.2 Radio-frequency spectrum.

information gained may be put to immediate effect to gain a tactical advantage on the battlefield; it may be used to picture the strategic scenario in peace time, in transition to war, or during a conflict. It may also be used to devise countermeasures to avoid a direct threat or to deny communications to an enemy. It must also be observed that such tactics are deployed by all sides in a conflict – in other words, the listeners are themselves being listened to.

The drive for intelligence is derived from a continuous need to be one step ahead of any potential adversary at all times – in peacetime, in transition to war, during actual conflict and in post-war peacekeeping operations. A typical cycle of intelligence is shown in Figure 6.3.

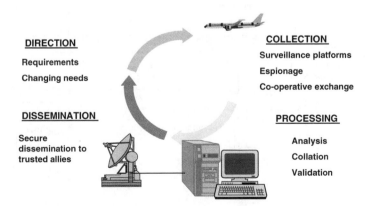

Figure 6.3 Intelligence cycle.

The cycle of intelligence begins with a requirement to gather information on a particular scenario. This may be tactical – the observation of a conflict – or it may be strategic – observation of a potential adversary's build-up of forces, their disposition and strength, and to identify new assets in the enemy inventory.

This requirement leads to a set of orders to collect information. This may be by means that include land, sea, air and space platforms, and is backed up by background information and espionage. Nations will also exchange information, although usually selectively. Raw information is analysed to identify new information or changes from previous intelligence. It is collated with other sources and with historical data. It is validated for accuracy and reliability by comparison with other intelligence and by other sources. It has now become 'intelligence' and is disseminated by secure means to trusted users.

Tactical users will make use of the intelligence to modify their battle plans and tactics. The intelligence may result in changes to the original requirement, and political situations may result in changing needs. Thus, new direction will be provided to the collectors of information.

As well as obtaining intelligence, military forces use electronic warfare actively to evade detection and to pursue aggressive attacks on enemy radar-guided weapons. Figure 6.4 illustrates some aspects of electronic warfare broken down into major subdivisions which will be described below.

In addition to all other forces in the electronic warfare (EW) scenario, the Air Forces play their own role. Figure 6.5 shows the high-flying EW aircraft gathering and analysing signals, and the low-flying tactical EW aircraft accompanying strike forces to counteract enemy defences. The high-flying surveillance platform is equipped with a range of sensors and receivers to cover the broad range of emitting systems on the ground and in the air. A vast amount of data is collected and analysed in real time to provide information of use to forces on the ground, and to provide a basis for intelligence to be used in the longer term. The low-

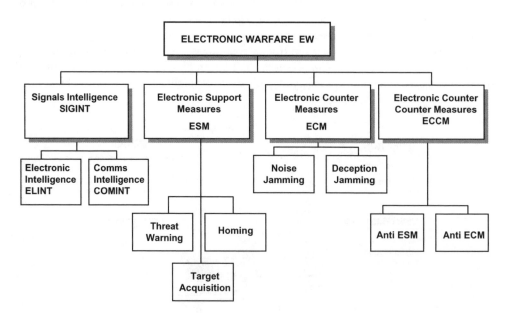

Figure 6.4   Electronic warfare elements.

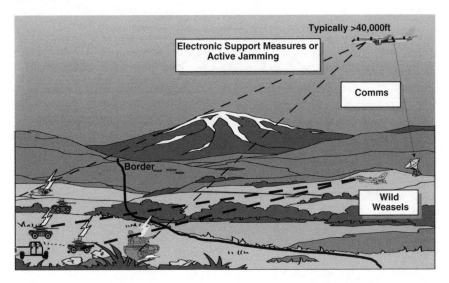

Figure 6.5 EW airborne roles.

flying aircraft is often equipped with a more selective range of sensors to identify and attack specific targets.

The aircraft type and the sensors and mission systems selected for these aircraft are determined by the requirement to perform strategic or tactical electronic warfare, and to obtain the appropriate intelligence. This requirement can be derived from analysis of a top-level national requirement such as 'defence of the realm'. This can be progressively decomposed or broken down into subsets of requirements which lead to the definition of a particular role (Price, 2005).

6.2 Signals Intelligence (SIGINT)

Intelligence is collected from a number of different sources to form a strategic picture. These sources include:

- ELINT or electronic intelligence;
- COMINT or communications intelligence.

Confirmation of electronic warfare intelligence is usually performed by comparison with local information collection and photographic evidence, including:

- HUMINT or human intelligence;
- IMINT or image intelligence;
- PHOTINT or photographic intelligence.

The first two items on this list are often gathered by high-flying EW aircraft on long duration patrols, usually flying a patrol on the friendly side of a border and beyond missile

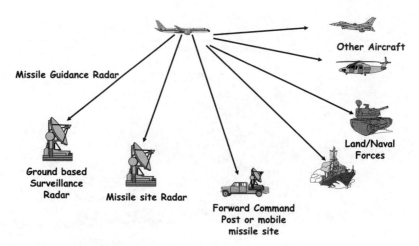

Figure 6.6 Users of radar systems.

engagement range. The aircraft is often a converted commercial type providing accommodation for a flight crew and a mission crew of operators able to detect, locate and identify sources of radio-frequency emissions at very long ranges. Their task is a combination of routine gathering and identification together with an ability to spot new or unusual emitters or patterns of use. This they perform with their own knowledge and experience and by using databases of known emitter characteristics. This is reinforced by the ability to communicate with external agencies to obtain further information.

The intelligence obtained from analysis of this electronic information is complemented by human intelligence in the field and by photographic intelligence which is used to confirm the existence and precise locations and types of target.

6.2.1 Electronic Intelligence (ELINT)

Figure 6.6 shows some examples of radar emitters, or systems operating in the radar bands that are likely to be of interest to ELINT collection aircraft. These include:

- Ground-based surveillance radars scanning borders looking for airborne or land-based intruders and forming a defensive security screen;
- Missile site or anti-aircraft artillery (AAA) radars scanning for threats and preparing to lock on and to track targets for directing defensive weapons;
- Forward command post radars providing advanced and localised warning of intrusion in order to direct local defences;
- Land and naval forces operating their own radar systems for detection and target tracking;
- Other fixed-wing aircraft and helicopters operating with their own characteristic radar types.

The ELINT system must provide a wide area coverage, preferably as near to spherical as possible with few 'shadows' as may be caused by wing tips, fin or fuselage masking. Figure 6.7 illustrates the functions to be performed by an ELINT system. The antennas are

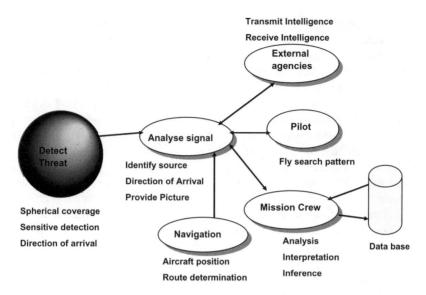

Figure 6.7   Functional overview of an ELINT system.

located on the aircraft to provide suitable coverage of the scenario to be monitored and detect an arriving signal and its direction of arrival (DoA). The signal is analysed to identify the source and its DoA, and to scan intelligence received from other sources to try to confirm the signal source. This is fused with the aircraft navigational data so that a picture can be provided showing the source relative to the ELINT aircraft. The crew will interpret the information and provide the information to other operators. An example system block diagram is shown in Figure 6.8.

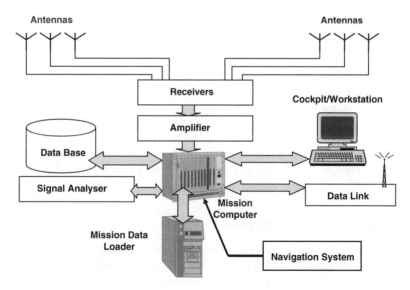

Figure 6.8   Typical ELINT block diagram.

This system receives signals via a number of antennas situated on the aircraft to provide maximum spherical coverage. These antennas are connected to preamplifiers or amplifiers by appropriate cables. This may include the routing of low-loss coaxial cables from wing-tip antennas to fuselage-installed receivers and amplifiers. The signals are processed by the mission computer to add labels or colour for ease of identification.

Operators are able to ask for further signal analysis to extract key signal characteristics, and may also ask for comparisons to be made with similar signals held in a database. With this analysis it is possible to identify the type of transmitter, which may enable identification of the type of installation or vehicle that made the original transmission.

An aircraft with a large number of operators can process many signals and is able to build up a picture of emitters over a wide area. Each operator will deal with signals from a particular band, logging each signal on receipt. The operator's workstation is equipped with a roll-ball and keyboard, or a touch screen, which allows the operator to annotate the signals, call up analysis or database checks and to store signals. The tactical commander is able to retrieve the received and processed signals and build up a composite picture.

The operators and the pilots work as a team to capture the best possible picture of the signal environment. Data link communication allows ground stations or operational commanders to join the team and to use other sources of intelligence to direct a specific search. The identified emitter remains in the real-time display, tagged as friendly or hostile with its characteristics.

The system can be used to identify radar signals from many sources, including:

- Fixed ground or airfield radar;
- Mobile missile battery radar;
- Ship radar;
- Aircraft radar;
- Missile radar;
- Submarine radar.

Skilled analysis and comparison with the intelligence database entries enables users to identify threat types by radar type, vehicle class and sometimes individual vehicle, especially ships where the number of high-value assets is small.

There will be highly classified threats that will need specific antennas and analysis techniques for identification. This situation arises as national security agencies develop new transmitters using different bands or different countermeasure techniques to avoid detection. This will always be a continuous activity in electronic warfare.

The database of historical intelligence, the flight plan and tactics and the collected intelligence would be of value to an enemy if the aircraft were to be forced to land or if it were destroyed. The data storage devices must not be captured intact, and for this reason are usually fitted with an explosive charge to ensure complete destruction.

6.2.2 Communications Intelligence (COMINT)

Figure 6.9 depicts some examples of users of communications. These users employ bands that are mandated for peacetime use such as VHF and UHF for air traffic control or shipping lanes, as well as satellite communications and data link for long-range encrypted data communications. All forces use a variety of frequency bands. Also shown in this figure is the

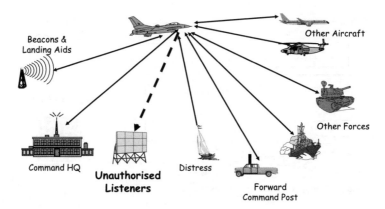

Figure 6.9   Users of communications.

unauthorised listener – the electronic warfare listeners of all participants in a conflict, as well as those agencies not directly involved but who want to gather more intelligence. It should be noted that this activity also takes place in peacetime and may include listening to friends as well as enemies.

Communications intelligence (COMINT) is gathered by scanning the normal communications frequency bands and locking on to detected transmissions. In peacetime it may be possible to receive in clear speech, but this is extremely unlikely in times of tension or during conflict. However, a great deal of intelligence can be obtained from the following characteristics of communications activity:

- The location of individual transmitters;
- The locations of groups of transmitters and the numbers in the groups;
- The frequency of the transmission carrier;

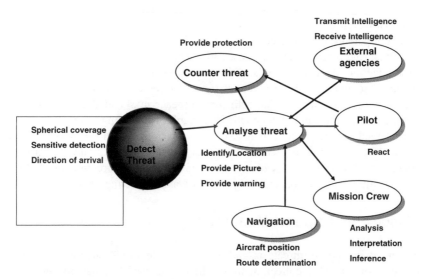

Figure 6.10   Functional view of a COMINT system.

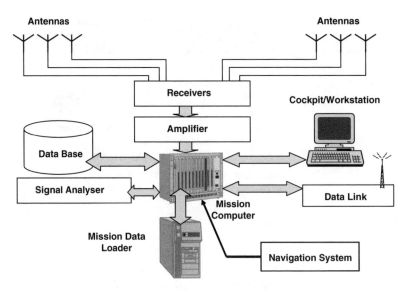

Figure 6.11 Typical COMINT block diagram.

- The style of the operator;
- The relative intensity of messages;
- Intervals between message groups;
- Periods of silence;
- Periods of activity, especially sudden or unusual activity;
- Overall pattern of communications during various states of force readiness.

For these reasons the information-gathering aircraft attempts to obtain a position fix on a transmitter and records the activity for later analysis. Depending upon the communications frequency, it is not always possible to obtain an accurate fix on a particular platform within a task group. However, observation over a period of time allows an overall picture to be built up regarding a potential foe's force structure and intended electronic order of battle (EoB).

The antennas and receivers are optimised for reception over a broad band with high-gain, high-noise rejection. An example COMINT system block diagram is shown in Figure 6.11. Over a period of several years a potential foe's electronic communications are collected, analysed and catalogued in order that both normal (peacetime) and abnormal (high-readiness states) may be recognised and understood. This enables responding forces to respond in kind by elevating readiness states and, if necessary, imposing restrictions of critical emissions. The overall effect is akin to an electronic form of 'cat and mouse', with neither side hoping inadvertently to disclose their readiness state or possible future intentions. Large-scale peacetime training exercises provide major intelligence-gathering opportunities.

6.3 Electronic Support Measures

Information on immediate threats is gathered by an electronic support Measures (ESM) system. This consists of a collection of sensitive antennas designed to detect signals in different frequency bands. The antennas are often grouped in a wing-tip pod. This allows a wide angle of view without obscuration by the fuselage, and also enables a fix on the signal

Figure 6.12 Wing-tip ESM pod installation – Nimrod MR2.

source to obtain an accurate direction of arrival (DoA) of the signal. Figure 6.12 shows the ESM pods mounted on the Nimrod MR 2 wing tips.

An effective ESM system rapidly identifies the signal band and location and determines the signal characteristics depending upon a number of discriminators. A signal analyser then examines the signal characteristics to identify the type of transmitter and the level of threat posed. Even the most cursory of analysis can establish whether the emitter is associated with surveillance, target tracking or target engagement. This analysis can compare the signal with known emitter characteristics obtained from an intelligence database or threat library and known signal types confirmed and new emissions identified and categorised. Every signal identification is logged with date, time and intercept coordinates, along with the known or suspected platform type, and the results are stored. A typical block diagram is shown in Figure 6.13.

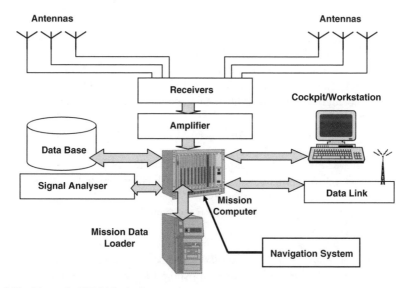

Figure 6.13 Example ESM block diagram.

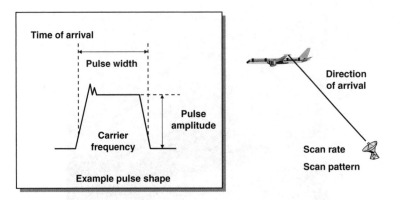

Figure 6.14   Important ESM parameters.

Signals received by the electronic support measures system may in some cases be analysed instantaneously to produce an identity for the transmitter of each signal received. Figure 6.14 shows some parameters of a signal that are essential for the understanding of the type of transmitter producing the signal. The nature of the pulse shape is used to determine the particular type of transmitter. The scan rate and the pattern of the scan also provide invaluable information about the mode of the transmitter. It is possible to detect the antennas changing from scanning mode to lock-on to tracking and hence determine the threat that the transmitting station poses.

As well as providing threat information, ESM is used by maritime and battlefield surveillance aircraft as a passive or listening sensor which adds important information to other sensors. It is especially useful when tracking submarines, where the use of the aircraft radar would be a source of intelligence to the submarine commander.

The salient signal characteristics or discriminators identified during the ESM collection and identification process include the following:

1. *Signal frequency.* Owing to the RF atmospheric propagation and transmission character-istics, the operating frequency is the first indicator of radar type as all RF emitters have to compete in the same physical world.
2. *Blip/scan ratio.* Examination of the blip/scan ratio will give preliminary indications of scan rate, sector scan width and possibly radar/emitter beamwidth.
3. *Scan rate.* The higher the scan rate, then generally the more likely is the threat of engagement.
4. *Scan pattern.* Search, track, track-while-scan (TWS) and ground-mapping (GM) modes will exhibit particular characteristics.
5. *Signal modulation.* Pulse, pulse compression, pulsed Doppler (PD), a continuous wave (CW) and other more sophisticated forms of modulation are indicative of the emitter mode(s) of operation and likely threat level.
6. *Pulse repetition frequency (PRF).* High PRF associated with a tracking mode signifies an imminent engagement.

The combination of analysis of all these modes of operation and when they are employed either singly or in combination is vital to establishing the likely capabilities and intentions of a threat platform, especially when used in combination with other intelligence information.

ESM may be employed at a strategic intelligence-gathering level using an AWACS or MPA aircraft to build the overall intelligence picture and electronic order of battle (EOB). Alternatively, such information may be gathered and utilised at a tactical level using radar warning receivers (RWR), whereby information is gathered and used at the strike platform level to enable strike aircraft to avoid the most heavily defended enemy complexes during the mission.

6.4 Electronic Countermeasures and Counter-countermeasures

Electronic countermeasures (ECM) and electronic counter-countermeasures (ECCM) take the form of interfering or deceiving the enemy's radio and radar systems in order to negate their use or, worst of all, compromise their performance. On occasions it is difficult to distinguish between 'chicken' and 'egg' as so many issues are considered during the design phase and then hastily need to be re-evaluated once a real conflict begins.

Therefore, the authors have chosen to consider these issues together rather than separately, as it is indeed a rapidly evolving process. The deployment of EW and successes and failures are invariably and rapidly recognised as the conflict develops and as both sides are inclined to receive untimely and unexpected unpleasant surprises. In some cases this is due to an inexact appreciation of the capabilities of the foe or where ELINT has not been able to provide the complete picture. Also, given the frailties of humankind, there is also a tendency to 'lose the recipe' and to relearn the hard way the lessons derived from a previous conflict.

These countermeasures, or 'jamming' as they are often loosely called, may be divided into two categories:

- Noise jamming;
- Deception jamming;

6.4.1 Noise Jamming

Active noise jamming is often performed by identifying an enemy detection system and broadcasting white noise at high power levels. For communication systems, noise jamming could employ the broadcast of music or other audio features designed to deny the use of the particular service. This effectively swamps the input circuitry of detection systems and prevents it from operating.

The effectiveness of a jamming system depends on a number of aspects of the system, for example:

- Transmitter power output;
- Transmission line losses between the transmitter and the radiating antenna;
- Antenna gain in the direction of the receiver to be jammed;
- Transmitter bandwidth.

The amount of energy delivered into a target transmitter depends on similar aspects of the target such as:

- Incoming jamming power;
- Receiver bandwidth;
- Antenna gain;
- Radar cross-sectional area of the aircraft being masked.

In order to be effective, the jamming transmitter must be able to emit sufficient power in the bandwidth of the target receiver to mask its intended signal or to simulate a deceptive signal realistically.

Most jamming transmitters are designed to operate over the range of frequencies expected, and, as has been shown above, this is extremely wide given the range of communications devices, search radar and guidance radar types to be found on the modern battlefield. In the case of radar, the bandwidth covered is often in the range 2–18 GHz. The jammer is most effective if it can be designed to target a specific frequency range or type of threat, in which case the power output is concentrated into a narrow spectrum. Given that a jammer must operate against a wide range of emitters, its power must be spread over an increased spectrum (Figure 6.15).

Figure 6.15 portrays a frequency spectrum in which four targets exist; the jammer has the task of nullifying each of the four targets. Three different techniques are shown:

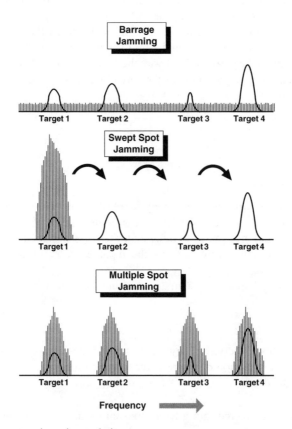

Figure 6.15   Rudimentary jamming techniques.

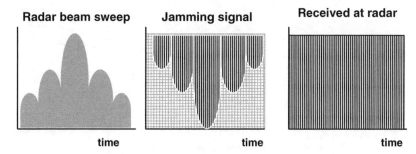

Figure 6.16 Inverse noise gain jamming.

1. *Barrage jamming*. In this example, jamming power is spread across the entire spectrum encompassing the targets. This results in a very low jamming power density (W/MHz) to the point that none of the targets is adversely affected.
2. *Swept spot jamming*. Swept spot jamming concentrates sufficient power in a narrow bandwidth to negate each target. The jammer switches to each of the targets in turn but is only present for a low-duty cycle. This may suffice if the target receiver saturation and automatic gain control capabilities are modest but will not suffice for higher-performance systems.
3. *Multiple-Spot Jamming*. Multiple-spot jamming divides the energy between the targets, effectively jamming them in parallel rather than sequentially. This requires a more sophisticated jamming transmitter.

The foregoing explanation is in itself very superficial. In reality the radar is unlikely to be transmitting continuously on a fixed frequency; modern radars have a considerable degree of frequency agility and can often change frequency and even signal modulation on a pulse-by-pulse basis. This makes noise jamming more difficult to achieve effectively.

These techniques are rudimentary and are not particularly effective except against the most primitive equipment and more sophisticated techniques may be employed.

Figure 6.16 illustrates the principle of inverse noise gain jamming. The target signal is analysed and a pattern of noise is generated on time that complements the original incoming signal. This results in a return signal received at the target radar that is a continuous noise pattern, thereby masking the return from the aircraft skin. With high power this can be used to swamp the return, thus denying the enemy range information. With even higher powers it is possible to enter the sidelobes of the threat radar to deny angle information.

6.4.1.1 Burnthrough

Burnthrough range is the range at which the strength of the radar echo becomes greater than that of the jamming noise. The radar return is proportional to $1/R^4$ since it must travel to the target and return to the host radar. The jamming signal only travels in one direction, and is thus proportional to $1/R^2$. The more closely an aircraft approaches the victim radar source, the more likely is the radar signal to break through the jamming noise (see Figure 6.17 which illustrates the principle).

In Figure 6.17 a plot is shown comparing the received power (dB) against range in nautical miles, and the effect of $1/R^2$ and $1/R^4$ for jammer and radar respectively can be clearly seen.

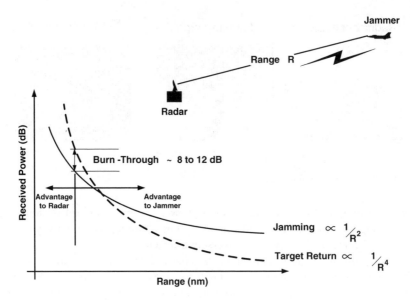

Figure 6.17 Effect of burnthrough.

However, at some point close to the radar, the target return signal will exceed the jamming signal by a suitable margin and the radar will prevail. The threshold associated with burnthrough is generally assumed to be of the order of 8–12 dB. At ranges greater than this the jammer has the advantage.

This balance depends upon a multitude of factors including the relative performances of jammer and radar transmitter and receivers, the antenna gain and sidelobe characteristics, the aspect of the engagement, etc. A radar antenna with low gain or poor sidelobe performance will be vulnerable to clutter, as already described in Chapter 4, and noise jamming is in effect a man-made form of clutter. Conversely, the higher the performance of a radar and the better the ability to discriminate against clutter, the more robust it will be in a jamming scenario.

Another significant disadvantage of noise jamming as a countermeasure extends to the jamming platform itself. By virtue of transmitting relatively high power, the jammer itself becomes a beacon whereupon the foe can use the jamming emissions as a source of guidance. Hence, many modern systems have a home-on-jam (HOJ) mode to enable the jammer itself to be attacked while radiating.

6.4.2 Deception Jamming

Deception jamming employs more sophisticated techniques to negate the performance of the radar. If subtly employed, the radar and radar operator may not realise that countermeasures are being used. Some typical techniques used to break the radar-tracking loops previously described in Chapter 4 are:

1. *False target generation.* If the modulating characteristics of the target radar are known, it is possible to transmit pulses that will appear as multiple targets in the victim radar.

Hence, by using the jamming transmitter with diligence and transmitting replica pulses after a time delay, these false, multiple, spurious targets will appear in subsequent radar range sweeps. An intelligent radar operator should realise that his radar is being deceived but may have a problem in deciding which of the multiple returns is the correct one.

2. *Range gate stealing*. This is a variation on the technique described above where one false pulse is generated that appears in the victim's radar at the same range as the jammer. It is then possible to capture the range gate with the artificial pulse; in particular, if the false pulse appears to be stronger than the original in the victim receiver, it is possible to 'steal' the range gate by progressively altering the false range. If desired, the range gate may be left on a false value or moved off to coincide with clutter, whereupon the target lock will be lost.

3. *Angle track breaking*. Similarly, there are ways of breaking the angle track mechanism, especially if the tracking mechanism of the victim radar is well understood. For example, in a conscan radar, angle track may be broken if the jamming signal is modulated at a frequency that approaches that of the conscan modulation frequency of the subject radar. This presupposes that the angle tracking method and conscan rate are known, which may not be the case in a wartime situation. Other simple ways of angle deception include terrain bounce, cross-eye, cross-polarisation and double cross.

4. *Velocity gate stealing*. This is similar to range gate stealing except that the incident signal is re-radiated back to the victim radar, initially without modification. Progressively the re-radiated signal is amplified and masks the original Doppler component upon which the velocity gate is centred. The deceiving radar may then steal the velocity gate in a similar manner to the range gate stealer described above.

Modern radars are inherently resistant – although not impervious – to jamming owing to a range of features inherent in the design. These characteristics are as follows:

- Low antenna sidelobes;
- Wide dynamic range with fast-acting automatic gain control (AGC);
- Constant false alarm rate (CFAR) reduction;
- Sidelobe blanking.

When these features are employed together with a range of other technology advances that evolved throughout the late 1980s and early 1990s, including greatly increased RF bandwidth, sensor fusion and the application of artificial intelligence techniques, then significant advances may be achieved. These developments have all contributed towards greatly enhanced radar performance. These techniques are outside the scope of this book and in many cases are classified.

6.4.3 Deployment of the Jamming Platform

The airborne jamming assets may deployed in two possible ways:

1. Self-screening platforms with their own on-board EW suite. The complexity and intensity of the modern battlefield is such that most platforms carry their own protection suite, also sometimes referred to in US parlance as aircraft survivability equipment (ASE).

2. Escort or stand-off jammers with a specialised EW role. The escort jamming role has been provided in the past by aircraft such as the F-4 Wild Weasel and EA-6 Prowler. Recently, the F-16C/J has taken on this role for the US Air Force and the F-18E/F is under development to replace the Prowler in the near future with the EF-18G. Such aircraft may also perform a stand-off jamming role, although this may also be performed by platforms with lower performance.

In reality, a modern conflict depends much upon the blending and merging of both asset types, depending upon the nature of the engagement. Jamming assets also offer complementary assistance to stealth platforms where they are deployed as low observability is easier to maintain in an aggressive EW environment, which has proved to be the case in recent Kosovo and Iraq engagements.

Escort jammers that accompany the main force are often referred to in US Air Force parlance as Wild Weasel squadrons and comprise strike aircraft types modified to perform a dedicated EW support and suppression role. The task is to precede or accompany the strike force, selectively jamming and confusing enemy defence radar and communications. These aircraft may also be armed with anti-radiation missiles (ARMs) that use the threat radar beam to guide themselves to the radar.

Aircraft operating in support of an attack force may also station themselves in a stand-off position outside the range of ground defences while maintaining a patrol so that an appropriate noise jamming signal can be used to confuse defences. Care must be taken that the jamming supplements and does not diminish the effectiveness of the attacking force.

6.4.4   Low Probability of Intercept (LPI) Radar

All these countermeasures depend upon the detection of the victim radars' emissions and upon having some prior knowledge of the frequency of operation and modulation techniques employed. The most obvious counter of all is to avoid detection as far as possible by utilising low probability of intercept (LPI) techniques. LPI techniques must be designed into the radar at the outset and involve a number of trade-offs where increasingly sophisticated design (and cost) is balanced against a lower probability of interception. Some of the design considerations include the following:

1. A reduction in peak power and an increase in the period of integration will result in the same overall detection capability for reduced peak power.
2. An increase in receiver bandwidth using spread spectrum techniques and a reduction in peak power, effectively spreading the modulation data across a wider band, will make the task more difficult for the jammer.
3. The radar has a much higher gain than that of a radar warning receiver (RWR) antenna and, while potentially disadvantageous during transmit, it has significant advantages during receive. Balancing peak power against antenna gain can yield benefits, and the aim is usually to increase antenna gain while reducing peak power. For effective LPI radars a design aim is to achieve a main beam gain of +55 dB above the first sidelobes. Other considerations include a high-duty cycle reducing peak power, low receiver losses and a low receiver noise factor.

6.5 Defensive Aids

An aircraft operating in a hostile military environment needs to be equipped with measures for self-defence. The crew will have been briefed on the threats on their outward and return transits, as well as enemy defences in the area to be attacked or where an engagement is to take place. This will be based on intelligence and will be in accordance with the most up-to-date intelligence compilation.

However, during the mission, the pilot must be warned of real tactical threats to the mission, and must have the means to minimise their effectiveness. The most common threats to low-flying aircraft are:

- Small Arms fire;
- Radar-guided anti-aircraft artillery (AAA or triple-A).
- Shoulder-launched surface-to-air missiles (SAM);
- SAM from ground sites, vehicles or ships.

Appropriate countermeasures include a means of detecting the threat and luring the threat away from the aircraft or causing the missile to detonate prematurely or far enough away from the aircraft so that no damage is sustained. This combination of sensor and counter-measure is often referred to as a defensive aids subsystem or DASS, and often abbreviated to Def-Aids. Figure 6.18 shows an aircraft equipped with a set of threat detectors and countermeasure subsystems.

There is little that can be done by a defensive aids system to have a significant impact on small arms fire and AAA, although counters may be devised for AAA gun tracking systems. High-speed evasive manoeuvres on the low-level run in to the target and may be firing a gun at ground sites may be a palliative, but the risk of a hit remains. Most aircraft are designed to minimise the catastrophic effects of missile or shell fragments by physical separation of critical equipment and wiring to reduce the probability of common mode damage effects. For those weapons employing active sensing there are mechanisms for reducing their effectiveness.

To counter weapons or systems utilising some form of electronic system or guidance, a defensive aids subsystem may include any or all of the following subsystems, depending upon the role and the intensity of the threat:

Typhoon DASS pods
Sensors & Towed
Decoy

Figure 6.18 Example of an aircraft equipped with a DASS.

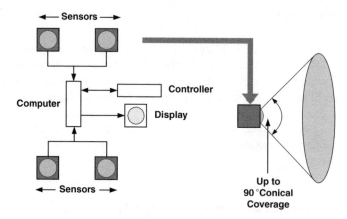

Figure 6.19   Functional layout of the radar warning receiver.

- Radar warning receiver;
- Missile warning receiver;
- Laser warning receiver;
- Countermeasure dispenser (CMD) – chaff or flares;
- Towed decoy.

On a military aircraft these systems will have the capability of interfacing with the aircraft/ mission avionics system by using MIL-STD-1553B data buses or other cost-effective commercial data bus equivalents.

6.5.1   Radar Warning Receiver

A typical radar warning receiver (RWR) is depicted in Figure 6.19. Sensors are located strategically around the peripheries of the aircraft – typically four sensors placed at the wing tips or sometimes at the top of the fin. The objective, as far as is possible, is to provide full-hemisphere horizontal coverage around the aircraft in order that the crew may detect and be alerted to potential RF threats. Each of these sensors may provide up to 90° conical coverage, although in some cases the angular reach may be less than this. A typical antenna used in this application would be a spiral antenna with an angular coverage of 75° but with a gain of ~10 dB.

   This figure should be compared with the 55 dB gain that would be the design point for a LPI radar – a difference of 45 dB or a factor of 32,000. This illustrates in part the disadvantage that the RWR faces while operating against a modern state-of-the-art radar. Other considerations such as the use of sophisticated spread spectrum modulation, radiated power management and advanced signal processing indicate why it is conceivably possible for a sophisticated AESA radar such as the AN/APG-80 as used on the F-22 to operate almost invisibly to some medium-capability RWR equipment.

   The quadrant-located spiral antennas detect and to some extent direction find (DF) any emissions within their respective area of coverage. Demodulated signals are analysed by the signal processor and categorised against a known threat library according to the following criteria:

- Frequency of operation;
- Modulation type;
- Signal strength;
- Direction of arrival.

In some cases an audio tone may be derived to provide the pilot or observer with audio cues – typically a tone equivalent to the PRF of the incoming radiation.

In early systems the processed outputs were displayed upon a plan position indicator (PPI) in a manner that depicted the angle of arrival according to a clock format with 12 o'clock dead ahead. In the late 1960s/early 1970s, when these systems were operationally deployed for the first time during the Vietnam War, this information would be presented on standard CRT green phosphor displays. Relative signal strength was shown by the length of the line from the centre of the clock, while the coding of the line into solid, dashed or dotted portrayal was indicative of the modulation type or possibly the band of operation. Early systems such as the air radio installation (ARI) 18228, as employed on the UK F-4K/M Phantom, used a hard-wired implementation to code specific threats and were therefore cumbersome to reprogramme.

The advent of digital processors now means that the threat library is coded in software allowing for rapid updates using a suitable software loading device. On a modern system, particularly since the advent of AMLCD colour displays, display symbology is much more likely to utilise stylised colour-coded symbology which is much more easily recognised by the pilot or radar operator, especially in a stressful combat environment.

Typical frequency coverage of a RWR system extends from 2 to 18 GHz and embraces a wide range of electronic threats across the RF spectrum. Modern RWR equipment offers a much more dynamic response to specific threats than was possible with the first-generation equipment.

6.5.2 Missile Warning Receiver

A missile warning receiver operates on a similar principle, except the missile warning systems (MWS) operate by detecting infrared (IR) or ultraviolet (UV) emissions during and following a missile launch. A typical system is portrayed in Figure 6.20.

Although the conical coverage of an IR/UV sensor may be as much as 110°, these systems often provide the option of expansion to six rather than four hemispherical sensors – see the F-35 example in Chapter 9. Apart from the sensors, in an overall sense the system works in a very similar fashion to the RWR above. Threat analysis is undertaken within a central signal processor/computing unit, and the results are output to a suitable tactical display. In modern systems this will be a colour tactical display.

An example of a missile warning system is shown in Figure 6.21.

6.5.3 Laser Warning Systems

A laser warning system again uses similar principles, except that the sensors are operating in the laser band. The example shown covers the 0.5–1.8 µm band and addresses the threat

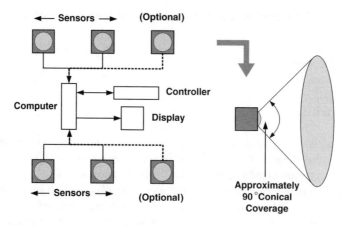

Figure 6.20   Missile warning receiver layout.

posed by the following lasers: doubled NdYAG, ruby, GaAs, NdYAG, Raman shifted NdYAG and exbium glass lasers. Angle of arrival (AoA) is claimed to be within 15° rms and sensor angles are 110°. (Figure 6.22).

6.5.4 Countermeasure Dispensers

The defensive aids will be equipped with a technique generator that interprets the threat and defines a suitable defensive response using the following:

- Chaff;
- Flares;
- Towed decoy.

Figure 6.21   Missile warning system (SAAB Avitron).

Figure 6.22 Typical laser warning system (SAAB Avitron).

6.5.4.1 Chaff and Flares

A chaff and flare dispenser is fitted to many aircraft so that appropriate mixes of chaff and flares can be selected and deployed to confuse missile seekers or defence radars. This can be done by providing alternative decoy targets for seekers, or by disguising the aircraft by changing its radar return so that operators cannot set up an aiming solution. Chaff and flares are usually deployed as 'last ditch' countermeasures against an incoming missile. Their effective protection zone is to the rear of the aircraft, and they offer no protection against missile engagements in the forward hemisphere.

Chaff consists of reflectively coated strips of plastic or metal foil. The strips are designed to a half-wavelength ($\lambda/2$) of a typical homing radar. The chaff can be dispensed in patterns or blooms to disguise the dispensing aircraft with a view to confusing a radar operator. When released into the turbulent airflow, chaff disperses rapidly (blooms) and for a brief period of time generates a very large, static radar image between the target aircraft and the threat radar, which is probably in tracking mode or CW illumination mode, providing radar guidance for an in-flight missile. If the timing is right, radar or missile lock may be broken.

Flares are a countermeasure against IR homing missiles. When deployed, a flare burns with an IR wavelength similar to that of the target aircraft IR signature. It works by initially appearing in the missile seeker head coincident with the target aircraft, but is left behind as the target aircraft performs evasive manoeuvres. Its thermal image is designed to be longer than the target aircraft and it then becomes the preferred IR target for the missile.

Timing of deployment is critical. Too soon and the divergence of target aircraft and flare will be detected and the flare ignored, too late and the missile will detonate on the flare and fragments may hit the target anyway.

Flare deployment can be used in a 'saturation' mode during periods of extremely high risk, where the target aircraft is in very close proximity to a missile launcher and has no time to manoeuvre if missile launch is detected, for example, transport aircraft carrying out low-level air drops or landing on captured airfield in hostile territory where MANPAD or Stinger IR missiles may be launched within a few hundred feet of the aircraft. In these circumstances the crew may choose to pre-empt target launch detection by the tactical deployment of multiple flares in the high-risk zone.

Figure 6.23   Example of flare dispensing.

Figure 6.23 shows an example of a C-130 Hercules deploying multiple flares.

6.5.4.2   Towed Decoy

The towed decoy is essentially a heat source that is towed behind the target aircraft on a cable. Located in a wing-tip pod, a wing-mounted pylon pod or deployed from inside the aircraft, the towed decoy is released and extended on a cable which restrains the device and provides a source of electrical power. The purpose is to cause infrared seekers in missiles to home on to the decoy rather than the jet-pipes of the towing aircraft. Any explosion should be sufficiently distant so as not to cause damage from the explosion or from missile fragments. The decoy can be rewound or it may be jettisoned by cutting the cable if there is a failure of the rewind mechanism. An example is shown in Figure 6.24.

Towed Decoys

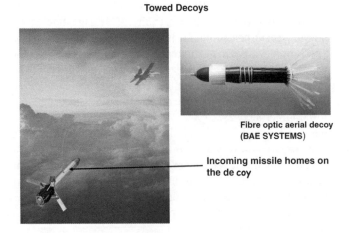

Figure 6.24   Example of a towed radar decoy.

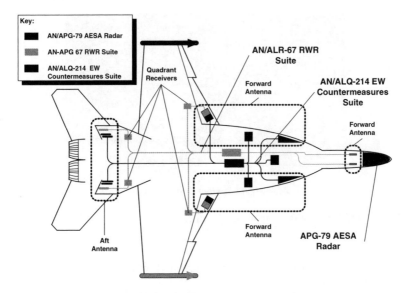

Figure 6.25 Simplified overview of F/A-18E/F countermeasures suite.

6.5.5 Integrated Defensive Aids Systems

In order to convey the complexity and extent of self-screening EW systems on modern combat aircraft, the example of the F/A-18E/F Super Hornet will be briefly analysed. This aircraft has the following countermeasures suite fit:

- AN/APG-79 AESA radar;
- AN/ALR-67 radar warning receiver;
- AN/ALQ-214 integrated defensive electronic countermeasures (IDECM);
- AN/ALE-47 countermeasures dispenser;
- ALE-50/55 towed decoy.

6.5.5.1 AN/APG-79 AESA Radar

The Raytheon APG-79 active electronically scanned array radar is an 1100-element radar that has all the advantages and flexibility inherent in this type of radar. In particular, flexibility of mode of operation, high scan rates, sophisticated modulation and signal processing and LPI features give the aircraft significant operating advantages in a hostile EW environment.

6.5.5.2 AN/ALR-67 Radar Warning Receiver

The RWR suite is an integrated suite comprising the following components:

- Countermeasures computer;
- Countermeasures receiver;
- Low-band integrated antenna;

ALR-67(V)3 System Components

Figure 6.26   AN/ALR-67 RWR components (Raytheon).

- 6 × integrated antenna detectors (two low band and four high band);
- 4 × quadrant receivers.

The countermeasure receiver receives inputs from the two low-band antennas and from the four high-band antennas via the respective quadrant receivers. The quadrant receivers provide preconditioning to reduce transmission losses between antenna and receiver. The receiver digitises and categorises the received signals and is able to handle a dense pulse environment while at the same time handling faint signals from distant threats.

The countermeasure computer incorporates a 32 bit machine with the application software encoded in Ada. The software structure enables complete reprogramming of the master threat file without any software changes. (Figure 6.26). The system weighs less than 100 lb – well under the normal weight of a system of this kind.

6.5.5.3   AN/ALQ-214 Integrated Defensive Electronic Countermeasures (IDECM)

The IDECM system is a radio-frequency countermeasures (RFCM) suite comprising the following units:

- Receiver;
- Processor;
- Signal conditioning amplifier;
- Modulator/techniques generator;
- Two optional plug-in transmitters may also be used.

The weight of the system, including the rack, is ∼168 lb. If the ALE-50 towed decoy option is included, a further 54 lb is added. The fibre-optic towed decoy actually transmits the jamming signal according to the top-level architecture shown in Figure 6.27 and the units shown in Figure 6.28.

After the interception of the incoming victim radar signal, the appropriate counter-measures are applied and the RF is converted to light energy for transmission down the fibre cable to the decoy. The light energy is converted to RF energy and amplified by the travelling wave tube (TWT) transmitter. The resulting jamming signal is transmitted to the target radar.

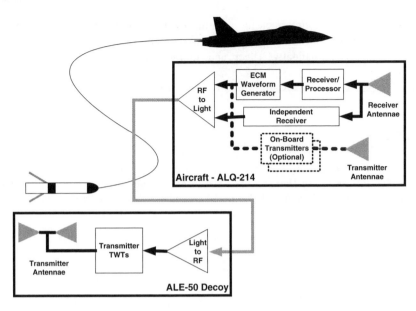

Figure 6.27 AN/ALQ-214 concept of operations.

6.5.5.4 AN/ALE-47 Countermeasure Dispenser

The AN/ALE-47 countermeasure dispenser – a successor to the ALE-39 – is able to dispense up to 60 expendables comprising chaff, flares or radar decoys. This is all achieved under computer control, enabling the pilot to achieve the optimum mix of expendables and deployment sequence for a given threat scenario.

The system has four main modes of operation:

1. *Automatic*. The countermeasure system evaluates the threat data from the on-board EW sensors and merges them with stored threat data to determine the optimum dispensed stores mix. The system automatically dispenses this countermeasure load.

Figure 6.28 AN/ALQ-214 units (BAE SYSTEMS).

2. *Semi-automatic*. The countermeasure system determines the optimum stores mix as for the automatic mode, but the crew activate deployment.
3. *Manual*. The crew manually select and initiate one of a number of preselected programmes.
4. *Bypass*. In the event of a system failure the crew can reconfigure the system in flight.

References

Lynch Jr, D. (2004) *Introduction to RF STEALTH*, SciTech Publishing inc.
Price, A. (2005) *Instruments of Darkness: The History of Electronic Warfare,* Greenhill Books.
Stimson, G.W. (1998) *Introduction to Airborne Radar*, 2nd edn, Scitech Publishing inc.

7 Communications and Identification

7.1 Definition of CNI

All military aircraft need certain computing sensing and computing resources apart from the mission sensors and weapons to enable them to complete their mission. These are:

1. *Communications*. The ability to be able to communicate by either voice or data link means with cooperative forces, be it wingmen in the same flight of aircraft, airborne command centre or troops on the ground.
2. *Navigation*. The military platform needs to be able to navigate with sufficient accuracy to a target, rendezvous point, waypoint, or initial point as dictated by the mission requirements.
3. *Identification*. The rules of engagement for a given theatre of operation will necessitate the classification and identification of a target before permission to engage is given.

The American military refer to this collection of resources as communications, navigation, identification (CNI).

Some of the CNI sensors such as air data, radar altimeters and inertial systems are autonomous to the platform, in other words the platform needs no other input from third-party sources. Others such as communications, radio navigation beacons and global navigation satellite systems (GNSSs) require the participation of other organisations to respond or the provision of a network of aids or a constellation of satellites to provide the navigational framework. Military platforms use a combination of all these sensors with the additional rider that, for certain covert stages of a mission, no emissions are made by the platform as radio silence – more correctly known as emission control (EMCON) – procedures are enforced.

Military Avionics Systems I. Moir and A. Seabridge
© 2006 John Wiley & Sons, Ltd

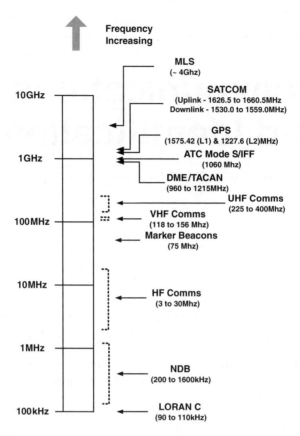

Figure 7.1   CNI RF spectrum.

7.1.1   RF Spectrum

The RF spectrum associated with the CNI functions is shown in Figure 7.1. The CNI spectrum covers a range of different equipment spanning almost five decades from 100 kHz to 4 GHz and comprising a range of functions as described below. Communications and identification are addressed within this chapter while navigation is discussed in Chapter 8. For ease of reference, the equipment is listed in ascending order of operational frequency:

1. Communications:

   - High-frequency (HF) communications;
   - Very high-frequency (VHF) communications;
   - Ultrahigh-frequency (UHF) communications;
   - Satellite communications (SATCOM);
   - Data links.

2. Identification:

- Air traffic control (ATC) mode S;
- Traffic collision and avoidance system (TCAS);
- Identification friend or foe (IFF).

With one or two exceptions, this equipment is all freely available for use by the civil community as well as by the military platform. All operational frequencies are published on aeronautical charts to ensure safe and successful integration and interoperability of all traffic within the wider airspace. There are a few exceptions, namely:

1. Civil traffic does not usually use the UHF communications band. Military users may also use UHF SATCOM which is not widely available.
2. Civil traffic would not ordinarily be equipped with TACAN.
3. Certain GPS codes offering more accurate navigation capabilities may be denied to the civil user community.
4. IFF is compatible with ATC modes S but not available to civil users.

7.1.2 Communications Control Systems

The control of the aircraft suite of communications systems, including internal communications, has become an increasingly complex task. This task has expanded as aircraft speeds and traffic density have increased and the breadth of communication types have expanded. The communications control function is increasingly being absorbed into the flight management function as the management of communication type, frequency selection and intended aircraft flight path become more interwoven. Now the flight management system can automatically select and tune the communications and navigation aids required for a particular flight leg, reducing crew workload and allowing the crew to concentrate more on managing the on-board systems.

7.2 RF Propagation

The number of antennas required on-board an aircraft to handle all the sensors, communications and navigation aids is considerable. The CNI aspects of RF systems integration on a fighter aircraft have already been described in Chapter 2.

Civil aircraft adopted for military applications also have a comprehensive CNI antenna complement. This is compounded by the fact that many of the key pieces of equipment may be replicated in duplicate or triplicate form. This is especially true of VHF, HF, VOR and DME equipments. Figure 7.2 shows typical antenna locations on a Boeing 777 aircraft; this is indicative of the installation on most civil aircraft operating today, particularly those operating transoceanic routes. Owing to their operating characteristics and transmission properties, many of these antennas have their own installation criteria. SATCOM antennas communicating with satellites will have the antennas mounted on the top of the aircraft so as to have the best coverage of the sky. ILS antennas associated with the approach and landing

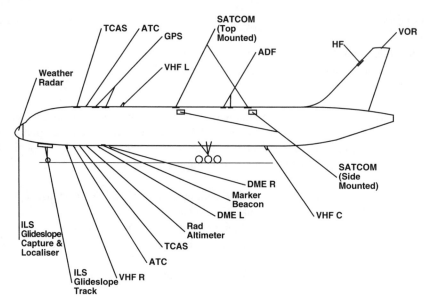

Figure 7.2   Typical aircraft CNI antenna (Boeing 777 example).

phase will be located on the forward, lower side of the fuselage. Others may require continuous coverage while the aircraft is manoeuvring and may have antennas located on both upper and lower parts of the aircraft; multiple installations are commonplace. In addition to these antennas, military aircraft will have additional communications fitted commensurate with their military role.

In aviation, communications between the aircraft and the ground (air traffic/local approach/ground handling) have historically been by means of voice communication. More recently, data link communications have been introduced owing to their higher data rates and in some cases superior operating characteristics. As will be seen, data links are becoming widely used in the HF, VHF and UHF bands for basic communications but also to provide some of the advanced reporting features required by FANS. In the military community, data links have a particular significance in relation to target reporting and the sharing of tactical and targeting information, as will be described in the section on network-centric operations. The most common methods of signal modulation are:

1. *Amplitude modulation (AM)*. This type of modulation concentrates the information being carried by the transmission in relatively narrow sidebands. AM communications are susceptible to noise and jamming.
2. *Frequency modulation (FM)*. FM modulation is more sophisticated and spreads the transmission across a wider frequency spectrum than AM, thereby reducing the vulnerability of the signal to interference and jamming. This technique is generically known as spread spectrum modulation and can be used in a number of ways using differing modulation techniques. The spread spectrum is a useful adjunct to low probability of intercept (LPI) systems where the intention is to make the task of an adversary detecting the signals more difficult.

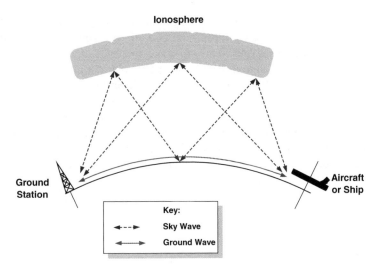

Figure 7.3 HF communications signal propagation.

There are many extremely sophisticated methods of signal modulation. In the military environment these are used to maximise the performance of the signal under adverse operating conditions while minimising the probability of intercept.

7.2.1 High Frequency

High frequency (HF) covers the communications band between 3 and 30 MHz and is a very common communications means for land, sea and air. The utilised band is HF SSB/AM over the frequency range 2.000–29.999 MHz using a 1 kHz (0.001 MHz) channel spacing. The primary advantage of HF communications is that this system offers communication beyond the line of sight. This method does, however, suffer from idiosyncrasies with regard to the means of signal propagation.

Figure 7.3 shows that there are two main means of propagation, known as the sky wave and the ground wave.

The sky wave method of propagation relies upon single- or multiple-path bounces between the earth and the ionosphere until the signal reaches its intended location. The behaviour of the ionosphere is itself greatly affected by radiation falling upon the earth, notably solar radiation. Times of high sunspot activity are known adversely to affect the ability of the ionosphere as a reflector. It may also be affected by the time of day and other atmospheric conditions. The sky wave as a means of propagation may therefore be severely degraded by a variety of conditions, occasionally to the point of being unusable.

The ground wave method of propagation relies upon the ability of the wave to follow the curvature of the earth until it reaches its intended destination. As for the sky wave, the ground wave may on occasions be adversely affected by atmospheric conditions. Therefore, on occasion, HF voice communications may be corrupted and prove unreliable, although HF data links are more resistant to these propagation upsets, as described below.

HF communications are one of the main methods of communicating over long ranges between air and ground during oceanic and wilderness crossings when there is no line of sight between the aircraft and ground communications stations. For reasons of availability, most long-range civil aircraft are equipped with two HF sets with an increasing tendency also to use HF data link if polar operations are contemplated.

HF data link (HFDL) offers an improvement over HF voice communications owing to the bit encoding inherent in a data link message format which permits the use of error-correcting codes. Furthermore, the use of more advanced modulation and frequency management techniques allows the data link to perform in propagation conditions where HF voice would be unusable or incomprehensible. An HFDL service is provided by ARINC using a number of ground stations. These ground stations provide coverage out to ∼2700 nm and on occasion provide coverage beyond that. Presently, HFDL ground stations are operating at the following locations (Figure 7.4):

1. Santa Cruz, Bolivia.
2. Reykjavik, Iceland.
3. Shannon, Ireland.
4. Auckland, New Zealand.
5. Krasnoyarsk, Russia.
6. Johannesburg, South Africa.
7. Hat Yai, Thailand.
8. Barrow, Alaska, USA.
9. Molokai, Hawaii, USA.
10. Riverhead, New York, USA.
11. San Francisco, California, USA.
12. Bahrain.
13. Gran Canaria, Canary Islands.

7.2.2 Very High Frequency

Voice communication at very high frequency (VHF) is probably the most heavily used method of communication used by civil aircraft, although ultrahigh frequency (UHF) is

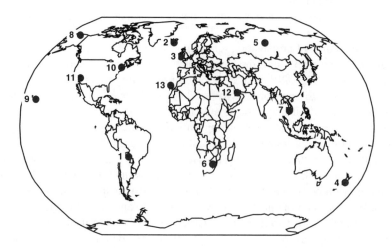

Figure 7.4 HF data link ground stations.

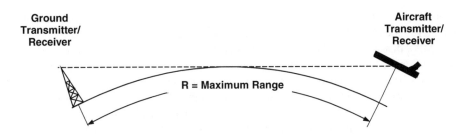

Figure 7.5 VHF signal propagation.

generally preferred for military use. The VHF band for aeronautical applications operates in the frequency range 118.000–135.975 MHz with a channel spacing in past decades of 25 kHz (0.025 MHz). In recent years, to overcome frequency congestion, and taking advantage of digital radio technology, channel spacing has been reduced to 8.33 kHz (0.00833 MHz) which permits 3 times more radio channels in the available spectrum. Some parts of the world are already operating on the tighter channel spacing – this will be discussed in more detail later in the chapter in the section on global air transport management (GATM).

The VHF band also experiences limitations in the method of propagation. Except in exceptional circumstances VHF signals will only propagate over line of sight. That is, the signal will only be detected by the receiver when it has line of sight or can 'see' the transmitter. VHF transmissions possess neither of the qualities of HF transmission and accordingly neither sky wave nor ground wave properties apply. This line-of-sight property is affected by the relative heights of the radio tower and aircraft. This characteristic applies to all radio transmissions greater than ~100 MHz, although the precise onset is determined by the meteorological conditions prevailing at the time (Figure 7.5).

The formula that determines the line-of-sight range for VHF transmissions and above is as follows:

$$R = 1.2.\sqrt{H_t} + 1.2.\sqrt{H_a}$$

where R is the range (nautical miles), H_t is the height of the transmission tower (ft) and H_a is the height of the aircraft (ft).

Therefore, for an aircraft flying at 35 000 ft, transmissions will generally be received by a 100 ft high radio tower if the aircraft is within a range of around 235 nautical miles.

Additionally, VHF and higher-frequency transmissions may be masked by terrain, by a range of mountains, for example. These line-of-sight limitations also apply to equipment operating in higher-frequency bands and mean that VHF communications and other equipment operating in the VHF band or above – such as the navigation aids VOR and DME – may not be used except over large land masses, and then only when there is adequate transmitter coverage. Most long-range aircraft have three pieces of VHF equipment, with one usually being assigned to ARINC communications and reporting system ACARS transmissions, although not necessarily dedicated to that purpose. The requirements for certifying the function of airborne VHF equipment are given in Advisory Circular AC 20-67B (1986), while RTCA DO-186 (1984) specifies the necessary minimum operational performance standards (MOPS). There are advanced techniques that may be used in sophisticated military equipment that mitigate against these fundamental limitations. Such systems are said to possess an over-the-horizon (OTH) capability.

A number of VHF data links (VHFDL) may be used, and these are discussed in more detail later in the chapter. ACARS is a specific variant of VHF communications operating on 131.55 MHz that utilises a data link rather than voice transmission. As will be seen during the discussion on future air navigation systems, data link rather than voice transmission will increasingly be used for air-to-ground, air-to-ground and air-to-air communications as higher data rates may be used while at the same reducing flight crew workload. ACARS is dedicated to downlinking operational data to the airline operational control centre. The initial leg is by using VHF communications to an appropriate ground receiver, thereafter the data may be routed via land-lines or microwave links to the airline operations centre. At this point it will be allowed access to the internal airline storage and management systems: operational, flight crew, maintenance, etc.

All aircraft and air traffic control centres maintain a listening watch on the international distress frequency: 121.5 MHz. In addition, military controllers maintain a listening watch on 243.0 MHz in the UHF band. This is because the UHF receiver could detect the harmonics of a civil VHF distress transmission and relay the appropriate details in an emergency (second harmonic of $121.5\,\text{MHz} \times 2 = 243.0\,\text{MHz}$; these are the international distress frequencies for VHF and UHF bands respectively).

7.2.3 Satellite Communications

Satellite communications provide a more reliable method of communications using the International Maritime Satellite Organisation (INMARSAT) satellite constellation which was originally developed for maritime use. Now satellite communications, abbreviated to SATCOM, form a useful component of aerospace communications. In addition there are dedicated and secure military satellite systems not addressed in this book for obvious reasons. In this publication, SATCOM is described as it uses similar principles of operation and is also used in conjuction with the global air transport management (GATM) described later.

The principles of operation of SATCOM are shown in Figure 7.6. The aircraft communicates via the INMARSAT constellation and remote ground earth station by means of C-band uplinks and downlinks to/from the ground stations and L-band links to/from the aircraft. In this way, communications are routed from the aircraft via the satellite to the ground station and on to the destination. Conversely, communications to the aircraft are routed in the reverse fashion. Therefore, provided the aircraft is within the area of coverage or footprint of a satellite, communication may be established.

The airborne SATCOM terminal transmits on frequencies in the range 1626.5–1660.5 MHz and receives messages on frequencies in the range 1530.0–1559.0 MHz. Upon power-up, the radio-frequency unit (RFU) scans a stored set of frequencies and locates the transmission of the appropriate satellite. The aircraft logs on to the ground earth station network so that any ground stations are able to locate the aircraft. Once logged on to the system, communications between the aircraft and any user may begin. The satellite to ground C-band uplink and downlink are invisible to the aircraft, as is the remainder of the earth support network.

The coverage offered by the INMARSAT constellation was a total of four satellites in 2001. Further satellites are planned to be launched in the near future. The INMARSAT satellites are placed in earth geostationary orbit above the equator in the locations shown in Figure 7.7.

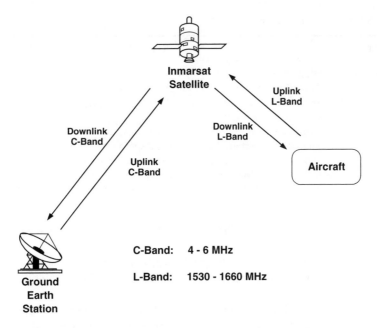

Figure 7.6   SATCOM principles of operation.

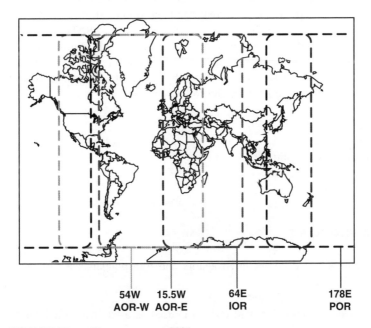

Figure 7.7   INMARSAT satellite coverage – 2001.

Table 7.1   SATCOM configurations

Configuration	Capabilities
Aero-H/H+	High gain. Aero-H offers a high-gain solution to provide a global capability and is used by long-range aircraft. Aero-H+ was an attempt to lower cost by using fewer satellite resources. Provides cockpit data, cockpit voice and passenger voice services
Aero-I	Intermediate gain. Aero-I offers similar services to Aero-H/H+ for medium- and short-range aircraft. Aero-I uses the spot beam service
Aero-C	Version that allows passengers to send and receive digital messages from a PC
Aero-M	Single-channel SATCOM capability for general aviation users

1. Two satellites are positioned over the Atlantic: AOR-W at 54° west and AOR-E at 15.5° west.
2. One satellite is positioned over the Indian Ocean: IOR at 64° east.
3. One satellite is positioned over the Pacific Ocean: POR at 178° east.

Blanket coverage is offered over the entire footprint of each of these satellites. In addition there is a spot beam mode that provides cover over most of the land mass residing under each satellite. This spot beam coverage is available to provide cover to lower-capability systems that do not require blanket oceanic coverage.

The geostationary nature of the satellites does impose some limitations. Owing to low grazing angles, coverage begins to degrade beyond 80° north and 80° south and fades completely beyond about 82°. Therefore, no coverage exists in the extreme polar regions, a fact assuming more prominence as airlines seek to expand northern polar routes. A second limitation may be posed by the performance of the on-board aircraft system in terms of antenna installation, and this is discussed shortly. Nevertheless, SATCOM is proving to be a very useful addition to the airborne communications suite and promises to be an important component as future air navigation system (FANS) procedures are developed.

A number of different systems are offered by SATCOM as described in Table 7.1. A SATCOM system typically comprises the following units:

- Satellite data unit (SDU);
- Radio-frequency unit (RFU);
- Amplifiers, diplexers/splitters;
- Low-gain antenna;
- High-gain antenna.

7.3 Transponders

There are a number of different interrogators and transponders used on military aircraft. These are as follows:

1. Distance measurement equipment (DME) is used as a navigation aid for both civil and military aircraft (see Chapter 8).

2. Tactical air navigation (TACAN) is used as a navigation aid for military aircraft solely (see Chapter 8).
3. ATC mode S is used both on military and civil aircraft, usually in association with the traffic collision avoidance system (TCAS). ATC mode S and TCAS are described below.
4. Automatic dependent surveillance – address mode (ADS/A) is used to support FANS and GATM developments during oceanic crossings using HF communications.
5. Automatic dependent surveillance – broadcast mode (ADS/B) is used to support FANS and GATM developments over land using VHF communications.
6. Identification friend or foe (IFF) is used by the military specifically for identification of threat aircraft. IFF is compatible with ATC mode S and works on the same frequencies of 1090 MHz (TX or interrogator) and 1090 MHz (RX or transponder) but carries additional identification codes specifically for military purposes.

DME and TACAN are described under the Communications and Navaids section while ATC mode S is described under the GATM section of Chapter 8.

7.3.1 Air Traffic Control (ATC) Transponder – Mode S

As a means to aid the identification of individual aircraft and to facilitate the safe passage of aircraft through controlled airspace, the ATC transponder allows ground surveillance radars to interrogate aircraft and decode data which enables correlation of a radar track with a specific aircraft. The principle of transponder operation is shown in Figure 7.8. A ground-based primary surveillance radar (PSR) will transmit radar energy and will be able to detect an aircraft by means of the reflected radar energy – termed the aircraft return. This will

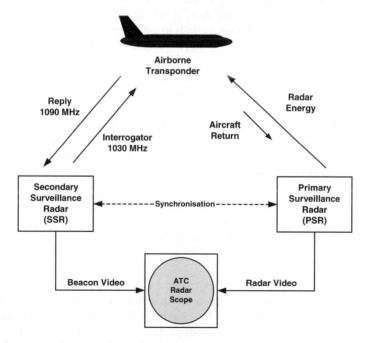

Figure 7.8 Principle of transponder operation.

enable the aircraft return to be displayed on an ATC console at a range and bearing commensurate with the aircraft position. Coincident with the primary radar operation, a secondary surveillance radar (SSR) will transmit a series of interrogation pulses that are received by the on-board aircraft transponder. The transponder aircraft replies with a different series of pulses which give information relating to the aircraft, normally aircraft identifier and altitude. If the PSR and SSR are synchronised, usually by being co-boresighted, then both the presented radar returns and the aircraft transponder information may be presented together on the ATC console. Therefore, the controller will have aircraft identification (e.g. BA 123) and altitude presented alongside the aircraft radar return, thereby greatly improving the controller's situational awareness.

The system is also known as identification friend or foe (IFF)/secondary surveillance radar (SSR), and this nomenclature is in common use in the military field. On-board the aircraft, the main elements are as listed below:

1. ATC transponder controller unit for setting modes and response codes.
2. A dedicated ATC transponder unit.
3. An ATC antenna unit with an optional second antenna. It is usual to utilise both upper and lower mounted antennas to prevent blanking effects as the aircraft manoeuvres.

The SSR interrogates the aircraft by means of a transmission on the dedicated frequency of 1030 MHz which contains the interrogation pulse sequence. The aircraft transponder replies on a dedicated frequency of 1090 MHz with a response that contains the reply pulse sequence with additional information suitably encoded in the pulse stream.

In its present form the ATC transponder allows aircraft identification – usually the airline call-sign – to be transmitted when using mode A. When Mode C is selected, the aircraft will respond with its identifier together with altitude information.

More recently, an additional mode – mode S or mode select – has been introduced with the intention of expanding this capability. In ATC mode S the SSR uses more sophisticated monopulse techniques that enable the aircraft azimuth bearing to be determined more quickly. Upon determining the address and location of the aircraft, it is entered into a roll call file. This, together with details of all the other aircraft detected within the interrogator's sphere of operation, forms a complete tally of all the aircraft in the vicinity. Each mode S reply contains a discrete 24 bit address identifier. This unique address, together with the fact that the interrogator knows where to expect the aircraft from its roll call file, enables a large number of aircraft to operate in a busy air traffic control environment (see section 7.3.2 for details of the traffic collision avoidance system).

ATC mode S has other features that enable it to provide the following:

• Air-to-air as well as air-to-ground communication;
• The ability of aircraft autonomously to determine the precise whereabouts of other aircraft in their vicinity.

Mode S is an improved conventional secondary radar operating at the same frequencies (1030/1090 MHz). Its 'selectivity' is based on unambiguous identification of each aircraft by unique 24 bit addresses. This acts as its technical telecommunications address, but does not replace the mode A code. There are also plans for recovery of the A and C codes via mode S.

Apart from this precise characterisation of the aircraft, mode S protects the data it transmits owing to the inclusion of several parity bits which means that up to 12 erroneous bits may be tolerated by the application of error detection and correction algorithms. For transmission, these parity bits are superimposed on those of the mode S address.

Finally, mode scan may be used to exchange longer, more varied data streams, which can even be completely unplanned. To do this, mode S transmissions between the station and the transponder use highly sophisticated 56 or 112 bit formats called frames. They fall into three main categories: 56 bit surveillance formats, 112 bit communication formats with a 56 bit data field, which are in fact 'extended' surveillance formats (uplink COMM-As and downlink COMM-Bs), and 112 bit communication formats with an 80 bit data field (uplink COMM-Ds and downlink COMM-Ds). This feature will be of use in facilitating the transmission and interchange of flight plans dynamically revised in flight which is one of the longer-term aims of FANS.

Mode S also has the capability of providing a range of data formats, from level 1 to level 4. These are categorised as follows:

1. Level 1. This is defined as the minimum capability mode S transponder. It has the capability of reply to mode S interrogations but has no data link capability. All the messages provided by level 1 are short (56 bit) messages.
2. Level 2. These transponders support all the features of the level 1 transponder with the addition of standard length data link word formats. This can entail the use of longer messages (112 bit). Some of the messages are used for TCAS air-to-air communication while others are utilised for air-to-ground and ground-to-air communication as part of the enhanced surveillance data access protocol system (DAPS) requirements.
3. Level 3. The level 3 transponders embrace the same functionality as level 2 with the additional ability to receive extended length messages (ELM) which comprise 16 segments of information, each containing a 112 bit message.
4. Level 4. Level 4 has the full functionality of level 3 with the capability of transmitting ELM messages of up to 16 segments of 112 bit word messages.

Originally it was envisaged that ATC mode S would be the primary contender to provide the CNS/ATM functionality by providing large block transfers of information. More recently it has been realised that VDL mode 4 might better serve this need, and levels 3 and 4 are no longer required.

When used together with TCAS, ATC mode S provides an important feature for FANS, that of automatic dependent surveillance – A (ADS-A). This capability will assist the safe passage of aircraft when operating in a direct routing mode.

7.3.2 Traffic Collision and Avoidance System

The traffic collision and avoidance system (TCAS) was developed in prototype form during the 1960s and 1970s to provide a surveillance and collision avoidance system to help aircraft avoid collisions. It was certified by the FAA in the 1980s and has been in widespread use in the United States in its initial form. The TCAS is based upon a beacon interrogator and operates in a similar fashion to the ground-based SSR already described. The system comprises two elements: a surveillance system and a collision avoidance system. The TCAS detects the range bearing and altitude of aircraft in the near proximity for display to the pilots.

The TCAS transmits a mode C interrogation search pattern for mode A and C transponder equipped aircraft and receives replies from all such equipped aircraft. In addition, the TCAS transmits one mode S interrogation for each mode S transponder equipped aircraft, receiving individual responses from each one. It will be recalled that mode A relates to range and bearing, while mode C relates to range, bearing and altitude and mode S to range, bearing and altitude with a unique mode S reply. The aircraft TCAS equipment comprises a radio transmitter and receiver, directional antennas, computer and flight deck display. Whenever another aircraft receives an interrogation it transmits a reply and the TCAS computer is able to determine the range depending upon the time taken to receive the reply. The directional antennas enable the bearing of the responding aircraft to be measured. The TCAS can track up to 30 aircraft but only display 25, the highest-priority targets being the ones that are displayed.

The TCAS is unable to detect aircraft that are not carrying an appropriately operating transponder or that have unserviceable equipment. A transponder is mandated if an aircraft flies above 10 000 ft or within 30 miles of major airports; consequently, all commercial aircraft and the great majority of corporate and general aviation aircraft are fitted with the equipment.

The TCAS exists in two forms: TCAS I and TCAS II. TCAS I indicates the range and bearing of aircraft within a selected range; usually 15–40 nm forward, 5–15 nm aft and 10–20 nm on each side. The system also warns of aircraft within ±8700 ft of the aircraft's own altitude.

The collision avoidance system element predicts the time to, and separation at, the intruder's closest point of approach. These calculations are undertaken using range, closure rate, altitude and vertical speed. Should the TCAS ascertain that certain safety boundaries will be violated, it will issue a traffic advisory (TA) to alert the crew that closing traffic is in the vicinity via the display of certain coloured symbols. Upon receiving a TA, the flight crew must visually identify the intruding aircraft and may alter their altitude by up to 300 ft. A TA will normally be advised between 20 and 48 s before the point of closest approach with a simple audio warning in the flight crew's headsets: 'TRAFFIC, TRAFFIC'. TCAS I does not offer any deconfliction solutions but does provide the crew with vital data in order that they may determine the best course of action.

TCAS II offers a more comprehensive capability with the provision of resolution advisories (RAs). TCAS II determines the relative motion of the two aircraft and determines an appropriate course of action. The system issues an RA via mode S, advising the pilots to execute the necessary manoeuvre to avoid the other aircraft. An RA will usually be issued when the point of closest approach is within 15 and 35 s, and the deconfliction symbology is displayed coincident with the appropriate warning.

A total of ten audio warnings may be issued. Examples are:

- 'CLIMB, CLIMB, CLIMB';
- 'DESCEND, DESCEND, DESCEND';
- 'REDUCE CLIMB, REDUCE CLIMB'.

Finally, when the situation is resolved: 'CLEAR OF CONFLICT'.

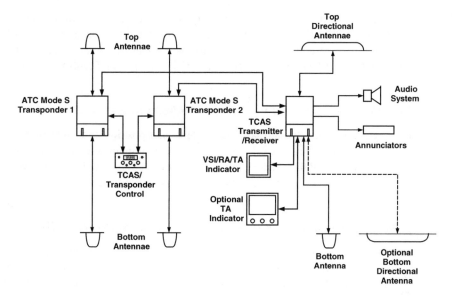

Figure 7.9 TCAS architecture showing related equipment and displays.

TCAS II clearly requires a high level of integration between the active equipment. Figure 7.9 shows the interrelationship between:

- TCAS transmitter/receiver;
- ATC mode S transponders;
- VSI display showing vertical guidance for TAs and RAs;
- Optional horizontal situational indicator for RAs that could be the navigation display;
- Audio system and annunciators;
- Antennas for ATC mode S and TCAS.

This is indicative of the level of integration required between ATC mode S transponders, TCAS, displays and annunciators. It should be noted that there are a variety of display options and the system shown does not represent the only TCAS option.

More recently, further changes have been introduced to TCAS II – known as TCAS II change 7. This introduces software changes and updated algorithms that alter some of the TCAS operating parameters. Specifically, change 7 includes the following features:

- Elimination of nuisance warnings;
- Improved RA performance in a multiaircraft environment;
- Modification of vertical thresholds to align with reduced vertical separation minima (RVSM) – see the section on global air transport management (GATM) and the civil equivalent future air navigation system (FANS);
- Modification of RA display symbology and aural annunciations.

The change 7 modifications became mandatory in Europe for aircraft with 30 seats or more from 31 March 2001 and for aircraft with more than 19 seats from 1 January 2000. The rest of the world will be following a different but broadly similar timescale for implementation. Change 7 is not mandated in the United States but it is expected that most aircraft will be equipped to that standard in any case. Further information can be found on AC-12955A, RTCA DO-181 DO-185 certification and performance requirements for TCAS II and mode S.

7.3.3 Automatic Dependent Surveillance – Address Mode (ADS-A)

ADS-A will be used to transmit the aircraft four-dimensional position and flight plan intent based upon GPS position during oceanic crossings. The communications media will be SATCOM or HF data link (HFDL). ADS-A requires the aircraft to be fitted with an FMS and CDU and with some means of displaying message alerts and annunciation.

7.3.4 Automatic Dependent Surveillance – Broadcast Mode (ADS-B)

ADS-B will be used to transmit four-dimensional position and flight plan intent based upon GPS position using line-of-sight VHF communications. Either mode S or digital VHF radio will be used to transmit the data. ADS-A requires a cockpit display of traffic information.

7.3.5 Identification Friend or Foe (IFF)

There are two ways in which IFF equipment may be used:

- Providing 360° coverage in order to be able to respond to interrogation and receive transponder returns from friendly aircraft in any direction. In this respect the operation is very similar to the airborne operation of ATC mode S when used in association with the TCAS.
- Used in association with a primary radar sensor in order to be able specifically to identify targets appearing within the radar scan. This operates in the same way as a ground surveillance radar interrogating aircraft in the vicinity of an airfield.

An example of the first type is the advanced IFF (AIFF) AN/APX-113(V) used on-board the F-16 aircraft but which is typical of equipment of this type (Figure 7.10). IFF equipment is sometimes referred to in a generic sense as IFF mark XII which relates the generic family of

Figure 7.10   IFF set AN/APX-113(V). (BAE SYSTEMS)

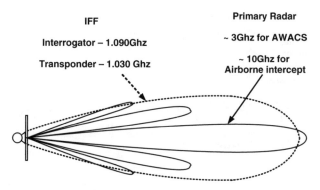

Figure 7.11 Co-boresighting of IFF interrogator with radar.

present IFF equipment. Equipment such as the AN/APX-113(V) has the following capabilities:

- Multiple antenna configurations – electronic or mechanical scan;
- MIL-STD-1553 interface to connect to the rest of the avionics system units;
- Ability to provide encryption capability;
- Provision of growth to accommodate future functional modes.

The use of the interrogator co-located and co-boresighted with the main radar creates a problem as illustrated in Figure 7.11. The primary radar will be operating at a higher frequency than the 1090/1030 MHz that the interrogator uses. An airborne early warning radar will be operating at ∼3 GHz, while an airborne intercept (AI) radar will be operating higher still at ∼10 GHz. For a given radar antenna size the beamwidth is inversely proportional to frequency, so, the higher the frequency, the more narrow is the beamwidth. The IFF beamwidth will therefore encompass the main beam and several sidelobes of the radar beam and may therefore be receiving returns from targets that are not of fundamental interest to the radar. This effect will be more pronounced for AI as opposed to AWACS radar.

7.4 Data Links

The use of voice was the original means of using RF communications. However, the use of speech has severe limitations; it is slow in terms of conveying information and prone to misunderstanding, whereas high bandwidth data links can delivery more information, if necessary incorporating error correction or encryption. In the avionics sense, typical data link users are portrayed in Figure 7.12.

Primary users include:

- Strategic airborne sensor platforms such as E-4, E-6, E-8, Global Hawk and satellites;
- Tactical airborne sensors and shooters – F-15, F-16, F-18, Harrier, Tornado, Eurofighter Typhoon and tactical UAVs/UCAVs among others;
- Shipborne sensors;
- Land forces.

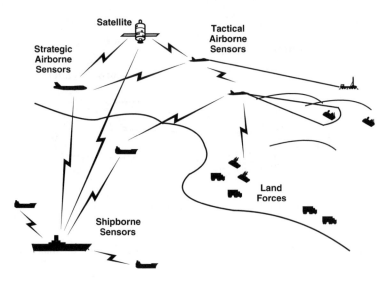

Figure 7.12 Typical data link users.

Many of the data links are limited to line-of-sight operation owing to the transmission characteristics of the RF frequencies being employed. However, the use of communications satellites to perform a relay function permits transmission of data over the horizon (OTH), thereby enabling intra- and intertheatre communications.

Typical data packages that may be delivered by data links include:

- Present position reporting;
- Surveillance;
- Aircraft survival, EW and intelligence information;
- Information management;
- Mission management;
- Status.

The primary data links used for communications between airborne platforms and space and surface platforms are as follows:

1. Link 16. This is the most commonly used avionics data link and is usually manifested in avionics systems as the joint tactical information distribution system (JTIDS). The JTIDS is also compatible with the US Navy data link satellite tactical data information link J (S-TADIL J). Link 16 operates in the UHF frequency band in the same frequency range as identification friend or foe (IFF), distance measurement equipment (DME) and TACAN, as described below.
2. Link 11. Certain strategic aircraft assets such as E-4 and Nimrod MRA4 associated with joint maritime operations also have the capability of operating with link 11 – a data link commonly used by naval forces.

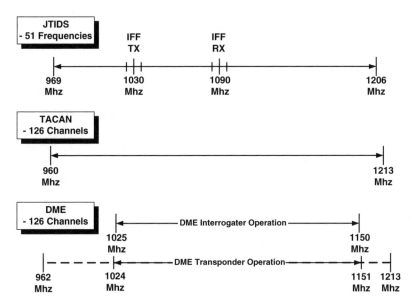

Figure 7.13   JTIDS frequency band.

7.4.1   JTIDS Operation

The characteristics of the JTIDS frequency band and how this is shared with other equipment is shown in Figure 7.13. The JTIDS characteristics are as follows:

1. Data are transmitted in the UHF band between 969 and 1206 MHz.
2. Frequency-hopping techniques are employed to provide ECM jam-resistant properties.
3. A total of 51 channels are provided at 3 MHz spacing.

JTIDS transmissions are constrained to avoid interfering with the IFF frequencies at 1030 MHz (TX) and 1090 MHz (RX), and JTIDS is not employed within ±20 MHz of these frequencies. The problem of mutual equipment interference is one that has to be frequently faced on highly integrated military avionic platforms.

Integration with the host aircraft avionics system usually takes the form shown in Figure 7.14. The JTIDS terminal and associated antenna are shown on the right of the diagram. The equipment shown in this particular case is the URC-138 terminal, a typical example of which is shown in the inset. Such a terminal – compatible with tanker/transport, fighter and helicopter environments will have the following capabilities/characteristics:

Data rate	28.8–238 kbps
Weight	40 lbs
Dimensions	12.5 in deep
	7.5 in high
	10 in wide
	(equivalent to 8ATR)
Power	750 W

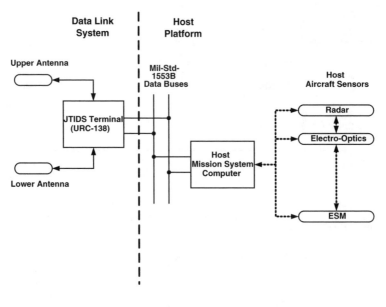

JTIDS Terminal (Rockwell Collins)

Figure 7.14 JTIDS integration with the avionics system.

The JTIDS terminal typically interfaces with the host aircraft mission system computer via MIL-STD-1553B data buses. The host mission systems computer is connected in turn to the aircraft platform sensors embracing radar, electrooptics, ESM/EW and other operational sensors depending upon the host platform sensor fit.

Such a system will in most cases include secure voice capabilities and the ability to transmit encrypted data. Clearly, in a real battlefield scenario there is a need to share classified information between some but certainly not all of the participants. With the data that many participants will utilise in a communications systems used in a military environment, therefore, there needs to be the capability of separating secure/classified data from the data that are 'in the clear' or open to all participants.

7.4.2   Other Data Links

Apart from JTIDS which specifically operates in the UHF band, other transmission techniques may be used to communicate between military platforms. These are:

- SATCOM;
- HF data links;
- Local cooperative data links.

SATCOM and HF data links or HFDL are already used extensively by the maritime and civil aviation communities. The same transmission capabilities are open to the military community, except that in many cases data protection/encryption may be required depending upon the sensitivity of the message content. Therefore, many military communications systems are designed to include an encryption/decryption device at the front end – between the processor and transmit/receive elements. By using suitable encryption 'keys', the necessary levels of encryption may be achieved depending upon the sensitivity of the message content.

Aircraft such as the F-22 Raptor use a local cooperative data link to aid in the data sharing and coordination of a group of aircraft embarked upon a shared mission. As outlined in Chapter 2, on the F-22 it was the intention to utilise two phased array related cooperative data links. These are:

- A common high-band data link (CHBDL) or in-flight data link (IFDL) operating at around 10 GHz and utilising three antenna locations to pass data between adjacent aircraft;
- A cooperative engagement capability (CEC) using similar antenna configurations.

It is not clear whether either or both of these facilities have been included in the final aircraft configuration.

7.5   Network-centric Operations

Network-centric operations are becoming the latest 'force multiplier' element of modern airborne warfare in the same way as air-to-air refuelling and the availability of digital signal processing have been in the past. The use of high-bandwidth digital communications, together with sophisticated signal processing capabilities and the high bandwidth of internal platform interconnective buses and highways, have enabled the data interchange between a variety of sensor and weapons platforms to ascend to much higher levels. The command and control (C^2) structure, sensor and weapon delivery platforms have become integrated at a level that would previously have been unimaginable. This connectivity, allied with the capability of ultrahigh-resolution sensors, enables target and threat data to be shared at all levels of the force structure with unprecedented speed and fidelity.

The nature of network-centric operations may be appreciated by reference to Figure 7.15. This figure illustrates three tiers of interconnected centres or nodes, each of which are interconnected at the three respective levels and which are also interconnected between levels or layers by specific network nodes. In the figure, interconnecting nodes between layers are shown in black while supporting nodes within a layer are portrayed in white.

The network comprises three layers, which in descending order of importance/authority are as follows:

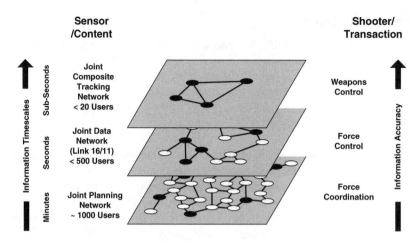

Figure 7.15   Nature of network-centric operations.

1. *Weapons control layer.* This embraces a limited number of participants – 20 or fewer – all operating at the strategic level, all interconnected within the layer and also connected to subordinate layers to implement force control. It is at this level that the rules of engagement (RoE) for a particular task execution will be decided and implemented. Airborne platforms in this category may include platforms such as AWACS or similar airborne assets.
2. *Force control layer.* This layer exercises control over the force structure, implementing the RoE and engaging targets on the basis of sensor and tactical information exchange. This joint data network will be typified by link 11 and link 16 users exchanging data at the tactical level and deciding the priorities according to different target types depending upon geographical and time currency of intelligence and target data. This layer may include up to 500 users including strategic and tactical force assets. Aircraft platforms may include maritime reconnaissance and fighter aircraft.
3. *Force coordination layer.* This layer embraces a joint planning network invoking force coordination at the local or theatre level and exercising force coordination to achieve maximum force effect or to avoid 'blue-on-blue' fratricide engagements. In the aviation context this may include fighter aircraft, attack helicopters, UAVs and forward air controllers (FAC). This network may extend across 1000 users or contributors.

It is noteworthy that the nature of the information changes as it migrates from the lowest to the highest level within this hierarchy according to the simple tabulation below:

Layer	Information timecales	Information accuracy
Weapons control	Subseconds	High accuracy
		^
Force control	Seconds	^
		^
Force coordination	Minutes	^
		Low accuracy

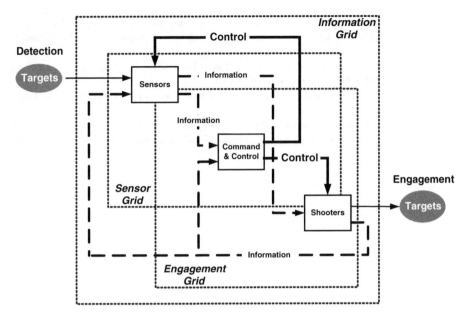

Figure 7.16   Sensor/shooter information grid.

The doctrines associated with network-centric operations have many proponents, especially in the United States. One of the most celebrated and vocal proponents is USN Vice Admiral Arthur K. Cebrowski together with J.J. Garstka (1998).

Key elements to the operation of a network-centric operation relate to the information grid interrelationship between the 'sensors' and 'shooters' involving the command and control element (Figure 7.16). This depicts the information flow and command links that exist between the detection of a target on the left to the engagement of the target on the right. It embraces the overlapping nature of information and engagement grids that determine the process by which information is processed between sensor, command and control and shooter to ensure that the necessary information is provided to the command function in order that a target may be correctly assessed, command and control may be exercised and battle damage assessment may be accomplished.

The high-bandwidth communications available for intelligence and target data interchange between these functional entities mean that radar video or electrooptic images may be exchanged in near real time. Therefore, the decision time to identify, categorise and authorise target engagement has reportedly decreased from hours (1991 Gulf War) to tens of minutes (Afghanistan War), with the aim of reducing this to a matter of minutes in future conflicts.

In the avionics environment this information exchange is achieved by using data links using a series of transmission means as described elsewhere in this chapter. On-board the airborne platform the availability of high-bandwidth fibre-optic or fibre-channel communications as described in Chapter 2, Technology and Architectures.

References

Advisory circular AC 20-67B, Airborne VHF communications installations, 16 January 1986.
Advisory circular AC 20-131A, Air worthiness approval of traffic alert and collision avoidance systems (TCAS II) and mode S transponders, 29 March 1993.

Advisory circular 129-55A, Air carrier operational approval and use of TCAS II, 27 August 1993.

Cebrowski, A.K. and Garstka, J.J. (1998) Network-centric warfare: its origin and future. *Proceedings of Naval Institute*, January.

RTCA DO-181, Minimum operational performance standards for air traffic control radar beacon system/ mode select (ATCRBS/mode S) airborne equipment.

RTCA DO-185, Minimum operational performance standards for traffic alert and collision avoidance systems (TCAS) airborne equipment.

RTCA DO-186, Minimum operational performance standards (MOPS) for radio communications equipment operating with the radio frequency range 117.975 to 137.000 MHz, dated 20 January 1984.

8 Navigation

8.1 Navigation Principles

8.1.1 Introduction

Navigation has been an ever-present component of humankind's exploitation of the capability of flight. While the principles of navigation have not changed since the early days of sail, the increased speed of flight, particularly with the advent of the jet age, has placed an increased emphasis upon accurate navigation. The increasingly busy skies, together with rapid technology developments, have emphasised the need for higher-accuracy navigation and the means to accomplish it. Navigation is no longer a matter of merely getting from A to B safely, it is about doing this in a fuel-efficient manner, keeping to tight airline schedules, and avoiding other air traffic – commercial, general aviation, leisure and military. Navigation of military aircraft has to comply with the same regulations as civil traffic when operating in controlled airspace. Platforms adopted from civil aircraft will retain the civil navigation systems as described in the companion volume 'Civil Avionics Systems' (Moir and Seabridge, 2003), some of which are described here for ease of reference. More than likely, legacy military platforms will be fitted with a bespoke system meeting most but possibly not all the latest requirements specified for controlled airspace and may on occasion need to operate with certain limitations until the necessary upgrades are embodied.

Outside controlled airspace in operational theatres the navigational accuracy will be determined by the accuracy provided by the platform mission, weapons system and possibly the weapons being carried. Operational mission navigation constraints include the optimisation of routing to avoid hazardous surface-to-air missile (SAM) and anti-aircraft artillery (AAA). Routing may also take into account the need to maximise the aircraft stealth or low-observability characteristics. Therefore, while military aircraft need to adopt all the necessary features to enable them to operate safely alongside civil aircraft in today's crowded airspace, they also add a further layer of complexity to the mission management function.

This section summarises some of the modern methods of navigation, leading to more detailed descriptions of how each technique operates. A later section in the chapter relates to

Military Avionics Systems   I. Moir and A. Seabridge
© 2006 John Wiley & Sons, Ltd

future air navigation system (FANS) requirements, also known within military circles as global air transport management (GATM).

The main methods of navigation as practised today may be summarised and simplified as follows:

- Classic dead-reckoning navigation using air data and magnetic, together with Doppler or LORAN-C;
- Radio navigation using navigation aids – ground-based radio-frequency beacons and airborne receiving and processing equipment;
- Barometric inertial navigation using a combination of air data and inertial navigations (IN) or Doppler;
- Satellite navigation using a global navigation satellite system (GNSS), more usually a global positioning system (GPS);
- Multiple-sensor navigation using a combination of all the above.

The more recent the pedigree of the aircraft platform, the more advanced the navigational capabilities are likely to be. However, it is common for many legacy platforms to be retrofitted with inertial and GNSS navigation, the accuracy of which far outshines the capability of the original system.

8.1.2 Basic Navigation

The basic navigation parameters are shown in Figure 8.1 and may be briefly summarised as follows:

1. An aircraft will be flying at a certain height or altitude relative to a barometric datum (barometric altitude) or terrain (radar altitude).
2. The aircraft may be moving with velocity components in the aircraft X (V_x), Y (V_y) and Z (V_z) axes. Its speed through the air may be characterised as indicated airspeed (IAS) or Mach number (M). Its speed relative to the ground is determined by true airspeed (TAS) in still air conditions.
3. The aircraft will be flying on a certain heading; however, the prevailing wind speed and direction will modify this to the aircraft track. The aircraft track represents the aircraft path across the terrain and will lead to the destination or next waypoint of the aircraft. Wind speed and direction will modify the aircraft speed over the ground to ground speed.
4. The aircraft heading will be defined by a bearing to magnetic (compass) north or to true north relating to earth-related geographic coordinates.
5. The aircraft will be flying from its present position – defined by latitude and longitude to a waypoint also characterised by latitude and longitude.
6. A series of flight legs – defined by way points – will determine the aircraft designated flight path from the departure airfield to the destination airfield.

As has already briefly been described, there are sensors and navigation techniques that may be used solely or in combination to navigate the aircraft throughout the mission.

The relationship of the different axis sets is shown in Figure 8.2. These may be characterised as follows:

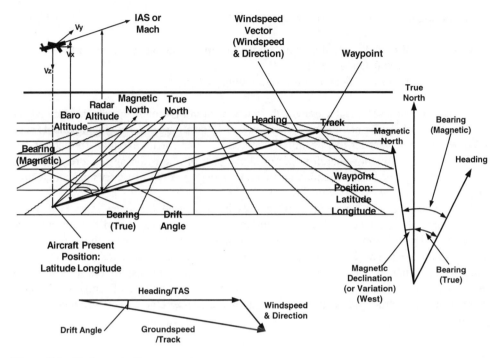

Figure 8.1 Basic navigation parameters.

1. Earth datum set. As shown in Figure 8.2, the earth axis reference set comprises the orthogonal set E_x, E_y, E_z, where:

- E_x represents true north;
- E_y represents east;
- E_z represents the local gravity vector.

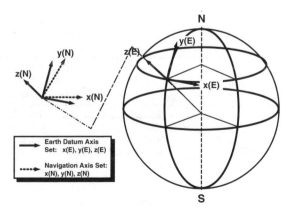

Figure 8.2 Earth-related coordinates.

2. The orthogonal aircraft axis set where:

- A_x is the aircraft longitudinal axis (corresponding to the aircraft heading);
- A_y is the aircraft lateral axis;
- A_z is the aircraft vertical axis (corresponding to E_z).

For navigation purposes, the accuracy with which the aircraft attitude may be determined is a key variable for Doppler navigation systems in which the Doppler velocity components need to be resolved into aircraft axes. Similarly, attitude is used for IN axis transformations.

The navigation function therefore performs the task of manoeuvring the aircraft from a known starting point to the intended destination, using a variety of sensors and navigation aids.

The classic method of navigation which has been in used for many years is to use a combination of magnetic and inertial directional gyros used together with airspeed information derived from the air data sensors to navigate in accordance with the parameters shown in Figure 8.1. This is subject to errors in both the heading system and the effects of en-route winds which can cause along-track and across-track errors. In the 1930s it was recognised that the use of radio beacons and navigation aids could significantly reduce these errors by providing the flight crew with navigation assistance related to precise points on the ground.

8.2 Radio Navigation

For many years the primary means of navigation over land, at least in continental Europe and the North American continent, was by means of radio navigation routes defined by VHF omniranging/distance measuring equipment (VOR/DME) beacons as shown in Figure 8.3. By arranging the location of these beacons at major navigation or crossing points, and in some cases airfields, it was possible to construct an entire airway network that could be used by the flight crew to define the aircraft flight from take-off to touchdown. Other radio frequency aids include distance measuring equipment (DME) and non-distance beacons (NDB). The operation of the radio navigation and approach aids is described elsewhere in this chapter.

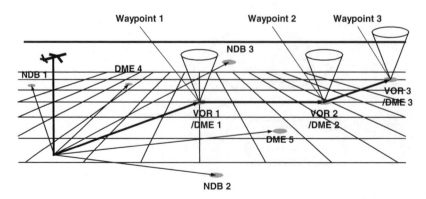

Figure 8.3 Radio navigation using VOR, DME and automatic direction finding (ADF).

Figure 8.3 shows:

1. Three VOR/DME beacon pairs: VOR 1/DME 1, VOR 2/DME 2 and VOR 3/DME 3 which define waypoints 1 to 3. These beacons represent the intended waypoints 1, 2 and 3 as the aircraft proceeds down the intended flight plan route – most likely an identified airway. When correctly tuned, the VOR/DME pairs succesively present the flight crew with bearing to and distance from the next waypoint.
2. Off-route DME beacons, DME 4 and DME 5, may be used as additional means to locate the aircraft position by means of the DME fix obtained where the two DME 4 and DME 5 range circles intersect. As will be seen, DME/DME fixes are a key attribute in the modern navigation system.
3. Off-route NDB beacons may be used as an additional means to determine the aircraft position by obtaining a cross-fix from the intersection of the bearings from NBD 1 and NDB 2. These bearings are derived using the aircraft ADF system.
4. In the military context: as well as these beacons TACAN and VORTAC beacons may be used specifically. TACAN has the particular advantage that it may be used in an offset mode where a navigational point may be specified in terms of a range and bearing offset from the TACAN beacon itself. TACAN also has certain features whereby it may be used to determine the range and bearing to other formations, eg a tanker aircraft, thereby facilitating airborne rendezvous operations.

Thus, in addition to using navigation information from the 'paired' VOR/DME or TACAN beacons that define the main navigation route, position fix, cross-fix, range or bearing information may also be derived from DME or NDB beacons in the vicinity of the planned route by using automatic direction-finding techniques. As has already been described in Chapter 7, a major limitation of the radio beacon navigation technique results from line-of-sight propagation limitations at the frequencies at which both VOR and DME operate. As well as the line-of-sight and terrain-masking deficiencies, the reliability and accuracy of the radio beacons can also be severely affected by electrical storms. Over longer ranges, LORAN-C could be used if fitted.

Owing to the line-of-sight limitations of these radio beacons, these navigation techniques were only usable overland where the beacon coverage was sufficiently comprehensive or for close off-shore routes where the beacons could be relied upon.

8.2.1 Oceanic Crossings

In 1969 the requirements were already specified for self-contained long-range commercial navigation by advisory circular AC 121-13. The appropriate document specified that self-contained navigation systems should be capable of maintaining a maximum error of ± 20 nm across track and ± 25 nm along track for 95% of the flights completed. Two systems were addressed in the specification: one using Doppler radar and the other using an inertial navigation system (INS).

In June 1977, the North Atlantic (NAT) minimum navigation performance specifications (MNPS) were altered to reflect the improved navigation sensors – see advisory circular AC 120-33. This defined the separation requirements for long-range navigation over the North Atlantic. The lateral separation was reduced from 120 to 60 nm while retaining the previous vertical separation of 2000 ft. Statistical limits were specified as to how long an aircraft was allowed to spend 30 nm off-track and between 50 and 70 nm off-track – the latter actually representing an overlap with an adjacent track. The standard deviation of lateral track errors was specified as 6.3 nm.

The Doppler radar system was specified as being an acceptable navigation means applying within certain geographical boundaries. Eastern and western entry points or 'gateways' were specified as entry and departure points into and out of the North Atlantic area. These gateways were identified as a number of specific named NDB or VOR beacons on both sides of the ocean. The North Atlantic transit area was specified as being the oceanic area bounded by the eastern and western gateways and lying between the latitude of 35°N and 65°N. By the standards of the allowable navigation routes available to today's aviators, this represented a very restricted envelope.

The aircraft equipment requirements were also carefully specified:

- Dual Doppler and computer systems;
- Dual polar path compasses;
- ADF;
- VOR;
- One LORAN receiver capable of being operated from either pilot's station.

8.3 Inertial Navigation Fundamentals

The availability of inertial navigation systems (INS) to the military aviation community during the early 1960s added another dimension to the navigation equation. Now flight crew were able to navigate by autonomous means using an on-board INS with inertial sensors. By aligning the platform to earth-referenced coordinates and present position during initialisation, it was now possible to fly for long distances without relying upon LORAN, VOR/DME or TACAN beacons. Waypoints could be specified in terms of latitude and longitude as arbitrary points on the globe, more suited to the aircraft's intended flight path rather than a specific geographic feature or point in a radio beacon network (Figure 8.4). The operation of inertial platforms is described in detail in section 8.8.

The specifications in force at this time also offered an INS solution to North Atlantic crossings as well as the dual-Doppler solution previously described. The inertial solution required serviceable dual INS and associated computers to be able to undertake the crossing.

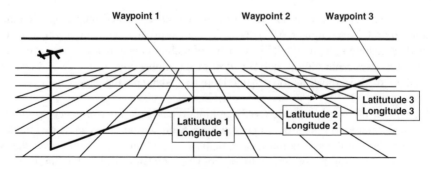

Figure 8.4 Fundamentals of inertial navigation.

There were also limitations on the latitudes at which the ground alignment could be performed – 76° north or south – as attaining satisfactory alignment becomes progressively more difficult the nearer to the poles the INS becomes.

Advisory circular AC 25-4 set forth requirements for operating an INS as a sole means of navigation for a significant portion of the flight. These requirements may be summarised as follows:

1. The ability to provide the following functions:

   - Valid ground alignment at all latitudes appropriate for the intended use of the INS;
   - The display of alignment status to the crew;
   - Provision of the present position of the aircraft in suitable coordinates: usually latitude from +90° (north) to −90° (south) and longitude from +180° (east) to −180° (west);
   - Provision of information on destinations or waypoints;
   - Provision of data to acquire and maintain the desired track and the ability to determine deviation from the desired track (across-track error);
   - Provision of information needed to determine the estimated time of arrival (ETA).

2. The ability to comply with the following requirements:

   - ±20 nm across track and ±25 nm along track;
   - Maintainenance of this accuracy on a 95% probability basis at representative speeds and altitudes and over the desired latitude range;
   - The capacity to compare the INS position with visual fixes or by using LORAN, TACAN, VOR, DME or ground radar (air traffic control).

3. The provision of a memory or in-flight alignment means. Alternatively, the provision of a separate electrical power source – usually a dedicated stand-alone battery – able to support the INS with full capability for at least 5 min in the event of an interruption of the normal power supply.

8.4 Satellite Navigation

The foregoing techniques were prevalent from the 1960s to the 1990s when satellite navigation became commonly available. The use of global navigation satellite systems (GNSS), to use the generic name, offers a cheap and accurate navigational means to anyone who possesses a suitable receiver. Although the former Soviet Union developed a system called GLONASS, it is the US global positioning system (GPS) that is the most widely used. The principles of satellite navigation using GPS will be described in detail later in the chapter.

GPS receivers may be provided for the airborne equipment in a number of ways:

1. Stand-alone GPS receivers, most likely to be used for GPS upgrades to an existing system. These are multichannel (typically, 12-channel) global navigation satellite system (GNSS) receivers – the B777 utilises this approach.
2. GPS receivers integrated into a multifunction receiver unit called a multimode receiver (MMR). Here, the GPS receiver function is integrated into one LRU along with VOR and ILS receivers.

8.4.1   *Differential GPS*

One way of overcoming the problems of selective availability is to employ a technique called differential GPS (DGPS). Differential techniques involve the transmission of a corrected message derived from users located on the ground. The correction information is sent to the user who can apply the corrections and reduce the satellite ranging error. The two main techniques are:

1. Local-area DGPS. The corrections are derived locally at a ground reference site. As the position of the site is accurately known, the satellite inaccuracies can be determined and transmitted locally to the user, in this case by line-of-sight VHF data link. The local-area DGPS system under development in the United States is called the local-area augmentation system (LAAS) and is described below.
2. Wide-area DGPS. The wide-area correction technique involves networks of data collection ground stations. Information is collected at several ground stations which are usually located more than 500 miles apart. The correction information derived by each station is transmitted to a central location where the satellite corrections are determined. Corrections are sent to the user by geostationary satellites or other appropriate means. The wide-area augmentation system (WAAS) being developed in the United States is outlined below.

Note that differential techniques may be applied to any satellite system. For GPS the basic accuracy without selective availability is about ± 100 m as opposed to ± 8 m when the full system is available. The DGPS developments under way in the United States are intended to improve the accuracy available to civil users.

8.4.2   *Wide-area Augmentation System (WAAS)*

The operation of WAAS, shown in Figure 8.5, is described as follows:

1. WAAS is a safety-critical system that augments basic GPS and will be deployed in the contiguous United States, Hawaii, Alaska and parts of Canada.
2. WAAS has multiple wide-area reference stations which are precisely surveyed and monitor the outputs from the GPS constellation.
3. These reference stations are linked to wide-area master stations where corrections are calculated and the system integrity assessed. Correction messages are uplinked to geostationary earth orbit (GEO) satellites that transmit the corrected data on the communications L1 band to aircraft flying within the WAAS area of coverage. Effectively, the GEO satellites act as surrogate GPS satellites.
4. WAAS improves the GPS accuracy to around ± 7 m which is a considerable improvement over the 'raw' signal. This level of accuracy is sufficient for Cat I approach guidance.

Some problems were experienced in initial system tests during 2000. Commissioning of the WAAS requires extensive testing to ensure integrity levels, accuracy, etc., and the system was finally commissioned during 2004.

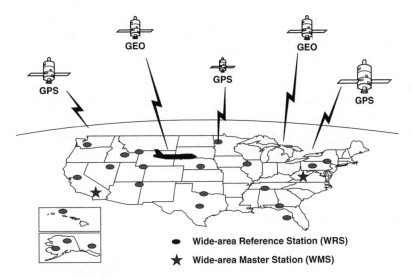

Figure 8.5   Wide-area augmentation system.

8.4.3   Local-area Augmentation System (LAAS)

The operation of LAAS is shown in Figure 8.6 and described below:

1. LAAS is intended to complement WAAS but at a local level.
2. LAAS works on similar principles except that local reference stations transmit correction data direct to user aircraft on VHF. As such, the LAAS coverage is restricted by VHF line-of-sight and terrain-masking limitations.

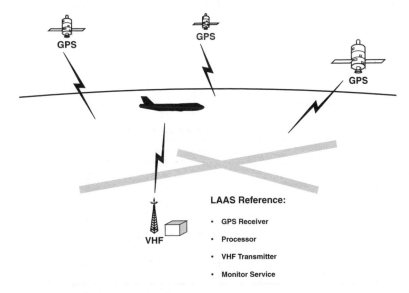

Figure 8.6   Local-area augmentation system.

3. LAAS improves the GPS accuracy to about ± 1 m, close to the higher GPS level of accuracy. This level of accuracy is sufficient to permit Cat II and Cat III approaches which are described more fully in Chapter 9.

Implementation is as before, with final deployment in 2006, although these timescales are apt to slip, as has the implementation of WAAS. According to present plans it is expected that LAAS will be deployed at up to 143 airfields throughout the United States. It was the anticipation of LAAS implementation that caused the United States to modify its stance upon the implementation of MLS as an approach aid successor to ILS, the space-based GPS system being seen as more flexible than ground-based MLS.

8.5   Integrated Navigation

Integrated navigation, as the name suggests, employs all the features and systems described so far. An integrated navigation solution using a multisensor approach blends the performance of all the navigation techniques already described together with GPS to form a totally integrated system. In this case the benefits of the GPS and IN derived data are blended to provide more accurate data fusion, in the same way as barometric and IN data are fused (Figure 8.7).

   Such an integrated system is a precursor to the introduction of the advanced navigation capabilities that will comprise the future air navigation system (FANS). FANS is designed to make more efficient use of the existing airspace such that future air traffic increases may be accommodated. Some elements of FANS have already been implemented, others will take several years to attain maturity. A key prerequisite to achieving a multisensor system is the installation of a high-grade flight management system (FMS) to perform the integration of all the necessary functions and provide a suitable interface with the flight crew.

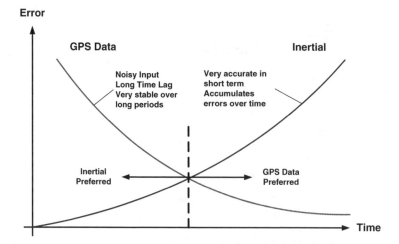

Figure 8.7   Integrated GPS and inertial navigation.

8.5.1 Sensor Usage – Phases of Flight

When assessing which navigation sensors to use for various phases of flight, the navigation accuracy, equipment availability and reliability and operational constraints all need to be taken into account. The advent of GPS with its worldwide coverage at high levels of accuracy has given a tremendous impetus to the navigation capabilities of modern aircraft. However, system integrity concerns have meant that the certification authorities have stopped short of relying solely on GPS.

Bearing in mind physical and radio propagation factors, and the relative traffic densities for various phases of flight, a number of requirements are specified for the use of GPS. Advisory circular AC 90-94 specifies the considerations that apply for the use of GPS as a sole or supplementary method of navigation.

These considerations apply for the following phases of flight:

1. Oceanic en route. Operation over long oceanic routes means that the aircraft will be denied the availability of most of the line-of-sight radio navigation aids such as NDB, VOR, TACAN, etc. LORAN-C may be available in some circumstances. The aircraft will need to depend upon an approved primary long-range method of navigation. For most modern transport aircraft that means equipping the aircraft with a dual- or triple-channel INS or ADIRS. Supplementary means such as GPS may be used to update the primary method of navigation. Aircraft using GPS under instrument flight rules (IFR) must be equipped with another approved long-range navigation system: GPS is not certified as a primary and sole means of navigation. Certain categories of GPS equipment may be used as one of the approved long-range navigation means where two systems are required. The availability of a functioning receiver autonomous integrity monitor (RAIM) capability is also important owing to the impact that this has upon GPS integrity. Providing RAIM is available, the flight crew need not actively monitor the alternative long-range navigation system.
2. Domestic en route. Once overland, most of the conventional navigation aids may be available, unless the aircraft is transiting a wilderness area such as Siberia. For the most part, NDB, VOR, TACAN and LORAN-C will be operational and available to supplement GPS. These ground-based systems do not have to be used to monitor GPS unless RAIM failure occurs. Within the United States, Alaska, Hawaii and surrounding coastal waters, IFR operation may be met with independent NDB, VOR, TACAN or LORAN-C equipment. This may not necessarily be the case outside the US NAS.
3. Terminal. GPS IFR operations for the terminal phases of flight should be conducted as for normal RNAV operations using the standard procedures:

   • Standard instrument departures (SIDs);
   • Standard terminal arrival routes (STARs);
   • Standard instrument approach procedure (SIAP).

   The normal ground-based equipment appropriate to the phase of flight must be available, out, as before, it does not need to be used to monitor GPS unless RAIM fails.
4. Approach. In the United States an approach overlay programme has been introduced by the FAA to facilitate the introduction of instrument approaches using GPS. The key features of the GPS overlay programme are described below.

8.5.2 GPS Overlay Programme

The GPS overlay programme allows pilots to use GPS equipment to fly existing VOR, VOR/DME, NDB, NDB/DME, TACAN and RNAV non-precision instrument approach procedures. This facility only applies in US airspace and was introduced in February 1994. The approach aid appropriate to the type of approach being flown must be available for use, but need not be monitored provided RAIM is available. In April 1994, 'phase III' approaches introduced the first GPS specific approaches with GPS specifically included in the title. For these approaches the traditional avionics need not be available – either ground-based or airborne equipment – provided RAIM is available. For aircraft fitted with GPS without a RAIM capability these navigation aids must be available.

8.5.3 Categories of GPS Receiver

The different types of GPS are mandated in technical standing order (TSO) C-129a. This categorises the different GPS receivers by three major classes:

1. Class A. This equipment incorporates a GPS receiver and the navigation capability to support it.
2. Class B. This consists of GPS equipment providing data to an integrated navigation system such as a flight management system or an integrated multisensor navigation system.
3. Class C. This includes equipment comprising GPS sensors which provide data to an autopilot or flight director in order to reduced flight technical errors.

This classification therefore categorises the GPS equipment types according to function. TSO C-129a also specifies which class of equipment may be used for the typical flight phases described above. It also specifies whether the RAIM function is to be provided by the GPS or the integrated system.

8.6 Flight Management System

It is clear from the foregoing description of the aircraft navigation functions that navigation is a complex task and becoming more so all the while. FMS functionality has increased rapidly over the last decade, and many more enhancements are in prospect as the future features required by FANS are added. A typical FMS will embrace dual computers and dual Multifunction control and display units (MCDUs) as shown in Figure 8.8. In military bomber and transport aircraft the system implementation is likely to be in the form portrayed in this figure. For military fighter aircraft the functions will be similar but embedded in the avionics system navigation computers and mission computers as appropriate.

Figure 8.8 is key to depicting the integration of the navigation functions described above. System sensor inputs, usually in dual-redundant form for reasons of availability and integrity, are shown on the left. These are:

- Dual INS/IRS;
- Dual navigation sensors: VOR/DME, DME/DME, etc;

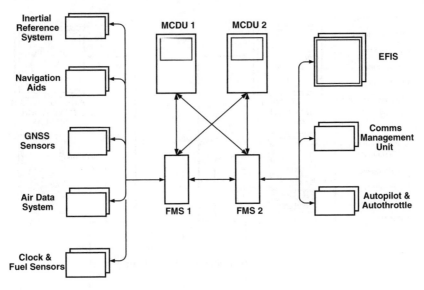

Figure 8.8 Typical flight management system (FMS).

- Dual GNSS sensors – usually GPS;
- Dual air data sensors;
- Dual inputs from on-board sensors relating to fuel on-board and time.

These inputs are used by the FMS to perform the necessary navigation calculations and provide information to the flight crew via a range of display units:

- Electronic flight instrument system (EFIS);
- Communications control system;
- Interface with the autopilot/flight director system to provide the flight crew with flight direction or automatic flight control in a number of predefined modes.

The FMS–crew interface is shown in Figure 8.9. The key interface with the flight crew is via the following displays:

1. Captain's and first officer's navigation displays (ND), part of the electronic flight instrument system (EFIS). The navigation displays may show information in a variety of different ways.
2. Control and display units 1 and 2, part of the FMS. The CDUs both display information and act as a means for the flight crew manually to enter data.

The FMS computers perform all the necessary computations and show the appropriate navigation parameters on the appropriate display. The navigation displays show the navigation and steering information necessary to fly the intended route. These are colour displays and can operate in a number of different formats, depending upon the phase of flight.

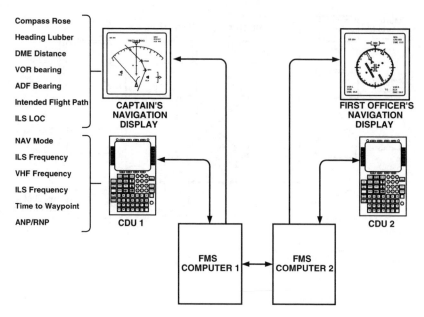

Figure 8.9   FMS control and display interface.

8.6.1 FMS CDU

The FMS CDU is the key flight crew interface with the navigation system, allowing the flight crew to enter data as well as having vital navigation information displayed. A typical FMS CDU is shown in Figure 8.10. The CDU has a small screen on which alpha-numeric information is displayed, in contrast to the pictorial information displayed on the EFIS navigation displays. This screen is a cathode ray tube (CRT) monochrome display in early systems; later systems use colour active matrix liquid crystal display (AMLCD) (see Chapter 11). The tactile keyboard has alpha-numeric keys in order to allow manual entry of navigation data, perhaps inserting final alterations to the flight plan, as well as various function keys by which specific navigation modes may be selected. The line keys at the sides of the display are soft keys that allow the flight crew to enter a menu-driven system of subdisplays to access more detailed information. On many aircraft the CDU is used to portray maintenance status and to execute test procedures using the soft keys and the menu-driven feature. Finally, there are various annunciator lights and a lighting control system.

Examples of the data displayed on the CDU are indicated in Figure 8.11. The CDU displays the following parameters using a menu-driven approach:

1. An ETA waypoint window. This shows the estimated time of arrival (ETA) at the waypoint, in this case waypoint 15.
2. Early/late timing information. This represents the earliest and latest times the aircraft can reach the waypoint given its performance characteristics.
3. Information on the runway – an ILS approach to runway 27.
4. Wind information for the approach – wind bearing 290.

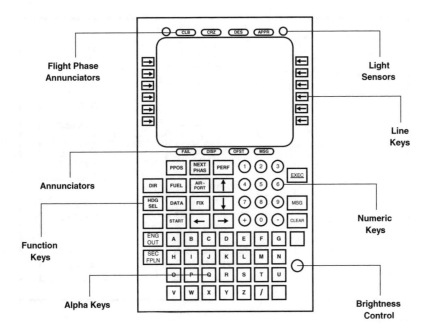

Figure 8.10 Typical FMS control and display unit.

5. Information on the navigation aids being used: VOR, DME and ILS/LOC.
6. An ANP/RNP window. This compares the actual navigation performance (ANP) of the system against the required navigation performance (RNP) for the flight phase and navigation guidance being flown. In this case the ANP is 0.15 nm against a RNP of 0.3 nm and the system is operating well within limits.

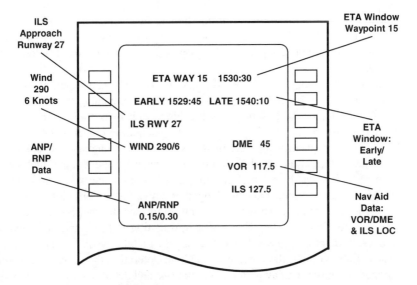

Figure 8.11 Typical FMS CDU display data.

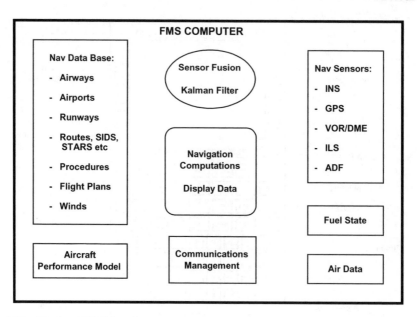

Figure 8.12 Top-level FMS functions.

8.6.2 FMS Functions

The functions of the FMS at a top level are shown in Figure 8.12. This diagram gives an overview of the functions performed by the FMS computers. These may be summarised as follows:

1. Navigation computations and display data. All the necessary navigation computations are undertaken to derive the navigation or guidance information according to the phase of flight and the sensors utilised. This information is displayed on the EFIS navigation display or the FMS CDU. Flight director and steering commands are sent to the autopilot for the flight director with the pilot in the loop or for the engagement automatic flight control modes.

2. Navigation sensors. INS, GPS, VOR, ILS, ADF, TACAN and other navigation aids provide dual-sensor information to be used for various navigation modes.

3. Air data. The ADCs or ADIRS provides the FMS with high-grade corrected air data parameters and attitude information for use in the navigation computations.

4. Fuel state. The fuel quantity measurement system and the engine-mounted fuel flow-meters provide information on the aircraft fuel quantity and engine fuel flow. The calculation of fuel use and total fuel consumption is used to derive aircraft and engine performance during the flight. When used together with a full aircraft performance model, optimum flight guidance may be derived which minimises the fuel consumed.

5. Sensor fusion and Kalman filter. The sensor information is fused and validated against other sources to determine the validity and degree of fidelity of the data. By using a sophisticated Kalman filter, the computer is able to determine the accuracy and integrity of the navigation sensor and navigation computations and determine the actual navigation performance (ANP) of the system in real time.

6. Communications management. The system passes information to the communication control system regarding the communication and navigation aid channel selections that have been initiated by the FMS in accordance with the requirements of the flight plan.
7. Navigation database. The navigation base contains a wide range of data that are relevant to the flight legs and routes the aircraft may expect to use. This database will include the normal flight plan information for standard routes that the aircraft will fly together with normal diversions. It will be regularly updated and maintained. A comprehensive list of these items includes:

- Airways;
- Airports – approach and departure information, airport and runway lighting, obstructions, limitations, airport layout, gates, etc;
- Runways including approach data, approach aids, category of approach (Cat I or Cat II/III) and decision altitudes;
- Routes, clearance altitudes, SIDS, STARS and other defined navigation data;
- Procedures including notification of short-term airspace restrictions or special requirements;
- Flight plans with standard diversions.
- Wind data – forecast winds and actual winds derived throughout flight.

8. Aircraft performance model. The inclusion of a full performance model adds to the systems ability to compute four-dimensional (x, y, z, time) flight profiles and at the same time make optimum use of the aircraft energy to optimise fuel use.

The FMS provides the essential integration of all of these functions to ensure that the overall function of controlling the navigation of the aircraft is attained. As may be imagined, this does not merely include steering information to direct the aircraft from waypoint to waypoint. The FMS also controls the tuning of all the appropriate aircraft receivers to navigation beacons and communications frequencies via the communications control units and many other functions besides. The flight plan that resides within the FMS memory will be programmed for the entire route profile, for all eventualities, including emergencies. More advanced capabilities include three-dimensional navigation and the ability to adjust the aircraft speed to reach a waypoint within a very small time window (typically ± 6 s). The various levels of performance and sophistication are summarised in Table 8.1. Military aircraft such as the Boeing multirole maritime aircraft (MMA) will be fitted with an FMS developed to provide all these features for a civil operator. The FMS will incorporate additional specific modes of operation to facilitate performance of the mission, e.g. flying mission attack profiles at low level.

The FMS capabilities will be examined in a little more detail.

8.6.3 LNAV

Lateral navigation or LNAV relates to the ability of the aircraft to navigate in two dimensions, in other words, the lateral plane. LNAV was the first navigation feature to be implemented and involved navigating aircraft to their intended destination without any other considerations. LNAV comprises two major implementations:

Table 8.1 Summary of FMS capabilities

Function	Capability
LNAV	The ability to navigate laterally in two dimensions
VNAV	The ability to navigate laterally in two dimensions plus the ability to navigate in the vertical plane. When combined with LNAV, this provides three-dimensional navigation
Four-dimensional navigation	The ability to navigate in three-dimensions plus the addition of time constraints for the satisfaction of time of arrival at a waypoint
Full performance based navigation	The capability of four-dimensional navigation together with the addition of an aircraft specific performance model. By using cost indexing techniques, full account may be made of the aircraft performance in real time during flight, allowing optimum use of fuel and aircraft energy to achieve the necessary flight path
Future air navigation system (FANS) or global air transport management (GATM)	The combination of the full performance model together with all the advantages that FANS or GATM will confer, eventually enabling the concept of 'free flight'

- Airway navigation;
- Area navigation or RNAV.

8.6.4 Airway Navigation

Airway navigation is defined by a predetermined set of airways which are based primarily on VOR stations, although some use NDB stations. In the United States these airways are further categorised depending upon the height of the airway:

1. Airways based on VOR from 120 ft above the surface to 18 000 ft above mean sea level (MSL) carry the V prefix and are called *victor airways*.
2. Airways using VOR from 18 000 ft MSL to 45 000 ft MSL are referred to as *jet routes*.

Each VOR used in the route system is called either a terminal VOR or a low- or high-altitude en-route VOR. Terminal VORs are used in the terminal area to support approach and departure procedures and are usable up to ~25 nm; they are not to be used for en-route navigation. Low-altitude en-route VORs have service volumes out to a range of 40 nm and are used up to 18 000 ft on victor airways. High-altitude VORs support navigation on jet routes and their service volume may extend to a range of ~200 nm from the ground station. Clearly, the range of the VOR beacons is limited by line-of-sight propagation considerations, as has already been described.

Airway width is determined by the navigation system performance and depends upon error in the ground station equipment, airborne receiver and display system.

8.6.5 Area Navigation

Many aircraft possess an area navigation (RNAV) capability. The on-board navigation together with the FMS can navigate along a flight path containing a series of waypoints that are not defined by the airways. Navigation in these situations is not confined to VOR beacons but may use a combination of VOR, DME, LORAN-C, GPS and/or INS. Random routes have the advantage that they may be more direct than the airway system and also that they tend geographically to disperse the aircraft away from the airway route structure. Advisory circular AC 90-45A defines the regulations that apply to aircraft flying two-dimensional RNAV in IFR conditions in the US national airspace system. Advisory circular AC 90-45A is the original guidance on the use of RNAV within the United States, while advisory circular AC 20-130A is a more recent publication on navigation or FMS systems using multiple sensors – including GPS – and is probably more relevant to the sophisticated FMS in use today.

8.6.6 VNAV

Following on from the LNAV and RNAV capabilities, vertical navigation (VNAV) procedures were developed to provide three-dimensional guidance. Present VNAV systems use barometric altitude, as it will be recalled that the GPS satellite geometry does not generally provide accurate information in the vertical direction. Whereas DGPS systems such as WAAS will address and overcome this issue, these systems will not be available for some time. Advisory circular AC 90-97 provides guidance for the use of VNAV guidance in association with RNAV instrument approaches with a VNAV decision altitude (DA). One disadvantage of using barometric means to provide the VNAV guidance function is the non-standard nature of the atmosphere. Therefore, VNAV approaches embrace a temperature limit below which the use of VNAV decision height is not permitted. If the temperature on a particular day falls below this limit, then the flight crew must instead respect the published LNAV minimum decision altitude (MDA).

8.6.7 Four-Dimensional Navigation

The combination of LNAV and VNAV provides a three-dimensional navigation capability. However, in a busy air traffic management situation the element of time is equally important. A typical modern FMS will have the capability to calculate the ETA to a specific waypoint and ensure that the aircraft passes through that point in space within ±6 s of the desired time. Furthermore, calculations can be made in response to an air traffic control enquiry as to when the aircraft can reach an upcoming waypoint. By using information regarding the aircraft performance envelope, the FMS can perform calculations that determine the earliest and the latest possible time within which the aircraft can reach the waypoint. The ability to determine this time window can be of great use in helping the air traffic controller to maintain steady traffic flow during periods of high air traffic density.

8.6.8 Full Performance Based Navigation

If the FMS contains a full performance model provided by the aircraft manufacturer, then even more detailed calculations may be performed. By using the aircraft velocity and other dynamic parameters, it is possible to compute the performance of the aircraft over very small time increments. By using this technique, and provided that the sensor data are sufficiently accurate, the future dynamic behaviour of the aircraft may be accurately predicted. Using this feature, and knowing the four-dimensional trajectory and gate speeds that are detailed in the flight plan, the aircraft can calculate the optimum trajectory to meet all these requirements while conserving energy and momentum and ensuring minimum fuel burn. When this capability is combined with the increasing flexibility that FANS will provide, further economies will be possible. Today, most FMS systems are being developed with these emerging requirements in mind such that future implementation will depend upon system software changes and upgrades rather than aircraft equipment or architecture modifications.

8.6.9   FMS Procedures

Although the foregoing explanations have concentrated on performance enhancements, the assistance that the FMS provides the flight crew in terms of procedural displays cannot be forgotten. Typical examples include:

• Standard instrument departure (SID);
• En-route procedures;
• Standard terminal arrival requirements (STAR);
• ILS approach.

Examples of these procedures are given in the companion volume (Moir and Seabridge, 2003).

8.6.10   Traffic Collision and Avoidance System (TCAS)

The TCAS combines the use of the ATC mode S transponder with additional computing and displays to provide warning of the proximity of other aircraft within the air traffic control system. The operation of ATC mode S and the TCAS is described in detail in Chapter 7.

8.6.11   GPWS and EGPWS

While the TCAS is designed to prevent air-to-air collisions, the ground proximity warning system (GPWS) is intended to prevent unintentional flight into the ground. Controlled flight into terrain (CFIT) is the cause of many accidents. The term describes conditions where the crew are in control of the aircraft, but, owing to a misplaced sense of situational awareness, they are unaware that they are about to crash into the terrain. The GPWS takes data from various sources and generates a series of audio warnings when a hazardous situation is developing.

  The terrain awareness and warning system (TAWS) embraces the overall concept of providing the flight crew with prediction of a potential controlled flight into terrain. The new term is a generic one since the ground proximity warning system (GPWS) and enhanced GPWS became associated mainly with the Allied Signal (now Honeywell) implementation. The latest manifestation is designed to provide the crew with an improved situational awareness compared with previous systems. The FAA is presently in the process of specifying that turbine-equipped aircraft with six seats or more will be required to be equipped with a TAWS by 2003. Advisory circular AC 25-23 addresses the airworthiness requirements associated with the TAWS.

The GPWS/TAWS uses radar altimeter information together with other information relating to the aircraft flight path. Warnings are generated when the following scenarios are unfolding:

- Flight below the specified descent angle during an instrument approach;
- Excessive bank angle at low altitude;
- Excessive descent rate;
- Insufficient terrain clearance;
- Inadvertent descent after take-off;
- Excessive closure rate to terrain – the aircraft is descending too quickly or approaching higher terrain.

Inputs are taken from a variety of aircraft sensors and compared with a number of algorithms that define the safe envelope within which the aircraft is flying. When key aircraft dynamic parameters deviate from the values defined by the appropriate guidance algorithms, appropriate warnings are generated.

The installation of GPWS equipment for all airliners flying in US airspace was mandated by the FAA in 1974, since when the number of CFIT accidents has dramatically decreased.

More recently, enhanced versions have become available. The EGPWS offers a much greater situational awareness to the flight crew as more quantitative information is provided, together with earlier warning of the situation arising. It uses a worldwide terrain database which is compared with the present position and altitude of the aircraft. Within the terrain database the earth's surface is divided into a grid matrix with a specific altitude assigned to each square within the grid representing the terrain at that point.

The aircraft intended flight path and manoeuvre envelope for the prevailing flight conditions are compared with the terrain matrix and the result is graded according to the proximity of the terrain, as shown in Figure 8.13:

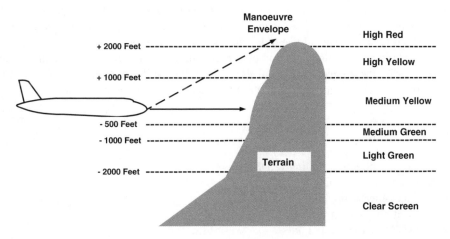

Figure 8.13 Principle of operation of the EGPWS (TAWS).

Terrain responses are graded as follows:

- No display for terrain more than 2000 ft below the aircraft;
- Light-green dot pattern for terrain between 1000 and 2000 ft below the aircraft;
- Medium-green dot pattern for terrain between 500 and 1000 ft below the aircraft;
- Medium-yellow dot pattern for terrain between 1000 ft above and 500 ft below the aircraft;
- Heavy-yellow display for terrain between 1000 and 2000 ft above the aircraft;
- Heavy-red display for terrain more than 2000 ft above the aircraft.

This type of portrayal using coloured imagery is very similar to that for the weather radar and is usually shown on the navigation display. It is far more informative than the audio warnings, given by earlier versions of GPWS. The EGPWS also gives audio warnings, but much earlier than those given by the GPWS. The earlier warnings, together with the quantitative colour display, give the flight crew a much better overall situational awareness in respect of terrain and more time to react positively to their predicament than did previous systems.

8.7 Navigation Aids

As aviation began to expand in the 1930s, the first radio navigation systems were developed. Initially, these were installed at the new growing US airports, and it is interesting to note that the last of these early systems was decommissioned as recently as 1979.

One of the most prominent was the 'radio range' system developed in Italy by Bellini and Tosi, which was conceived as early as 1907. The operation of the Bellini–Tosi system relied upon the transmission of morse characters A (dot-dash) and N (dash-dot) in four evenly spaced orthogonal directions. When flying the correct course, the A and N characters combined to produce a humming noise which the pilot could detect in his earphones. Deviation from the desired course would result in either the A or N characters becoming most dominant, signifying the need for corrective action by turning left or right as appropriate.

Following WWII, the International Civil Aviation Organisation (ICAO) produced international standards that led to the definition of the very high-frequency omnirange (VOR) system which is in widespread use today and is described below.

The use of radio navigation aids is important to military aircraft as much of their operation involves sharing the airspace with civil users. Therefore, military aircraft, especially those adopted from a civil aircraft platform, will utilise a similar suite on navigation aids. Typical aids are:

1. Navigation aids:

   - Automatic direction finding (ADF);
   - VHF omnirange (VOR);
   - Distance-measuring equipment (DME);
   - Tactical air navigation (TACAN);
   - VOR/TACAN (VORTAC);
   - Long range navigation (LORAN).

2. Landing aids:

- Instrument landing system (ILS);
- Microwave landing system (MLS).

8.7.1 Automatic Direction Finding

Automatic direction finding (ADF) involves the use of a loop direction finding technique to establish the bearing to a radiating source. This might be to a VHF beacon or a non-distance beacon (NDB) operating in the 200–1600 kHz band. Non-directional beacons, in particular, are the most prolific and widely spread beacons in use today. The aircraft ADF system comprises integral sense and loop antennas which establish the bearing of the NBD station to which the ADF receiver is tuned. The bearing is shown on the radio magnetic indicator (RMI) in the analogue cockpit of a 'classic' aircraft or more likely on the electronic flight instrument system (EFIS), as appropriate. ADF is used by surveillance aircraft such as MPA on an air sea rescue mission to home on to a personal locator beacon used by downed airmen or installed in life rafts.

8.7.2 Very High-frequency Omnirange (VOR)

The VOR system was accepted as standard by the United States in 1946 and later adopted by the International Civil Aviation Organisation (ICAO) as an international standard. The system provides a widely used set of radio beacons operating in the VHF frequency band over the range 108–117.95 MHz with a 100 kHz spacing. Each beacon emits a morse code modulated tone which may be provided to the flight crew for the purposes of beacon identification.

The ground station radiates a cardioid pattern which rotates at 30 r/min generating a 30 Hz modulation at the aircraft receiver. The ground station also radiates an omnidirectional signal which is frequency modulated with a 30 Hz reference tone. The phase difference between the two tones varies directly with the bearing of the aircraft. At the high frequencies at which VHF operates there are no sky wave effects and the system performance is relatively consistent. VOR has the disadvantage that it can be severely disrupted by adverse weather – particularly by electrical storms – and as such it cannot be used as a primary means of navigation for a civil aircraft.

Overland in the North American continent and Europe, VOR beacons are widely situated to provide an overall coverage of beacons. Usually these are arranged to coincide with major airway waypoints and intersections in conjunction with DME stations – see below – such that the aircraft may navigate for the entire flight using the extensive route/beacon structure. By virtue of the transmissions within the VHF band, these beacons are subject to the line-of-sight and terrain-masking limitations of VHF communications. Advisory circular AC 00-31A lays out a method for complying with the airworthiness rules for VOR/DME/TACAN.

8.7.3 Distance-measuring Equipment (DME)

Distance-measuring equipment (DME) is a method of pulse ranging used in the 960–1215 MHz band to determine the distance of the aircraft from a designated ground station.

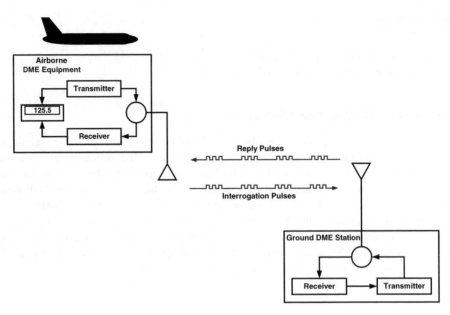

Figure 8.14   DME principle of operation.

The aircraft equipment interrogates a ground-based beacon and, upon the receipt of retransmitted pulses, unique to the on-board equipment, is able to determine the range to the DME beacon (Figure 8.14). DME beacons are able to service requests from a large number of aircraft simultaneously but are generally understood to have the capacity to handle ~200 aircraft at once. Specified DME accuracy is ±3% or ±0.5 nm, whichever is the greater (advisory circular AC 00-31A).

DME and TACAN beacons are paired with ILS/VOR beacons throughout the airway route structure in accordance with the table set out in Appendix 3 of advisory circular AC 00-31A. This is organised such that aircraft can navigate the airways by having a combination of VOR bearing and DME distance to the next beacon in the airway route structure. A more recent development – scanning DME – allows the airborne equipment rapidly to scan a number of DME beacons, thereby achieving greater accuracy by taking the best estimate of a number of distance readings. This combination of VOR/DME navigation aids has served the aviation community well in the United States and Europe for many years, but it does depend upon establishing and maintaining a beacon structure across the land mass or continent being covered. New developments in third-world countries are more likely to skip this approach in favour of a global positioning system (GPS), as described later in the chapter.

8.7.4   TACAN

Tactical air navigation (TACAN) is military omnibearing and distance-measuring equipment with similar techniques for distance measurement as DME. The bearing information is accomplished by amplitude modulation achieved within the beacon which imposes 15 and 135 Hz modulated patterns and transmits this data together with 15 and 135 Hz reference pulses. The airborne equipment is therefore able to measure distance using DME interrogation techniques while using the modulated data to establish bearing.

TACAN beacons operate in the frequency band 960–1215 MHz as opposed to the 108–118 MHz used by DME. This means that the beacons are smaller, making them suitable for shipborne and mobile tactical use. Some airborne equipment have the ability to offset to a point remote from the beacon which facilitates recovery to an airfield when the TACAN beacon is not co-located. TACAN is reportedly accurate to within ±1% in azimuth and ±0.1 nm in range, so it offers accuracy improvements over VOR/DME.

TACAN also has the ability to allow aircraft to home on to another aircraft, a feature that is used in air-to-air refuelling to enable aircraft to home on to the donor tanker.

8.7.5 VORTAC

As most military aircraft are equipped with TACAN, some countries provide VORTAC beacons which combine VOR and TACAN beacons. This allows interoperability of military and civil air traffic. Military operators use the TACAN beacon while civil operators use the VOR bearing and TACAN (DME) distance-measuring facilities. This is especially helpful for large military aircraft, such as transport or surveillance aircraft, since they are able to use civil air lanes and operational procedures during training or on transit between theatres of operations.

8.7.6 Hyberbolic Navigation Systems – LORAN-C

Hyperbolic navigation systems – of which long range navigation (LORAN) is the most noteworthy example – operate upon hyperbolic lines of position rather than circles or radial lines. Figure 8.15 illustrates the principle of operation of a hyperbolic system in a very elementary manner. This shows hyperbolic solid lines which represent points that are equidistant from the two stations. These points will have the same time difference between the arrival of signals from the blue-master and blue-slave stations (the term secondary station

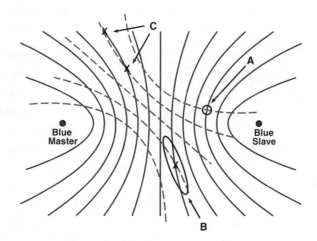

Figure 8.15 Principle of operation of a hyperbolic navigation system.

is probably a better and more accurate description). This in itself will not yield position, but, if a second pair of stations is used – angled approximately 45° to the first – shown as dashed lines, then position can be obtained. The relative positioning of the lines in this dual-chain example shows that three outcomes are possible:

1. At point A the lines cross at almost 90°, and this represents the most accurate fix.
2. At point B the lines cross at a much more acute angle and the result is a larger error ellipse.
3. At point C there are two possible solutions and an ambiguity exists that can only be resolved by using a further station.

LORAN-C is the hyperbolic navigation system in use today and was conceived in principle around the beginning of WWII. Worldwide coverage existed in 1996 and new facilities were being planned in the late 1990s. LORAN operates in the frequency band 90–110 kHZ as a pulsed system which enables the ground wave to be separated from the sky wave, the ground wave being preferred. A LORAN chain will comprise at least three stations, one being nominated as the master. The time difference of arrival between the master and slaves allows position to be determined. Each of the stations in a chain transmits unique identifiers which allow the chain to be identified. A typical example of a LORAN-C chain is shown in Figure 8.16 which shows the north-eastern US chain.

Within the defined area of coverage of the chain, LORAN-C will provide a user with a predictable absolute accuracy of 0.25 nm. A typical chain will have over 1000 nm operating range coverage. LORAN-C is also capable of relaying GPS positional error within the transmissions. LORAN-C is expected to remain in commission until at least 2008. Advisory circular AC 20-121A provides information to assist with the certification of LORAN-C navigation systems for use within the United States and Alaska.

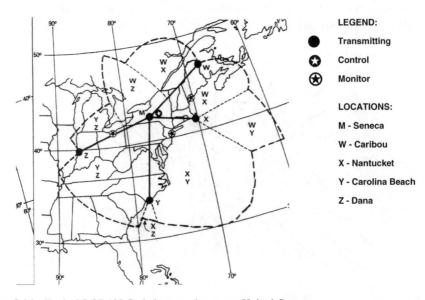

Figure 8.16   Typical LORAN-C chain – north-eastern United States.

8.7.7   Instrument Landing System

The instrument landing system (ILS) is an approach and landing aid that has been in widespread use since the 1960s and 1970s. The main elements of the ILS include:

1. A localiser antenna centred on the runway to provide lateral guidance. A total of 40 operating channels are available within the 108–112 MHz band. The localiser provides left and right lobe signals which are modulated by different frequencies (90 and 150 Hz) such that one signal or the other will dominate when the aircraft is off the runway centre-line. The beams are arranged such that the 90 Hz modulated signal will predominate when the aircraft is to the left, while the 150 Hz signal will be strongest to the right. The difference in signal is used to drive a cross-pointer deviation needle such that the pilot is instructed to 'fly right' when the 90 Hz signal is strongest and 'fly left' when the 150 Hz signal dominates. When the aircraft is on the centre-line, the cross-pointer deviation needle is positioned in the central position. This deviation signal is proportional to azimuth out to ±5° of the centre-line.

2. A glideslope antenna located beside the runway threshold to provide lateral guidance. Forty operating channels are available within the frequency band 329–335 MHz. As for the localiser, two beams are located such that the null position is aligned with the desired glideslope, usually set at a nominal 3°. In the case of the glideslope, the 150 Hz modulated signal predominates below the glideslope and the 90 Hz signal is stronger above. When the signals are balanced, the aircraft is correctly positioned on the glideslope and the glideslope deviation needle is positioned in a central position. As for the localiser needle, pilots are provided with 'fly up' or fly down' guidance to help them to acquire and maintain the glideslope (see Figure 8.17 for the general arrangement of the ILS). Figure 8.18 illustrates how guidance information is portrayed for the pilot according to the aircraft position relative to the desired approach path. On older aircraft this would

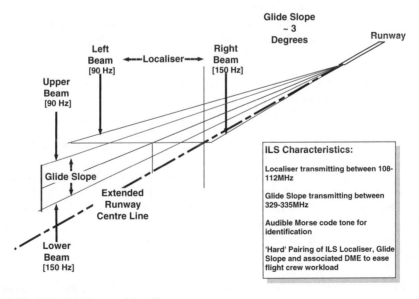

Figure 8.17   ILS glideslope and localiser.

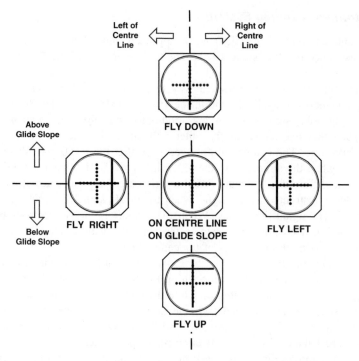

Figure 8.18   ILS guidance display.

be shown on the compass display, while on modern aircraft with digital cockpits this information is displayed on the primary flight display (PFD). The ILS localiser, glideslope and DME channels are paired such that only the localiser channel needs to be tuned for all three channels to be correctly aligned.

3. Marker beacons are located at various points down the approach path to give the pilot information as to what stage on the approach has been reached. These are the outer, middle and inner markers. Location of the marker beacons are:

- Outer marker approximately 4–7 nm from the runway threshold;
- Middle marker ~3000 ft from touchdown;
- Inner marker ~1000 ft from touchdown.

The high approach speeds of most modern aircraft render the inner marker almost superfluous and it is seldom used.

4. The marker beacons are all fan beams radiating on 75 MHz and provide different morse code modulation tones which can be heard through the pilot's headset. The layout of the marker beacons with respect to the runway is as shown in Figure 8.19. The beam pattern is ±40° along track and ±85° across track. The overall audio effect of the marker beacons is to convey an increasing sense of urgency to the pilot as the aircraft nears the runway threshhold.

A significant disadvantage of the ILS system is its susceptibility to beam distortion and multipath effects. This distortion can be caused by local terrain effects, large man-made

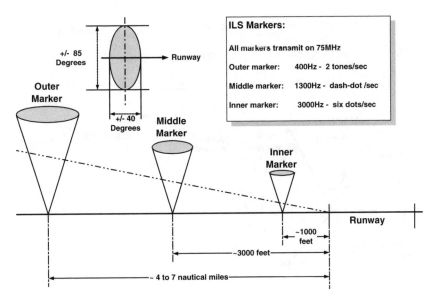

ILS Markers:

All markers transmit on 75MHz

Outer marker: 400Hz - 2 tones/sec

Middle marker: 1300Hz - dash-dot /sec

Inner marker: 3000Hz - six dots/sec

Figure 8.19   ILS approach markers.

structures or even taxiing aircraft which can cause unacceptable beam distortion, with the glideslope being the most sensitive. At times on busy airfields and during periods of limited visibility, this may preclude the movement of aircraft in sensitive areas, which in turn can lead to a reduction in airfield capacity. More recently, interference by high-power local FM radio stations has presented an additional problem, although this has been overcome by including improved discrimination circuits in the aircraft ILS receiver.

8.7.8   Microwave Landing System (MLS)

The microwave landing system (MLS) is an approach aid that was conceived to redress some of the shortcomings of the ILS. The specification of a time-reference scanning beam MLS was developed through the late 1970s/early 1980s, and a transition to the MLS was envisaged to begin in 1998. However, with the emergence of satellite systems such as the GPS there was also a realisation that both the ILS and MLS could be rendered obsolete when such systems reach maturity. In the event, the US civil community is embarking upon higher-accuracy developments of the basic GPS system: the wide-area augmentation system (WAAS) and local-area augmentation system (LAAS) have already been outlined. In Europe, the United Kingdom, the Netherlands and Denmark have embarked upon a modest programme of MLS installations at major airports.

The MLS operates in the frequency band 5031.0–5190.7 MHz and offers some 200 channels of operation. It has a wider field of view than the ILS, covering ±40° in azimuth and up to 20° in elevation, with 15° useful range coverage. Coverage is out to 20 nm for a normal approach and up to 7 nm for back azimuth/go-around. The co-location of a DME beacon permits three-dimensional positioning with regard to the runway, and the combination of higher data rates means that curved-arc approaches may be made, as opposed to the straightforward linear approach offered by the ILS. This offers advantages when operating

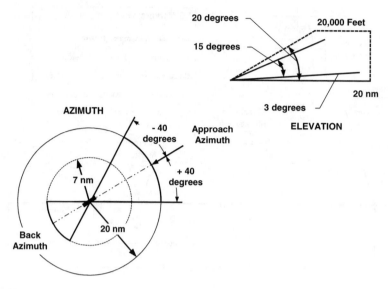

Figure 8.20   Microwave landing system coverage.

into airfields with confined approach geometry and tactical approaches favoured by the military. For safe operation during go-around, precision DME (P-DME) is required for a precise back azimuth signal.

A groundbased MLS installation comprises azimuth and elevation ground stations, each of which transmits angle and data functions which are frequency shift key (FSK) modulated and which are scanned within the volume of coverage already described. The MLS scanning function is characterised by narrow beam widths of around 1–2° scanning at high slew rates. Scanning rates are extremely high at 20 000 deg/s which provides data rates that are around 10 times greater than is necessary to control the aircraft. These high data rates are very useful in being able to reject spurious and unwanted effects due to multiple reflections, etc.

Typical coverage in azimuth and elevation for an MLS installation is shown in Figure 8.20.

8.8   Inertial Navigation

8.8.1   Principles of Operation

The availability of inertial navigation systems (INS) to the military aviation community during the early 1960s added another dimension to the navigation equation. Now flight crew were able to navigate by autonomous means using an on-board INS with inertial sensors. By aligning the platform to earth-referenced coordinates and present position during initialisation, it was now possible to fly for long distances without relying upon TACAN or VOR/DME beacons overland or hyperbolic navigation systems elsewhere. Waypoints could be specified in terms of latitude and longitude as arbitrary points on the globe, more suited to the aircraft's intended flight path rather than a specific geographic feature or point in a radio beacon network (Figure 8.13). This offered an enormous increase in operational capability as mission requirements could be defined and implemented using an indigenous on-board

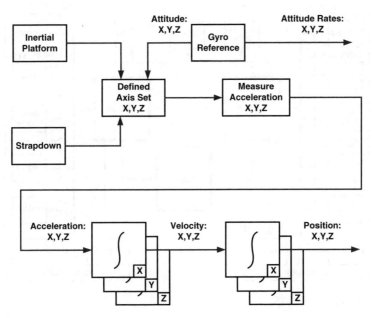

Figure 8.21 Principles of inertial navigation.

sensor with no obvious means of external detection except when the need arose to update the system.

The principles of inertial navigation depend upon the arrangement of inertial sensors such as gyroscopes and accelerometers in a predetermined orthogonal axis set. The gyroscopes may be used to define attitude or body position and rates.

The output from the accelerometer sensor set is integrated to provide velocities, and then integrated again to provide distance travelled (Figure 8.21). First in the military field and then in the commercial market place, inertial navigation systems (INS) became a preferred method for achieving long-range navigation such that by the 1960s the technology was well established.

The specifications in force at this time also offered an INS solution to North Atlantic crossings as well as the dual-Doppler solution previously described. The inertial solution required serviceable dual INS and associated computers to be able to undertake the crossing. There were also limitations on the latitudes at which the ground alignment could be performed – 76° north or south – as attaining satisfactory alignment becomes progressively more difficult the nearer to the poles the INS becomes.

For civil operators, accuracy requirements set forth requirements for operating an INS as a sole means of navigation for a significant portion of the flight. These requirements were described earlier in the chapter.

8.8.2 Stand-alone Inertial Navigation System

For reasons of both availability and accuracy, systems were developed with dual and triple INS installations. A typical triple INS installation of the type used by modern wide-body

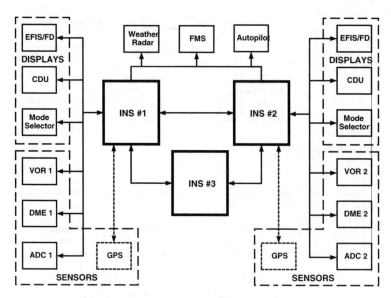

Figure 8.22 Typical triple INS system.

commercial transport aircraft is presented in Figure 8.22, showing three INS units integrated with the other major systems units. This type of system would be representative of an INS installation of a large aircraft before the availability of satellite sensors in the 1990s. By this time the gimballed IN platform would have been replaced by a more reliable strapdown system similar to the Litton LTN-92 system.

This integrated system comprised the following units:

1. Dual sensors:

   - VOR for bearing information;
   - DME for range information;
   - Air data computer (ADC) for air data;
   - Provision for a dual GPS interface.

2. Controls and displays:

   - Control and display unit (CDU);
   - Electronic flight instrument system/flight director (EFIS/FD);
   - Mode selector unit.

3. Other major systems receiving INS data for stabilisation or computation:

   - Weather or mission radar;
   - Flight management system (FMS);
   - Autopilot.

The weight of this system, which comprised three LTN-92 platforms with back-up battery power supplies, two CDUs and two mode selector units, was in the region of 234 lb.

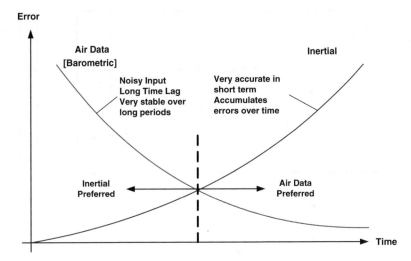

Figure 8.23 Sensor fusion of air data and inertial sensors.

By integrating the air data information with inertially derived flight information, the best features of barometric and inertial systems can be combined. Figure 8.23 illustrates the principle of sensor fusion where the short-term accuracy of inertial sensors is blended or fused with the long-term accuracy of air data or barometric sensors.

Means of taking external fixes were evolved so that longer-term inaccuracies could be corrected by updating the INS position during long flights. Some fighter aircraft systems such as Tornado also added a Doppler radar such that Doppler-derived data could be included in the navigation process. The availability of on-board digital computers enabled statistical Kalman filtering techniques to be used to calculate the best estimated position using all the sensors available.

The fundamental problem with the INS is the long-term and progressive accrual of navigation error as the flight proceeds, and, irrespective of the quality or type of the gyroscopes used, this fundamental problem remains.

8.8.3 Air Data and Inertial Reference Systems (ADIRS)

The system illustrated in Figure 8.22 utilises stand-alone ADCs; however, the use of ADCs in many new large aircraft systems was superceded by the introduction of air data modules (ADMs) in the late 1980s. The new integrated air data and inertial reference system (ADIRS) developed in the early 1980s combined the computation for air data and inertial parameters in one multichannel unit. As large civil plaftforms are increasingly adapted for use in the transport, air refuelling, anti-surface warfare (ASuW) or surveillance roles, this 'commercial' implementation is finding use in military applications.

Taking the B777 as an example, the primary unit is an air data and inertial reference unit (ADIRU) which provides the main source of air data and inertial information. This unit is supported by an attitude and heading reference system (AHARS) which on the B777 is

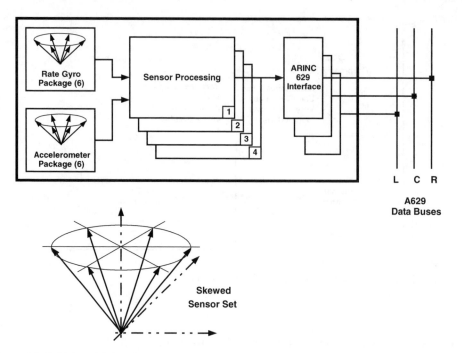

Figure 8.24 B777 ADIRU.

called the secondary attitude air data reference unit (SAARU). This provides secondary attitude and air data information should the primary source, the ADIRU, become totally unusable.

The B777 ADIRU is shown in Figure 8.24. There are six laser rate gyros (LRGs) and six accelerometers included in the unit. It can be seen that both sets of sensors are arranged in a hexad skew-redundant set in relation to an orthogonal axis set. This means that, by resolving the output of each of the six sensors in the direction of the axis set, each sensor is able to measure an element of the relevant inertial parameter – body rate or acceleration – in each axis. This provides a redundant multichannel sensor set with the prospect of achieving higher levels of accuracy by scaling and combining sensor outputs. Additionally, the output of erroneous sensors may be detected and 'voted out' by the remaining good sensors. This multiple-sensor arrangement greatly increases the availability of the ADIRU as the performance of the unit will degrade gracefully following the failure of one or more sensors. The ADIRU may still be used with an acceptable level of degradation until a replacement unit is available or the aircraft returns to base and only has to be replaced following the second failure of a like sensor (e.g. second rate or accelerometer sensor). By contrast, the failure of a sensor in an earlier three-sensor, orthogonally oriented set would lead to a sudden loss of the INS.

Coupled with the dual-hexagonal sensor arrangement, there are four independent lanes of computing within the ADIRU, each of which computes a wide range of navigation parameters:

North velocity	Corrected computed airspeed
East velocity	Corrected Mach number
Ground speed	Corrected total pressure
Latitude	Corrected static pressure
Longitude	CG longitudinal acceleration
Wind speed	CG lateral acceleration
Wind direction	CG normal acceleration
True heading	Flight path acceleration
Magnetic heading	Vertical speed
True track angle	Roll attitude
Magnetic track angle	Pitch attitude
Drift angle	Track angle rate
Flight path angle	Corrected angle of attack (AoA)
Inertial altitude	Roll attitude rate
Computed airspeed	Pitch attitude rate
Mach number	Heading rate
Altitude rate	Body yaw rate
Altitude	Body pitch rate
Total air temperature (TAT)	Body roll rate
Static air temperature	Body longitudinal acceleration
True airspeed	Body lateral acceleration
Static pressure (corrected)	Body normal acceleration
Impact pressure	

Finally, the ADIRU interfaces with the remainder of the aircraft systems by means of triple flight control ARINC 629 digital data buses: left, centre and right. The unit is provided with electrical power from a number of independent sources.

Further information on typical ADIRS implementations is given in the companion volume (Moir and Seabridge, 2003).

8.8.4 Inertial Platform Implementations

There are two methods by which the IN function may be achieved. These are as follows:

1. *Gyrostabilised platform.* In the gyrostabilised platform the sensing elements – gyroscopes and accelerometers – are placed on a platform that is itself stabilised to maintain a fixed position in space. This requires fine servomotors and mechanisms to maintain this stabilisation, and consequently gyrostabilised platforms are expensive to manufacture and tend to be unreliable. All of the earlier platforms were implemented in this fashion and many are still in service today.
2. *Strapdown or analytical platform.* The advent of digital computation and its application to avionics applications enabled the introduction of the strapdown or analytical platform. In this implementation the sensors are strapped directly on to the body of the vehicle and the necessary axis transformations to convert from the vehicle to space axes are performed numerically using digital computers. Originally this technique was used for military applications; today, virtually all IN platforms work in this way. The benefits of strapdown

Platform

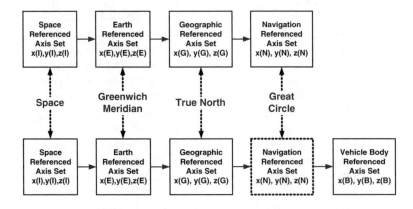

Strapdown or 'Analytical'

Figure 8.25 Inertial platform implementation.

platforms are that they are easier and cheaper to manufacture and are more reliable as they contain none of the servomotors and mechanisms that are a feature of the gyrostabilised platform. Consequently, they are more reliable by a factor of around 3.

Both the gyrostabilised platform and the strapdown platform need to undergo a series of axis transformations before they may be used for navigation on the surface of the earth (Figure 8.25). These axis transformations are as follows:

- Space axes to the Greenwich meridian;
- Greenwich meridian axes to true north;
- True north axes to great circle;
- Great circle axes to vehicle body (strapdown only).

These transformations are fully described in the following pages.

The initial space axes are shown in Figure 8.26. This initial reference shows a generic axis set: $X^{(I)}$, $Y^{(I)}$, $Z^{(I)}$ as a set of axes determined in space by the platform. For ease of reference it may be assumed that the inertial axis $Z^{(I)}$ coincides with the earth axes $Z^{(E)}$ at the outset.

8.8.5 Space Axes to the Greenwich Meridian

Given that $Z^{(I)}$ and $Z^{(E)}$ already coincide, the completion of the earth axis transformation is achieved by rotating the $X^{(I)}$ axis to coincide with the Greenwich meridian by rotating by Ω_t to $X^{(E)}$. The bold axes now represent the earth reference set $X^{(E)}$, $Y^{(E)}$, $Z^{(E)}$ (Figure 8.27).

8.8.6 Earth Axes to Geographic Axes

As it is unlikely that the navigation task will begin precisely aligned with the earth axes, account needs to be taken of where the platform is on the surface of the earth when powered

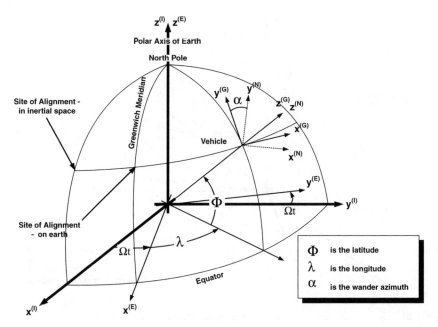

Figure 8.26 Initial space reference axis set $X^{(I)}$, $Y^{(I)}$, $Z^{(I)}$.

up. This process is also known as platform alignment and will be described separately shortly. The alignment process rotates the X axes from $X^{(E)}$ to $X^{(N)}$ by the angle of longitude. The process also aligns $Y^{(G)}$ such that it points north by rotating by the angle of latitude Φ. The outcome is that $Z^{(G)}$ is in line with the local earth vertical (Figure 8.28).

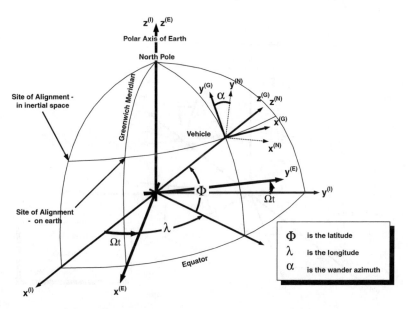

Figure 8.27 Space axes to the Greenwich meridian $X^{(E)}$, $Y^{(E)}$, $Z^{(E)}$.

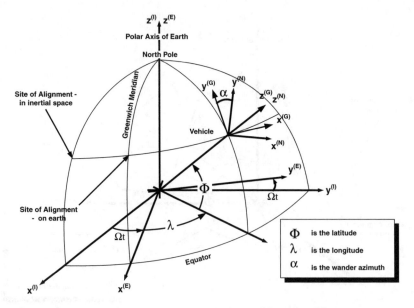

Figure 8.28   Greenwich meridian axes to geographic axes $X^{(G)}$, $Y^{(G)}$, $Z^{(G)}$.

8.8.7   Geographic to Great Circle (Navigation)

By aligning the platform at the wander angle α, the platform may be used to navigate a great circle route which represents the shortest possible path between two points on the surface of the globe (Figure 8.29).

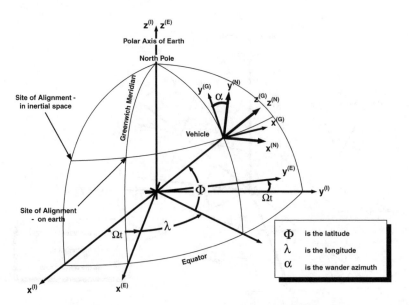

Figure 8.29   True north axes to great circle route (navigation) $X^{(N)}$, $Y^{(N)}$, $Z^{(N)}$.

8.8.8   Great Circle/Navigation Axes to Body Axes (Strapdown)

In the case of the analytical/strapdown platform, the navigation axes $X^{(N)}$, $Y^{(N)}$, $Z^{(N)}$ have to be realigned to the vehicle body axes $X^{(B)}$, $Y^{(B)}$, $Z^{(B)}$ respectively. This is achieved by executing the following rotations in turn:

- Rotating in the yaw axes by Ψ;
- Rotating in the pitch axis by θ;
- Rotating in the roll axis by Φ;

Refer to Figure 8.30.

8.8.9   Platform Alignment

The process of platform alignment follows the processes defined in Figure 8.31. The process is split into the following phases:

- Entry of the present position latitude and longitude of the platform into the INS.
- Platform levelling – both course and fine phases.
- Gyrocompass alignment.

During these processes the sensors are used to sense misalignment between the platform and the desired platform datum. The platform is inched towards these datums over a period of several minutes and, when the datums are reached, the platform is considered to be aligned and in a suitable state to be used for navigation. The platform attitudes for alignment correspond to the navigation axes $X^{(G)}$, $Y^{(G)}$, $Z^{(G)}$ described in Figure 8.31 above.

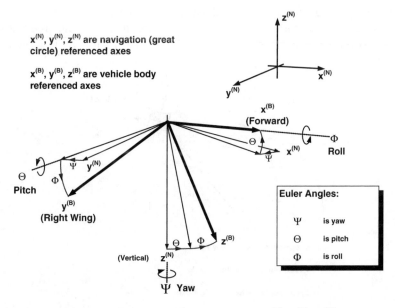

Figure 8.30   Great circle/navigation to body axes (strapdown) $X^{(B)}$, $Y^{(B)}$, $Z^{(B)}$.

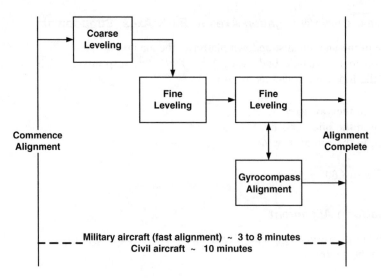

Figure 8.31   Process of platform alignment.

8.8.9.1 Platform Levelling

In the levelling phase the platform sensors and servomotors drive the platform to ensure that the Z direction corresponds to the local vertical at the point on the earth's surface where the platform is located. Now one of the three platform axes is correctly aligned (Figure 8.32).

8.8.9.2 Gyrocompass Alignment

Gyroscopic compassing uses a similar technique and commences once the fine alignment process is under way. In this case the Y axis is driven to align with true north. In an orthogonal axis set, when the Z axis corresponds to the local earth vertical and the Y axis

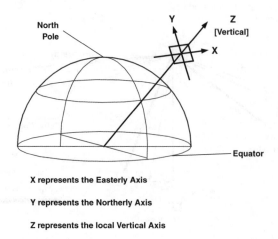

Figure 8.32   Aligned platform axes.

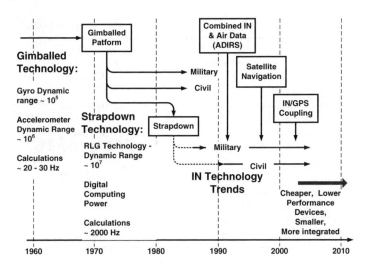

Figure 8.33 Historical development of inertial platforms.

corresponds to true north, the X axis corresponds to east and the platform is aligned and ready to perform navigation tasks.

The accuracy of navigation depends to some degree upon the accuracy of the alignment process, so, in general, but also within reason, the longer the alignment, the better is the accuracy. Accurate alignments are difficult above $\pm 70°$ north or south as greater inaccuracies are experienced.

8.8.10 Historical Perspective – Use of Inertial Platforms

The historical development of IN platforms is shown in Figure 8.33. Gimballed technology came into prominence in the 1960s, followed by strapdown in the early 1980s. In both cases the military avionics community was the first to exploit the technology. In the late 1980s and early 1990s the civil community developed the integratedADIRS concept which rapidly became adopted as the primary means of navigation. Meanwhile, the US military in particular were developing the satellite-based global positioning system (GPS). More recently, IN/GPS coupling has been adopted which enables the fusion of IN and GPS sensors in a similar fashion to baro-IN fusion. Both loosely coupled and tightly coupled implementations are commonly used.

Key attributes of the stabilised and strapdown platforms are as follows:

Gimballed platform	Gyro dynamic range $\sim$105
	Accelerometer dynamic range $\sim$106
	Calculations undertaken $\sim$20–30 Hz
Strapdown technology	Laser ring gyro (LRG) dynamic range $\sim$107
	Digital computing technology
	Calculations undertaken $\sim$2000 Hz

Table 8.2 Typical INS performance and physical characteristics

Parameter	Value
Navigation accuracy	0.8 nm/h
Velocity accuracy	2.5 ft/s rms
Pitch/roll accuracy	0.05° rms
Azimuth accuracy	0.05° rms
Alignment	3–8 min
Volume	4–8 ATR/MCU
Weight	20–30 lb
Power	30–150 W
Acceleration capability	30 g
Angular rate capability	400 deg/s
MTBF (fighter environment)	3500 h
MTBF (civil environment)	10 000 h

The future trend for IN/GPS products is to use cheaper, lower-performance IN sensors, smaller packages and with increasing integration.

The performance and physical characteristics of a typical LRG strapdown performance of 1996 vintage are summarised in Table 8.2.

8.9 Global Navigation Satellite Systems

8.9.1 Introduction to GNSS

Global navigation techniques came into being from the 1960s through to the 1990s when satellites became commonly available. The use of global navigation satellite systems (GNSS), to use the generic name, offers a cheap and accurate navigational means to anyone who possesses a suitable receiver. Although the former Soviet Union developed a system called GLONASS, it is the US ground positioning system (GPS) that is the most widely used. The European Community (EC) is developing a similar system called Gallileo which should enter service in the 2008–2010 timescale. A comparison of the three systems is given in Table 8.3.

GPS is a US satellite-based radio navigational, positioning and time transfer system operated by the Department of Defense (DoD) specifically for military users. The system provides highly accurate position and velocity information and precise time on a continuous global basis to an unlimited number of properly equipped users. The system is unaffected by weather and provides a worldwide common grid reference system based on the earth-fixed coordinate system. For its earth model, GPS uses the world geodetic system of 1984 (WGS-84) datum.

The Department of Defense declared initial operational capability (IOC) of the US GPS on 8 December 1993. The Federal Aviation Administration (FAA) has granted approval for US civil operators to use properly certified GPS equipment as a primary means of navigation in oceanic and certain remote areas. GPS equipment may also be used as a supplementary means of instrument flight rules (IFR) navigation for domestic en-route, terminal operations and certain instrument approaches.

Table 8.3 Comparison of global navigation satellite systems

GLONASS
Soviet Union launched 1982	24 satellites (only 10 in orbit in 2000)
	Three planes
	Inclination 64.8°
	Height 19 130 km

GPS
United States – early 1990s	24 satellites (29)
	Six planes
	Inclination 55°
	Height 20 180 km

Galileo
Europe – 2008–2010	30 satellites (27 + 3)
	Three planes
	Inclination 55°
	Height 23 616 km

8.9.2 Principles of Operation

The principles of satellite navigation using GPS are illustrated in Figure 8.34. GPS comprises three major components as characterised in the figure:

1. The control segment embraces the infrastructure of ground control stations, monitor stations and ground-based satellite dishes that exercise control over the system.

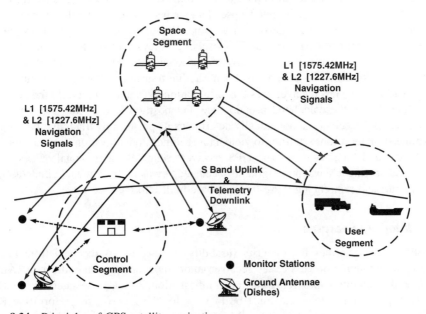

Figure 8.34 Principles of GPS satellite navigation.

2. The space segment includes the satellite constellation, presently around 25 satellites, that forms the basis of the network.
3. The user segment includes all the users: ships, trucks, automobiles, aircraft and hand-held sets. In fact, anyone in possession of a GPS receiver is part of the user segment.

The baseline satellite constellation downlinks data in two bands: L1 on 1575.42 MHz and L2 on 1227.60 MHz. A GPS modernisation programme recently announced will provide a second civil signal in the L2 band for satellites launching in 2003 onwards. In addition, a third civil signal, L5, will be provided on 1176.45 MHz on satellites to be launched in 2005 and beyond. Finally, extra signals for military users (Lm) will be included in the L1 and L2 bands for satellites launched in 2005 and beyond.

GPS operation is based on the concept of ranging and triangulation from a group or constellation of satellites in space which act as precise reference points. A GPS receiver measures distance from a satellite using the travel time of a radio signal. Each satellite transmits a specific code, called course/acquisition (CA), which contains information on the position of the satellite, the GPS system time and the health and accuracy of the transmitted data. Knowing the speed at which the signal travelled (approximately 186 000 miles/s) and the exact broadcast time, the distance travelled by the signal can be computed from the arrival time.

The GPS constellation of 24 satellites is designed so that a minimum of five are always observable by a user anywhere on earth. The receiver uses data from a minimum of four satellites above the mask angle (the lowest angle above the horizon at which it can use a satellite).

GPS receivers match the CA code of each satellite with an identical copy of the code contained in the receiver database. By shifting its copy of the satellite code in a matching process, and by comparing this shift with its internal clock, the receiver can calculate how long it took the signal to travel from the satellite to the receiver. The value derived from this method of computing distance is called a pseudorange because it is not a direct measurement of distance but a measurement derived from time. Pseudorange is subject to several error sources; for example, ionospheric and tropospheric delays and multipath. In addition to knowing the distance to a satellite, a receiver needs to know the exact position of the satellite in space; this is known as its ephemeris. Each satellite transmits information about its exact orbital location. The GPS receiver uses this information to establish precisely the position of the satellite. Using the calculated pseudorange and position information supplied by the satellite, the GPS receiver mathematically determines its position by triangulation. The GPS receiver needs at least four satellites to yield a three-dimensional position (latitude, longitude and altitude) and time solution. The GPS receiver computes navigational values such as distance and bearing to a waypoint, ground speed, etc., by using the known latitude/longitude of the aircraft and referencing these to a database built into the receiver.

8.9.3 Integrity Features

The GPS receiver verifies the integrity (usability) of the signals received from the GPS constellation through a process called receiver autonomous integrity monitoring (RAIM) to determine if a satellite is providing corrupted information. At least one satellite, in addition to those required for navigation, must be in view for the receiver to perform the RAIM function. Therefore, performance of the RAIM function needs a minimum of five satellites in view, or four satellites and a barometric altimeter (baro-aiding) to detect an integrity anomaly. For receivers capable of doing so, RAIM needs six satellites in view (or five satellites with baro-aiding) to isolate the corrupt satellite signal and remove it from the navigation solution.

RAIM messages vary somewhat between receivers; however, generally there are two types. One type indicates that there are insufficient satellites available to provide RAIM integrity monitoring. Another type indicates that the RAIM integrity monitor has detected a potential error that exceeds the limit for the current phase of flight. Without the RAIM capability, the pilot has no assurance of the accuracy of the GPS position. Areas exist where RAIM warnings apply and which can be predicted – especially at higher latitudes – and this represents one of the major shortcomings of GPS and the reason it cannot be used as a sole means of navigation.

8.9.4   GPS Satellite Geometry

The geometry of the GPS satellites favours accurate lateral fixes. However, because a number of the visible satellites may be low in the sky, determination of vertical position is less accurate. Baro-aiding is a method of augmenting the GPS integrity solution by using a non-satellite input source to refine the vertical (height) position estimate. GPS-derived altitude should not be relied upon to determine aircraft altitude since the vertical error can be quite large. To ensure that baro-aiding is available, the current altimeter setting must be entered into the receiver as described in the operating manual.

GPS offers two levels of service: the standard positioning service (SPS) and the precise positioning service (PPS). The SPS provides, to all users, horizontal positioning accuracy of 100 m or less with a probability of 95% 300 m with a probability of 99.99%. The PPS is more accurate than the SPS; however, this is intended to have a selective availability function limiting access to authorized US and allied military, federal government and civil users who can satisfy specific US requirements. At the moment, the selective availability feature is disabled, making the PPS capability available to all users pending the availability of differential GPS (DGPS) solutions to improve the SPS accuracy. This step has been taken pending the development of differential or augmented GPS systems which will provide high accuracy to civil users while preserving the accuracy and security that military users demand.

The basic accuracy without selective availability is about ± 100 m as opposed to ± 1 m when the full system is available. Developments are under way in the United States to improve the accuracy available to civil users. These are:

- The wide-area augmentation system (WAAS) to improve accuracy en route;
- The local-area augmentation system (LAAS) to improve terminal guidance.

8.10   Global Air Transport Management (GATM)

The rapidly increasing commercial air traffic density is leading to a pressing need to improve the air transport management (ATM) system by all available means and move on from the techniques and technologies that have served the industry for the last 40 years. This evolution will embrace the use of new technologies mixed with existing capabilities to offer improved air traffic management. The aims and objectives of ATM and a full description of the future air navigation system (FANS) may be found in Chapter 12 of the sister publication 'Civil Avionics Systems' (Moir and Seabridge, 2003). GATM is the military version of FANS and has to be compatible in all respects to enable the interoperability of civil and military aircraft within controlled airspace. This section provides a brief overview of some of the key features.

To this end, the air traffic control authorities, airline industry, regulatory authorities and airframe and equipment manufacturers are working to create the future air navigation system (FANS) to develop the necessary equipment and procedures. In order to be able to use controlled airspace on equal terms with commercial users, military platforms will need to embody GATM objectives.

The areas where improvements may be made relate to communications, navigation and surveillance, commonly referred to as CNS. The key attributes of these improvements may be briefly summarised as follows:

- *Communication.* The use of data links to increase data flow and permit the delivery of complex air traffic control clearances.
- *Navigation.* The use of GPS in conjunction with other navigational means to improve accuracy and allow closer spacing of aircraft.
- *Surveillance.* The use of data links to signal aircraft position and intent to the ground and other users.

These headings form a useful framework to examine the GATM improvements already made and those planned for the future.

8.10.1 Communications

The main elements of improvement in communications are:

- Air-to-ground VHF data link for domestic communications;
- Air-to-ground SATCOM communications for oceanic communications;
- High-Frequency data link (HFDL);
- 8.33 kHz VHF voice communications.

8.10.1.1 Air-to-Ground VHF Data Link

The emergence of data links as means of communications versus conventional voice communications has developed recently in the commercial community; they have long been used for military purposes.

Voice links have been used in the past for communications between the air traffic control system or ATM and the airline operational centre (AOC). The use of data links, controlled and monitored by the FMS or other suitable method on-board the aircraft, facilitates improved communication with the AOC and ATM systems. These data links may be implemented using one or more of the following:

- VHF communications;
- Mode S transponder;
- Satellite links.

Data link communications are being designed to provide more efficient communications for ATC and flight information services (FIS). Although these systems essentially replace voice communication, there will be a provision for voice back-up in the medium term.

Flight plan data, including aircraft position and intent in the form of future waypoints, arrival times, selected procedures, aircraft trajectory, destination airport and alternatives, will all be transferred to the ground systems for air traffic management. The data sent to the ground ATM system will aid the process of predicting a positional vector for each aircraft at a specific time. This information will aid the task of the ground controllers for validation or reclearance of the flight plan of an aircraft. Furthermore, use of the required time of arrival (RTA) feature will enable the air traffic controllers to reschedule aircraft profiles in order that conflicts do not arise.

For ATC flight service and surveillance, VHF data link (VDL) communications will be increasingly used for domestic communications. VHF communications are line-of-sight limited, as has already been explained. A number of options exist:

1. VDL mode 1. Compatible with existing ACARS transmitting at 2.4 kbps. This mode suffers from the disadvantage that it is character oriented.
2. VDL mode 2. Data only transmitted at 31.5 kpbs. As well as having a higher bandwidth, this protocol is bit rather than character oriented, making it 50–70% more efficient than the ACARS protocol. VDL mode 2 is able to support controller to pilot data link communications (CPDLC).
3. VDL mode 3. Simultaneous data and analogue voice communications using time division multiple-access (TDMA) techniques.
4. VDL mode 4. Used with the 1090 MHz signal of ATC mode S.

It is expected that the introduction of data link technology will benefit all users owing to a more efficient and less ambiguous nature of the messages passed. Significant improvements in dispatch delays and fuel savings are expected as these technologies reach maturity.

8.10.1.2 Air-to-Ground SATCOM Communications

SATCOM is a well proven data link that, as has already explained, is limited at very high latitudes in excess of about 82°. The SATCOM system is supported by the INMARSAT constellation already described earlier in the chapter.

8.10.1.3 HF Data Link

Modern technology enables HF data link transmissions to be more robust than HF voice and therefore less susceptible to the effects of the sunspot cycle. HF data link provides primary coverage out to 2700 nm and secondary coverage beyond that should propagation conditions be favourable. There is extensive cover by ground stations located in the northern hemisphere such that HF data link is a viable alternative to SATCOM for north polar transitions. Refer to the communications and Navaids description earlier.

8.10.1.4 8.33 kHz VHF Voice Communications

Conventional VHF voice channels are spaced at intervals of 25 kHz throughout the spectrum. A denser communications environment has resulted in the introduction of digital radios that permit spacing at 8.33 kHz, allowing three channels to be fitted in the spectrum where only one could be used previously. With effect from 7 October 1999, these radios have already been mandated in Europe for operation above 20 000 ft and will follow in the United States within a number of years; one of the difficulties in predicting the timescale is the vast number of radios that have to be replaced/retrofitted.

8.10.1.5 Protected ILS

Within Europe some ILS installations suffer interference from high-power FM local radio stations. Modifications have been mandated that introduce receiver changes to protect the ILS systems from this interference.

8.10.2 Navigation

A number of navigational improvements are envisaged:

1. Introduction of required navigation performance (RNP) and actual navigation performance (ANP) criteria. This defines absolute navigational performance requirements for various flight conditions and compares this with the actual performance the aircraft system is capable of providing.
2. Reduced vertical separation minima (RVSM).
3. Differential GPS (DGPS) enhancements:

   - WAAS – described earlier;
   - LAAS – described earlier.

4. Protected ILS.
5. Introduction of the microwave landing system (MLS) in Europe.

8.10.2.1 Area Navigation (RNAV)

Area navigation (RNAV) systems allow the aircraft to operate within any desired course within the coverage of station-referenced signals (VOR, DME) or within the limits of a self-contained system capability (IRS, GPS) or a combination of these. RNAV systems have a horizontal two-dimensional capability using one or more of the on-board navigational sensors to determine a flight path determined by navigation aids or waypoints referenced to latitude and longitude. In addition, the RNAV system provides guidance cues or tracking of the flight path. Many modern RNAV systems include a three-dimensional capability to define a vertical flight path based upon altimetry, and some include a full aircraft and engine performance model.

The performance of pre-RNAV systems has historically been defined according to the following criteria:

- Along-track error;
- Across-track error;
- Flight technical error (FTE).

The total navigation error is the root sum square (RSS) of these elements for a given navigation means or phase of flight.

The availability of the navigation capability is defined at 99.999%, and the integrity requirement for misleading navigation information is set at 99.9999%.

8.10.2.2 RNP RNAV and Actual Navigation Performance

The actual navigation performance (ANP) of the aircraft navigation system is represented by a circle that defines the accuracy of the aircraft navigation system for 95% of the time. The value of the ANP is derived by taking the value of all the navigation sensors and statistically weighing them against the other sensors. After a period of time a degree of confidence is established in which are the most accurate sensors and therefore the ANP value is established. The 95% probability circle is that which is compared with RNP to decide whether the navigation system performance is good enough for the route segment being flown. The ANP and RNP values are displayed on the FMS CDU such that the flight crew can readily check on the navigation system status. Should the ANP exceed the RNP value for a given route sector for any reason – for example owing to a critical navigation sensor failing – the crew are alerted to the fact that the system is not maintaining the accuracy necessary. This will result in the aircraft reverting to some lower-capability navigational means. In an approach guidance mode it may necessitate the crew executing a go-around and reinitiating the approach using a less accurate guidance means.

8.10.2.3 Required Navigation Performance (RNP)

The RNP defines the lateral track limits within which the ANP circle should be constrained for various phases of flight. The general requirements are as follows:

1. For oceanic crossings the RNP is ±12 nm, also referred to as RNP-12.
2. For en-route navigation the RNP is ±2 nm (RNP-2).
3. For terminal operations the RNP is ±1 nm (RNP-1).
4. For approach operations the RNP is ±0.3 (RNP-0.3).

Other specific RNP requirements may apply in certain geographical areas, e.g. RNP-4 and RNP-10 (Figure 8.35).

It is clear that this represents a more definitive way of specifying aircraft navigational performance, versus the type of leg being flown, than has previously been the case. Other more specific criteria exist: RNP-5 (also known as BRNAV or area navigation) has already been introduced in parts of the European airspace with the prospect that RNP-1 (also known as PRNAV or precision navigation) will be introduced in a few years. There are precision approaches in being – notably those in Juneau, Alaska – where RNP-0.15 is required for new precision approaches developed for mountainous terrain.

8.10.2.4 RNAV Standards within Europe

Two RNAV standards are being developed in Europe:

1. Basic RNAV (BRNAV). BRNAV was introduced in 1988 and is equivalent to RNP-5 for RNAV operations. Navigation may be accomplished by using the following means:

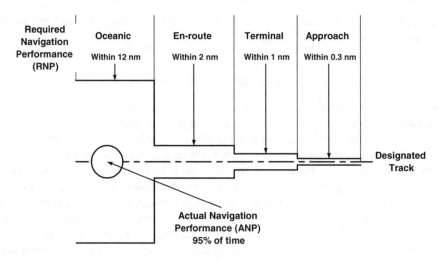

Figure 8.35   ANP versus RNP requirements.

- DME/DME;
- VOR/DME with a 62 nm VOR range limit;
- INS with radio updating or limited to 2 h since the last on-ground position update;
- LORAN-C with limitation.
- GPS with limitation.

Until 2005, primary sources of navigation will be DME/DME, VOR/DME and GPS. Advisory circular AC 90-96 on the approval of US operators and aircraft to operate under instrument flight rules (IFR) in European airspace designated for basic area navigation (BRNAV), 20 March 1998, approves the operation of US aircraft in European airspace under the application of existing advisory circulars.

2. Precision RNAV (PRNAV). PRNAV is intended to be introduced at some time in the future but not before 2005. PRNAV will invoke the use of navigation under RNP-1 accuracy requirements or better.

8.10.2.5   RVSM

One of the other ways of increasing traffic density is the introduction of the reduced separation vertical minima (RVSM) criteria. For many years aircraft have operated with a 2000 ft vertical separation at flight levels between FL290 and FL410. As traffic density has increased, this has proved to be a disadvantage for the busiest sections of airspace. Examination of the basic accuracy of altimetry indicated that there were no inherent technical reasons why this separation should not be reduced. Accordingly, RVSM was introduced to increase the available number of flight levels in this band and effectively permit greater traffic density. The principle is to introduce additional usable flight levels such that the flight level separation is 1000 ft throughout the band, as shown in Figure 8.36.

Originally a trial was mounted in 1997 to test the viability of the concept on specific flight levels – FL 340 and FL360 as shown in the figure. RVSM is now implemented throughout most of Europe from FL290 to FL410, introducing six new flight levels compared with

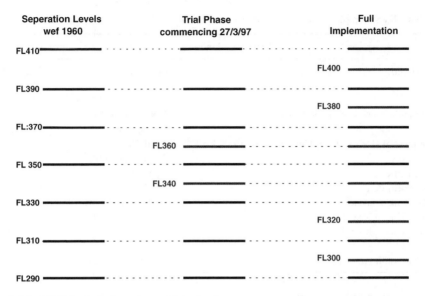

Figure 8.36   RVSM – insertion of new flight levels.

before. All the specified flight levels on the North Atlantic were implemented in 2001. Other regions in the globe will have RVSM selectively implemented to increase air traffic density according to Figure 8.36 and Table 8.4.

8.10.2.6   RVSM Implementation

At the time of writing, the plans for the worldwide implementation of RVSM are as shown in Table 8.4, and many have been implemented to plan. The Federal Aviation Authority (FAA) RVSM website lists the most recent schedule and level of implementation.

RVSM operation requires the aircraft to possess two independent means of measuring altitude and an autopilot with an accurate height hold capability. The operators of RVSM-equipped aircraft are not taken on trust: independent height monitoring stations survey aircraft passing overhead, measuring actual height compared with flight plan details, thereby assuring the performance of each aircraft and operator. RVSM implementation therefore embraces a watchdog function that ensures that all users are conforming to the RVSM accuracy and performance provisions.

8.10.2.7   Differential GPS Enhancements

DGPS enhancements are being developed for en-route and precision landings in the United States. The GPS enhancements – WAAS and LAAS implementations – in the United States have already been described earlier in the chapter and their introduction should lead to the following accuracies being achieved as a matter of course:

1. WAAS is anticipated to yield an accuracy of $\sim$7 m which will be sufficient for Cat I approaches.
2. LAAS is expected to provide enhanced accuracies of $\sim$1 m which will be sufficient for precision approaches Cat II and Cat III.

Table 8.4 RVSM implementation schedule – worldwide

RVSM status – Americas and Europe		
North Atlantic	March 1997	FL330-370
	October 1998	FL310-390
	24 January 2002	FL290-410
West Atlantic route system (WATRS)	1 November 2001	FL310-390
	24 January 2002	FL290-410
Europe tactical (UK, Ireland, Germany, Austria)	April 2001	FL290-410
Europe-wide	24 January 2002	FL290-410
South Atlantic	24 January 2002	FL290-410
Canada North domestic	April 2002	FL290-410
Canada South domestic	Coordinate with US domestic	
Domestic US – phase 1[a]	1 December 2004	FL350-390
Domestic US – phase 2	Late 2005–2006	FL 290-390 (FL410)
Caribbean/South America	RVSM group established	
RVSM status – Asia/Pacific		
Pacific	February 2000	FL290-390
	Tactical use	FL400-410
Australia	November 2001	FL290-410
Western Pacific/South China Sea	21 February 2002	Consult publications
Middle East	November 2003	TBD
Asia-Europe/South of Himalayas	November 2003	TBD

[a]DRVSM plan to be finalised not later than January 2002 on the basis of ATC simulation results and user inputs.

The introduction of DGPS technology is also envisaged for Europe and the Far East. In Europe there are two programmes in the planning stage that will enhance satellite navigation. The European Space Agency (ESA), the European Commission (EC) and the European organisation for the Safety of Air Navigation (Eurocontrol) are working together on the development of a global positioning and navigation satellite system (GNSS) plan. The GNSS programme is being carried out in two phases:

1. GNSS-1. This involves the development of the European geostationary navigation overlay system (EGNOS) which will augment the US GPS and Russian GLONASS systems.
2. GNSS-2. This involves the development of a second-generation satellite navigation system including the deployment of Europe's own satellite system – Galileo. At the time of writing the EU nations were wrangling about budget increases needed to fund the programme, so delay appears to be inevitable.

8.10.3 Surveillance

Surveillance enhancements include the following:

• TCAS II;
• ATC mode S;

- Automatic dependent surveillance A (ADS-A);
- Automatic dependent surveillance B (ADS-B).

The operation of TCAS and ATC mode S has already been described in Chapter 7, but their use in a FANS/GATM context will be briefly examined here.

When operating together with a mode S transponder and a stand-alone display or EFIS presentation, TCAS is able to monitor other aircraft in the vicinity by means of airborne interrogation and assessment of collision risk. TCAS II provides vertical avoidance manoeuvre advice by the use of RAs. TCAS II will soon be made mandatory for civil airliners – aircraft with a weight exceeding 15 000 kg or 30 or more seats – operating in Europe. This will be extended to aircraft exceeding 5700 kg or more than 10 seats, probably by 2005.

References

Moir, I. and Seabridge, A. (2003) *Civil Avionics Systems*. Professional Engineering Publicating/ American Institute of Aeronautics and Astronautics.

Advisory circular AC 121-13, Self-contained navigation systems (long range), 14 October 1969.

Advisory circular AC 120-33, Operational approval for airborne long-range navigation systems for flight within the North Atlantic minimum navigation performance specifications airspace, 24 June 1977.

Advisory circular AC 25-4, Inertial navigation systems (INS), 18 February 1966.

Advisory circular AC 90-94, Guidelines for using global positioning system equipment for IFR en-route and terminal operations and for non-precision approaches in the US national airspace system, 14 December 1994.

Technical standing order (TSO) C-129a, Airborne supplementary navigation equipment using global positioning system (GPS), 20 February 1996.

Advisory circular AC 90-45A, Approval of area navigation systems for use in the US national airspace system, 21 February 1975.

Advisory circular AC 20-130A, Airworthiness approval of navigation or flight management systems integrating multiple sensors, 14 June 1995.

Advisory circular AC 90-97, Use of barometric vertical navigation (VNAV) for instrument approach operations using decision altitude, 19 October 2000.

Advisory circular AC 25-23, Airworthiness criteria for the installation approval of a terrain awareness and warning system (TAWS) for Part 25 airplanes, 22 May 2000.

Advisory circular AC 00-31A, National aviation standard for the very high frequency omnidirectional radio range (VOR)/distance measuring euipment (DME)/tactical air navigation (TACAN) systems, 20 September 1982.

Advisory circular AC 20-121A, Airworthiness approval of LORAN-C navigation systems for use in the US national airspace systems (NAS) and Alaska, 24 August 1988.

Advisory circular AC 00-31A, National aviation standard for the very high frequency omnidirectional radio range(VOR)/distance measuring equipment (DME)/tactical air navigation systems, 20 September 1982.

Advisory circular AC 90-96, Approval of US operators and aircraft to operate under instrument flight rules (IFR) in European airspace, March 1998.

Federal Aviation Authority (FAA) RVSM website: www.faa.gov.ats/ato/rvsm1.htm

9 Weapons Carriage and Guidance

9.1 Introduction

Thus far the technologies associated with the overall integration of the avionics system and all the associated sensor technology have been described. In this chapter, exemplar platforms will be described along with the weapons they carry. By this means the concept of the total integration of avionics system, sensors and weapons will be outlined so that the reader can gain and understanding of the entire weapons system. To set a historical context and to illustrate the breadth of typical weapons systems, the following systems are described:

- F-16 mid-life Update and subsequent developments;
- AH-64 C/D Longbow Apache;
- Eurofighter Typhoon;
- F-22 Raptor;
- Nimrod MR4;
- F-35, formerly the joint strike fighter (JSF).

Figure 9.1 shows a comparative timescale for these developments. There are some interesting observations that may be made:

1. The relative ease of modifying an existing platform, especially where there are a large number of aircraft manufactured – the F-16 (>4000) and AH-64 (>1000) are good examples. The huge investment in developing and purchasing aircraft in these numbers means that there is an inducement to modifying existing platforms instead of buying new airframes. There is a limit to this, as eventually the aircraft will run out of fatigue life and will have to be replaced.

Military Avionics Systems   I. Moir and A. Seabridge
© 2006 John Wiley & Sons, Ltd

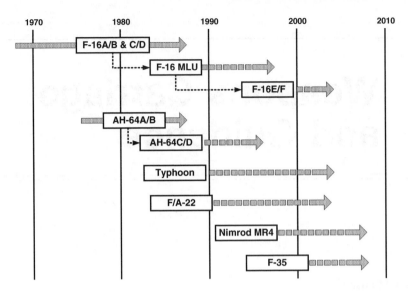

Figure 9.1   Comparative development programmes.

2. The Typhoon and Raptor have both suffered from political factors. Both platforms are very capable aircraft that were conceived as air superiority fighters during the latter stages of the Cold War and have lacked total political commitment since. The result has been in both cases a very long development phase of ~15 years leading to entry into service (EIS), and both aircraft are still evolving to adapt to deployment in air-to-ground roles.

As will be seen, all of these aircraft are able to carry combinations of weapons options best suited for air-to-air and air-to-ground roles in which they are deployed. In many cases, common weapons are capable of being carried and released from more than one platform type, and this offers economies of scale in each weapon manufacture, and interoperability between different nations and air forces – a point recognised in NATO some time ago. The key has lain with the standardisation of the weapons/aircraft mechanical and electrical interfaces. The latter has taken on increasing significance as 'smart weapons' have evolved that require information exchange with the platform and sensors.

The standard stores interface – MIL-STD-1760 – will also be described, as will the capabilities of common air-to-air and air-to-ground missiles that are used on the platforms listed above.

9.2   F-16 Fighting Falcon

The F-16 is a multirole fighter of which more than 4000 have been produced. The F-16 was selected by the US Air Force as a result of the lightweight fighter (LWF) contest in which the General Dynamics (now Lockheed Martin) YF-16 underwent a fly-off competition against the Northrop YF-17 and was successfully declared the winner in 1976. The F-16 was smaller than its contemporary, having a single engine versus two engines for the YF-17. The YF-16 was a more innovative design than the YF-17, which in many respects was derived from the Northrop F-5. One particularly innovative feature was that the aircraft was designed to be negatively aerodynamically stable, that is, with the aircraft centre of gravity (CG) behind or close to the aerodynamic centre of pressure (CP). This unstable design necessitated the provision of a quadruplex fly-by-wire (FBW) computer system which carried out the necessary flight control calculations and signalled the demands to the flight control actuators by electrical means rather than using the conventional lever and push rod configuration. This resulted in a safe, highly efficient design that provided a fast control response while also being programmed to provide the pilot with pleasant handling characteristics.

The F-16 is a classic example of a multirole aircraft serving in a variety of roles with a large number of air forces. The airframe has been grown to provide increased all-up weight, fuel and weapons loads; engine thrust has increased to accommodate the additional airframe mass without any loss in performance. Also interesting within the context of this book is the evolution of the avionics system to provide ever greater weapons systems capability within the aircraft. The F-16 MLU avionics systems architecture is depicted in a simplified form in Figure 9.2.

9.2.1 F-16 Evolution

The F-16 evolution over more than 20 years can be summarised as follows. The original F-16 was designed as a lightweight air-to-air day fighter. Air-to-ground responsibilities transformed the first production F-16s into multirole fighters. The empty weight of the block 10 F-16A is 15 600 lb. The empty weight of the block 50 is 19 200 lb. The A in F-16A refers to a block 10 to 20 single-seat aircraft. The B in F-16B refers to the two-seat version, hence the abbreviation A/B, C/D, etc. The block number is an important term in tracing the evolution of the F-16 and is particularly useful in tracking the evolving avionics systems configurations as the aircraft has developed.

The F-16A, a single-seat model, first flew in December 1976. The first operational F-16A was delivered in January 1979 to the 388th Tactical Fighter Wing at Hill Air Force Base, Utah. The F-16B, a two-seat model, has tandem cockpits that are about the same size as the single canopy in the A model. Its bubble canopy extends to cover the second cockpit. The various F-16 models may be summarised as follows:

1. *Block 1 and block 5* F-16s were manufactured through 1981 for the US Air Force and for four European air forces. Most block 1 and block 5 aircraft were upgraded to a block 10 standard in a programme called Pacer Loft in 1982.
2. *Block 10* aircraft (312 total) were built through 1980. The differences between these early F-16 versions are relatively minor. Block 15 aircraft represent the most numerous version of the more than 4000 F-16s manufactured. The transition from block 10 to block 15 resulted in two hard points added to the chin of the inlet. The larger horizontal tails, which grew in area by about 30% are the most noticeable difference between block 15 and previous F-16 versions.
3. *F-16C/D* aircraft, which are the single- and two-seat counterparts to the F-16A/B, incorporate the latest cockpit control and display technology. All F-16s delivered since November 1981 have built-in structural and wiring provisions and systems architecture that permits expansion of the multirole flexibility to perform precision strike, night attack and beyond-visual-range (BVR) interception missions. All active units and many Air National Guard and Air Force Reserve units have converted to the F-16C/D, which is deployed in a number of block variants.

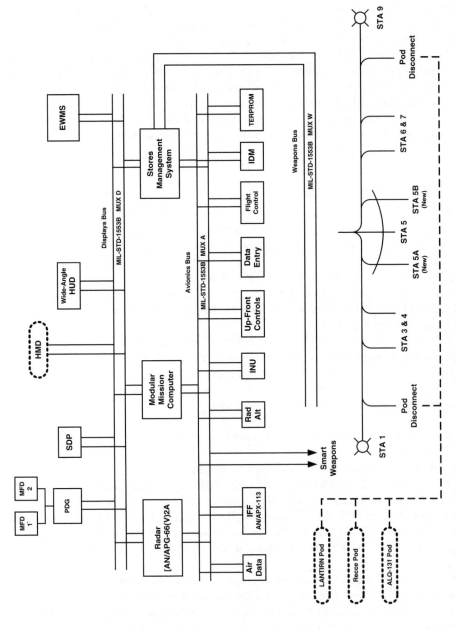

Figure 9.2 F-16 mid-life upgrade (MLU) – avionics system top-level architecture.

4. *Block 25* added the ability to carry AMRAAM to the F-16 as well as night/precision ground-attack capabilities and an improved radar, the Westinghouse (now Northrop-Grumman) AN/APG-68, with increased range, better resolution and more operating modes.
5. *Block 30/32*. This configuration adopted two new engines – block 30 designates a General Electric F110-GE-100 engine, and block 32 designates a Pratt & Whitney F100-PW-220 engine. Block 30/32 can carry the AGM-45 Shrike and the AGM-88A HARM, and, like the block 25, it can carry the AGM-65 Maverick.
6. *Block 40/42*. F-16CG/DG gained capabilities for navigation and precision attack in all weather conditions and at night with the LANTIRN pods and more extensive air-to-ground loads, including the GBU-10, GBU-12 and GBU-24 Paveway laser-guided bombs and the GBU-15. Block 40/42 production began in 1988 and ran through to 1995. Currently, the block 40s are being upgraded with several block 50 systems: the ALR-56M threat warning system, the ALE-47 advanced chaff/flare dispenser, an improved performance battery and Falcon UP structural upgrade.
7. *Block 50/52*. Equipped with a Northrop Grumman APG-68(V)7 radar and a General Electric F 110-GE-129 increased performance engine, the aircraft are also capable of using the Lockheed Martin low-altitude navigation and targeting infrared for night (LANTIRN) system (see Chapter 5). Technology enhancements include colour multi-functional displays and a programmable display generator, a new modular mission computer (MMC), a digital terrain system, a new colour video camera and colour triple-deck video recorder to record the pilot's head-up display view and an upgraded data transfer unit. In May 2000, the US Air Force certificated block 50/52 variant F-16s to carry the following:

- CBU-103/104/105 wind-corrected munitions dispenser;
- AGM-154 joint stand-off weapon (JSOW);
- GBU-31/32 joint direct attack munition (JDAM);
- Theatre airborne reconnaissance system.

Beginning in mid-2000, Lockheed-Martin began to deliver block 50/52 variants equipped with an on-board oxygen generation system (OBOGS) designed to replace the original liquid oxygen (LOX) system.
8. *Block 50D/52D*. Wild Weasel F-16CJ (CJ means block 50) comes in C-model (one-seat) and D-model (two-seat) versions. It is best recognized for its ability to carry the AGM-88 HARM and the AN/ASQ-213 HARM targeting system (HTS) in the suppression of enemy air defences (SEAD) mission relinquished by the F-4. This system allows HARM to be employed in the 'range-known' mode, providing longer-range engagements against specific targets. This specialised version of the F-16, which can also carry the ALQ-119 electronic jamming pod for self-protection, became the sole provider for Air Force SEAD missions when the F-4G Wild Weasel was retired from the Air Force inventory. Although F-18s and EA-6Bs are HARM capable, the F-16 provides the ability to use the HARM in its most effective mode.
9. *Block 60 or F-16E/F*. In May 1998 the UAE announced selection of the block 60 F-16 to be delivered between 2002 and 2004. The upgrade package consists of a range of modern systems including conformal fuel tanks for greater range, new cockpit displays, an internal infrared (IR) sensor suite, a new mission computer and other advanced features

including a new active electronically scanned array radar (APG-80) as described in Chapter 4 (see below under F-16 E/F).

9.2.2 F-16 Mid-life Update

Towards the end of the 1980s the US Air Force began to study an avionics upgrade for the F-16A/B and also attracted interest from four European NATO air forces that also operated the F-16. These air forces were those of the Netherlands, Belgium, Norway and Denmark. In the event, the end of the Cold War led the US Air Force to abandon their plans to adopt the mid-life update (MLU) as it was called, and their F-16A/Bs were retired instead. In all, a total of 343 NATO aircraft were modified as follows:

Air force	Number modified
Netherlands	136
Belgium	90
Denmark	61
Norway	56
Total	343

The MLU programme was undertaken by an industrial partnership formed from partners from the participating nations. The key technical improvements included:

1. Introduction of a new modular mission computer (MMC) to provide processing for weapons control, stores management and head-up display (HUD). Helmet-mounted display (HMD) features were also added. The MMC is a derivative of the common integrated processor (CIP) developed for the F-22 (see F-22 in this chapter) (Figure 9.3). This is a modular avionics rack populated by line replaceable modules (LRMs) as described in Chapter 2 under the JIAWG architecture. There are a total of 30 LRM slots provided in the MMC rack, of which 21 are populated; the remaining nine are available for growth. The use of LRMs allows the adoption of two-level maintenance, thereby eliminating intermediate-level (I-level) test equipment. The MMC replaces three LRUs

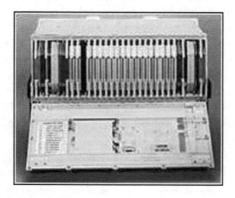

Figure 9.3   MLU modular mission computer.

from the previous architecture, reducing volume, weight and power consumption. The MMC has four functional areas:

- Data processing set (DPS) associated with weapons MUX bus control;
- Avionics display set (ADS) interfacing with the HUD;
- Avionics I/O (AIOS) interfacing with the avionics units;
- Power set (PS) controlling the power supplies and conditioning.

2. Improvements to the AN/APG-66 radar bring it to the AN/APG-66(V)2A standard. Improved operational modes include ten-target track-while-scan (TWS) capability, 64 : 1 Doppler beam sharpening (DBS) and enhanced air-to-air and ground-mapping modes. Target detection and tracking ranges have been improved by ~25%. Rationalisation of the digital signal processor (DSP) and radar processor (RP); these have been replaced by a single unit with accompanying improvements in weight, volume and power dissipation. Introduction of a radar MUX bus to integrate the radar units has provided further savings.
3. Advanced identification friend or foe (AIFF) AN/APX-113(V) incorporating both IFF interrogator and transponder functions.
4. Wide-angle HUD.
5. Multifunction displays (MFDs).
6. Data entry/cockpit interface set integrating communication and navigation functions.
7. Side stick controller and throttle grip to give hands-on throttle and stick (HOTAS) capability.
8. Improved data modem (IDM) to provide target data sharing capability using V/UHF and secure communications.
9. Digital terrain system using the terrain profile matching (TERPROM) principle.
10. Global positioning system (GPS).
11. Navigation/targeting pod provision – LANTIRN.

Many of these items were subsequently added to the US Air Force block 50 F-16s, but under a series of budget allocations and modification programmes.

The top-level architecture for the F-16 MLU is shown in Figure 9.2. The MMC performs the primary bus controller function for a total of five MIL-STD-1553B (MUX) buses: MUX A, MUX B, MUX C, MUX D and MUX W. The figure shows three of these MUX buses:

1. MUX A is the avionics bus that interconnects all the main avionics functions: air data, IFF, radar altimeter, inertial navigation unit (INU), up-front controls, data entry, flight control, IDM and TERPROM.
2. MUX D is the displays bus interfacing with the two MFDs, wide-angle HUD and HMD among others. The main avionics units – AN/APG-66 (V2) radar, MMC and stores management system (SMS) – straddle both MUX A and MUX D buses.
3. MUX W is the weapons bus controlled by the stores management system (SMS) and interfaces to the weapons stations by means of standard MIL-STD-1760/1553B interconnects.

There are a total of nine stores stations (STA) provided, as shown in Table 9.1. In addition, the two new chin-mounted stations – 5A and 5B – are also shown. As well as carrying

Table 9.1 F-16 weapons stations and stores configurations

	Left wing				Centre – fuselage					Right wing	
Role	STA 1 wing tip	STA 2	STA 3/3A	STA 4	STA 5A left chin	STA 5 centre-line	STA 5B right chin	STA 6	STA7/7A	STA 8	STA 9 wing tip
Defensive counterair	AMRAAM	AMRAAM	AIM-9	370 g tank				370 g tank	AIM-9	AMRAAM	AMRAAM
Interdiction 1	AMRAAM		GBU-24	370 g tank		LANTIRN		370 g tank	GBU-24		AMRAAM
Interdiction 2	AIM-9		AGM-65	370 g tank		ECM pod		370 g tank	AGM-65		AIM-9
SEAD	AIM-9		HARM	370 g tank		LANTIRN		370 g tank	HARM		AIM-9

Note: All these configurations can be augmented by the gun – an M-61A 20 mm rotary multibarrel cannon with 500 rounds.

weapons and fuel tanks on designated stations, the system is also capable of carrying the following pods for specific missions:

- LANTIRN pods – navigation and targeting;
- Reconnaissance pod;
- AN/ALQ-131 ECM pod.

The MLU/block 50/52 modifications have given an enormous mission capability far exceeding that of the original F-16 A/B configuration. All new F-16s being delivered are delivered to block 50/52 standard or block 60 – see below.

9.2.3 F-16 E/F (F-16 block 60)

The latest avionics upgrades to the F-16 were originally planned as block 60 as a continuation of the block numbering system. Block 60 – now officially termed the F-16E/F – has been ordered by the United Arab Emirates (UAE) and represents a totally new avionics system. Compared with the original F-16A/B, the E/F version has increased the maximum take-off weight by 50% (to 22 700 kg/50 000 lb) and increased engine thrust by 35% (to 32 500 lb/145 kN). In the F-16E/F, approximately 70% of the aircraft structure is new; additional conformal fuel tanks may be fitted to improve range and a new dorsal avionics bay has been introduced down the spine of the aircraft (Figure 9.4).

The avionics system for the F-16E/F offers a huge advance in capability over previous systems. The key attributes are:

1. Introduction of an active electronically scanned array (AESA) radar – AN/APG-80 – with a marked increase in radar performance and capabilities as described in Chapter 4. In certain circumstances, the increase in radar range offered by the AESA radar represents a 55% improvement over the block 50/52 radar (Chapter 4, Figure 4.29). The radar is cooled by a dedicated liquid cooling system with the radar ECS located in the dorsal spine equipment bay and the coolant lines run forward to the radar. The radar also uses COTS processor and fibre-channel (FC) technology to handle the high-bandwidth processing and data handling tasks.
2. The avionics systems integration is performed by a dual-redundant FC that is capable of transferring data at 1 Gbit/s. This provides much greater bandwidth than the 1553B MUX buses, allowing more data to be passed between aircraft systems. Avionics computing resources are based upon Power PC COTS technology, and the 1.3 million lines of software code have been written in C++. Existing/legacy software has been rewritten from the original Jovial or Ada languages into C++.
3. Introduction of an integrated FLIR and targeting system (IFTS) developed by Northrop Grumman and designated as the AN/AAQ-32. This system offers the navigation and targeting functions offered by the LANTIRN system but with separate navigation and targeting pods. The system comprises:

    - A passive navigation FLIR located on the upper fuselage just ahead of the cockpit;
    - A targeting FLIR minipod which looks similar to the XR Sniper pod being introduced on to Air National Guard (ANG) block 30s and US Air Force block 50s having a similar chisel nose; however, it is more likely to be based upon the Northrop Grumman

Figure 9.4   F-16E/F (Lockheed Martin).

and Rafael technology embodied in the Litening family of EO pods which is outlined in Chapter 5. (Table 5.3).

4. Improved Falcon Edge electronic warfare (EW) suite.
5. Improvements to the displays – introduction of three (5 × 7 in) colour MFDs and use of a 25° × 25° HUD.
6. Improvements to the flight control system to provide safety features such as automatic recovery from a deep stall and automatic ground collision avoidance. The flight control software has been rewritten in the commercial language C++.

The F-16 can carry a wide variety of weapons to satisfy a number of different roles (see Table 9.1 which displays typical weapons loads for four different missions). Table 9.1 is a simple portrayal of the weapons carried by the many air forces who operate the F-16. Many operators will have their own configurations of indigenously produced missiles and ECM pods. The wide range of weapons carried is a testimony to the flexibility of the aircraft.

9.3   AH-64 C/D Longbow Apache

The Army AH-64 C/D Longbow Apache is a formidable attack helicopter. The original Hughes AH-64 A/B Apache became a McDonnell Douglas Helicopter Company (MDHC) product before they in turn were taken over by Boeing.

9.3.1   Baseline System

The original AH-64A/B Apache was fitted with a state-of-the-art electrooptics suite that comprised the following capabilities:

- Day TV for use by day and low-level light conditions providing monochrome (black and white) imagery.
- FLIR-viewed imagery, real world and magnified during day, night and adverse weather.
- Direct vision optics (DVO) viewing real world in colour and with magnified images during day and low-light conditions.

These features are integrated in the integrated AN/AAQ-11 target acquisition and designator sight (TADS) and AN/ASQ-170 pilot's night-vision sight (PNVS) into the combined TADS/PNVS described in Chapter 5.

The US Army procured a total of 824 of the A/B variants of Apache, and several overseas countries also purchased this baseline variant of the aircraft. By the late 1980s, studies were under way to upgrade the aircraft, and this led to the AH-64C/D Longbow variant. This programme entails the remanufacture of the original 64A/B version and converts it to the later variant. The US Army intend to procure a total of 227 helicopters to the 64D variant fitted with the fire control radar (FCR) and 531 to the 64C variant which does not have the FCR but includes all the other avionics modifications.

The United Kingdom has also procured 67 of the Longbow Apache as the WAH-64, and these have entered service with the British Army. Other foreign nations that have purchased or have shown an interest in the Longbow Apache are: the Netherlands – 30; Singapore – 8, with an option for 20 more; Japanese Defence Force – 12, with a requirement for up to 60; Israel – 12 converted AH-64As plus nine new-build; Egypt – 36 rebuilt AH-64As; and Kuwait – 16.

9.3.2   Longbow Apache

The key elements of the AH-64C/D upgraded avionics systems are:

1. Mast-mounted fire control radar (FCR) with the designation AN/APG-78, more commonly known as the Longbow fire control radar. The FCR is a multimode millimetric wave (MMW) radar and provides the pilot with the capability of detecting, classifying and prioritising stationary and moving targets on the ground and in the air. The radar has four modes of operation:

   - *Air-targeting mode (ATM)*: this detects, classifies and prioritises fixed and rotary wing threats;

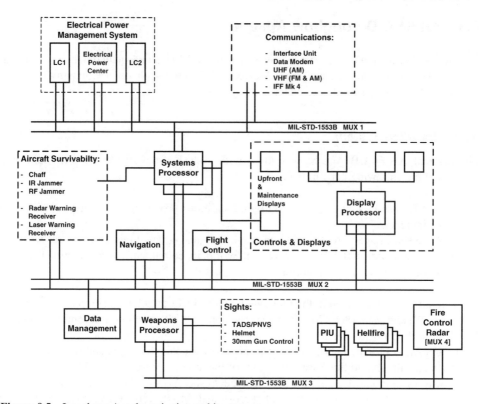

Figure 9.5 Longbour Apache avionics architecture.

- *Ground-targeting mode (GTM)*: this detects, classifies and prioritises ground and air targets;
- *Terrain-profiling mode (TPM)*: this provides obstacle detection and adverse weather pilotage aids to the Longbow.
- *Built-in test (BIT) mode*: this monitors radar performance in flight and isolates electronic failures before and during maintenance.

2. Introduction of a new sensor, the radio-frequency interferometer (RFI) – AN/APR-48A. The RFI is capable of monitoring a broad spectrum and identifying threat radars and also when threat emitters are tracking the helicopter. The system has a stored library of threats.
3. Improved navigation using a dual-channel enhanced global positioning system (GPS)/ inertial navigation system (INS) or EGI.
4. Improved Doppler velocity rate sensor.
5. The UK WAH-64 has a slightly different fit in terms of radios and IFF. There will be a new EW equipment fit called the helicopter integrated defensive aids system (HIDAS) including:

   - AN/AAR-57 common missile warning system (CWMS);
   - AN/APR-48 RFI (as for AH-64C/D);
   - Sky Guardian 2000 radar warning receiver (RWR);

- Type 1223 laser warning system;
- IR jamming system: either the AN/ALQ-144 as already fitted to the AH-64 or the AN/AAQ-24 Nemesis directed IR countermeasures (DIRCM) system fitted to C-130 gunships.

The new avionics is hosted in extended forward avionic bays (EFABs) each side of the cockpit. The EFABs are conditioned by means of two closed-cycle environmental control systems. The Apache is flown by a two-man crew: a pilot and a copilot gunner (CPG). The new mission system introduced by Longbow allows the crew member to use the FCR to seek air targets while the CPG uses the TADS/PNVS system to identify ground targets.

The Longbow Apache avionics system is shown in Figure 9.6. It can be seen that this federated architecture using 1553B MUX buses bears some similarity to the F-16 MLU architecture. There are four MUX buses:

1. MUX bus 1 integrates the electrical system and the communications suite with two systems processors. The communications suite includes a communication interface unit, a data modem, UHF (AM), VHF (AM and FM), IFF and secure communications.
2. MUX bus 2 is similar to an avionics bus in function, interfacing the following functions with the systems processors:

- Navigation;
- Flight control;
- Display processors, which in turn drive the display suite;
- Weapons processors;
- Data management;
- Aircraft survivability equipment (ASE) including:

– radar warning: AN/APR-39A (V)

– laser warning: AN/AVR-2A

Mast Mounted Assembly

Radio Frequency interferometer

RF Missile

RF Seeking Guidance

Figure 9.6 Fire control radar, RFI and RF-guided Hellfire. (Lockhead Martin)

– radar jammer: AN/ALQ-136(V)

– IR jammer: AN/ALQ-144(V).

3. MUX bus 3 is a weapons bus integrating the weapons processors, the pylon interface unit (PIU), Hellfire missiles and the AN/APG-78 fire control radar.
4. MUX bus 4 is a bus dedicated to integrating the fire control radar units.

See Figure 9.6 which depicts:

- The AN/APG-78 radar mast assembly;
- The AN/APR-48A;
- RF-guided AGM-114 Hellfire missile.

9.3.3 Modernisation of TADS/PNVS

Towards the end of the 1990s a programme was initiated to modify and upgrade the TADS/PNVS. The M-TADS and M-PNVS programmes (also known as Arrowhead) upgrade the system by replacing six of the eight existing system line replaceable units (LRUS). as well as introducing improved technology with an improved performance, Arrowhead draws upon in-service experience to make the system truly two-level maintenance capable. The LRUs are also being designed so that in many cases LRMs may be replaced instead of the entire LRU.

The external appearance of the M-TADS/PNVS is similar to the original TADS/PNVS apart from the PNVS shroud which now houses two sensors: a new FLIR and an in-built image intensifier (effectively an integrated 18 mm NVG tube). The pilot will be able to use either FLIR or an image intensifier, depending upon which is best for the conditions that prevail. Ultimately there are plans completely to fuse the output of both sensors (Figure 9.7).

The new FLIR will have an improved 4×480 detector, as opposed to the original 1×180, and therefore image quality will improve. This performance improvement has a particular bearing in avoiding wires at low level and in adverse weather. Automatic boresighting (harmonisation) is incorporated, with the process taking about 2 min after being initiated by the CPG. A number of other improvements including a new multitarget

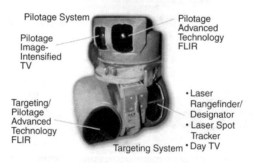

Figure 9.7 M-TADS/PNVS – Arrowhead. (Lockhead Martin)

tracker system (capable of tracking up to six targets) and extended-range algorithms will enable target tracking at ranges around twice that of the existing system.

The M-TADS/PNVS is delivered as a field retrofit kit that can be installed in about 8 h. Deliveries commenced in 2004.

9.3.4 Weapons

The Apache is capable of carrying the following weapons:

1. Up to 16 AGM-114 Hellfire missiles. The latest AGM-114 K has a dual-seeker capability: pulsed radar or semi-active laser guidance and a range in excess of 8 km (5 miles).
2. Hydra 70 rocket pod carrying up to 1970 mm folding-fin aerial rocket (FFAR) unguided rockets.
3. M 320 30 mm 'chain gun' located under the chin of the aircraft and capable of carrying up to 1200 rounds of ammunition. The chain gun can be slaved to the pilot/CPG's helmet sight.

The Hellfire missiles and pylon interface units (PIUs) are interfaced with the weapons MUX bus as shown in Figure 9.5. Four missiles can be accommodated on each M229 missile launcher (Figure 9.8).

Although there have been test firings of other missiles, including the AIM-9 Sidewinder, AIM-28 air-to-air stinger (both heat-seeking missiles) and the British laser-guided Starstreak, none of these has been deployed for operational service. (see Figure 9.9 for typical weapons loads of 16 × Hellfire missiles and a mixed Hellfire missiles (eight) plus rockets load).

9.4 Eurofighter Typhoon

The Eurofighter Typhoon had its origin in an operational need in the 1980s to provide an air superiority fighter for the European NATO air forces to counter the threat of the Soviet Union and Warsaw Pact countries. After the decision of France not to participate, the project became

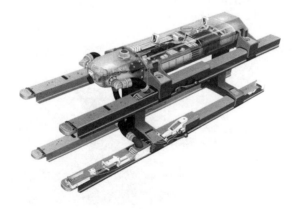

Figure 9.8 Apache M229 missile launcher. (Lockhead Martin)

Figure 9.9 Typical Longbow Apache weapons.

a four-nation joint project with the participating partners being Britain, Germany, Italy and Spain. The project had to accommodate national as well as common equipment fits and both single-seat and two-seat versions, the latter for training rather than operational purposes.

The Typhoon configuration had previously been very successfully demonstrated using a single-aircraft flight demonstrator called the experimental aircraft programme (EAP) which first flew in 1986. This aircraft demonstrated cardinal-point technologies including colour multifunction displays, an integrated utilities management system (UMS), the first of its type incidentally to fly anywhere in the world, and a digital fly-by-wire system to control the highly unstable aircraft. At the time it was the first aircraft flying in Europe with extensive use of MIL-STD-1553 buses. This aircraft flight demonstrator was funded by the UK Ministry of Defence (MOD), together with UK Industry, and with some help from German and Italian Industry. It flew for around 2 years, gathering vital data about the aircraft dynamics and the interaction of the new digital systems, and proved to be a highly successful venture gaining valuable experience that would be used during the design of the Typhoon (Figure 9.10). The aircraft is now at Loughborough University in the United Kingdom.

The aircraft is now in service with the air forces of the four participating nations. The avionics architecture is shown in Figure 9.11. The full capabilities of the system may be best explained by describing the following elements:

Figure 9.10 Eurofighter Typhoon (Eurofighter GmbH).

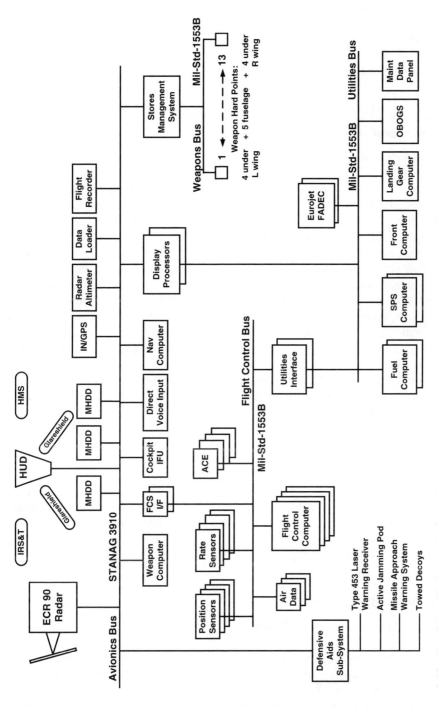

Figure 9.11 Eurofighter Typhoon avionics architecture.

- Sensors and navigation;
- Displays and controls;
- Flight control;
- Utilities control;
- Systems integration;
- Survival/countermeasures;
- Weapons.

9.4.1 Sensors and Navigation

The Typhoon sensors include the following:

1. Captor radar. This is an X-band (8–12 GHz) radar multimode pulse Doppler radar. A track-while-scan (TWS) mode can track, identify and prioritise up to 20 targets simultaneously. Air-to-ground modes include a ground moving target indication (GMTI), spot mapping and surface ranging. A synthetic aperture radar (SAR) mode has the capability of high resolution for specific mapping purposes. Sophisticated frequency analysis techniques provide a non-cooperative target recognition capability where the signal returned from a target aircraft may be analysed and its signature recognised as being from a particular aircraft type. At some stage, Typhoon may be retrofitted with a European AESA radar with technology developed jointly from the United Kingdom, Germany and France AMSAR programme. A small demonstration array has been tested, and a full-scale array of 1000 or more elements is being flown on a test bed aircraft.
2. Infrared search and track (IRST). This is a second-generation IRST system called PIRATE and was described in Chapter 5. It provides passive IR detection in the MWIR (3–5 μm) and LWIR (8–11 μm) bands.
3. IFF interrogator and transponder. An IFF interrogator and mode S transponder compatible with the NATO IFF Mk 12 standard.
4. FLIR targeting pod. The aircraft will have the ability to carry a contemporary FLIR targeting pod, as yet this capability is not operational.
5. Dual INS/GPS. A laser-rate gyro-based INS together with GPS provides better navigational accuracy within several metres. A terrain avoidance warning system (TAWS) based upon TERPROM working with the INS/GPS and covert radio altimeter allows passive low-level navigation and terrain avoidance.
6. Air data. Triplex air data sources provide high integrity data to the FBW system.

9.4.2 Displays and Controls

The displays and controls include the following:

1. HOTAS capability providing hands-on throttle and stick control of sensors, weapon control and communications and cursor control. A total of 24 selector buttons are provided (12 on each control).
2. Direct voice input (DVI) with 200 commands and a response time of 200 ms. A 95% recognition capability is claimed.

3. Wide-angle HUD with a $35° \times 25°$ FOV.
4. Three multifunction head-down displays (MHDDs) using colour AMLCD technology. Any of the displays – usually the centre display – can show a moving map using digital terrain data to portray the position of the aircraft. If necessary, the target and threat scenario may be overlaid, providing the pilot with complete tactical awareness.
5. Helmet-mounted sighting system (HMSS) with an HMD providing a binocular system with up to $40°$ FOV.

9.4.3 Flight Control

The FBW is a full-authority active control technology (ACT) digital system to provide carefree handling of the aircraft using all-moving foreplanes mounted near the nose, wing trailing edge elevons, leading edge slats, rudder and airbrake. The system has quadruplex digital flight control computers, each containing eight Motorola 68020 processors and specially designed ASICS to achieve the necessary levels of safety. The flight control computers, sensors and flight control actuators are connected using a MIL-STD-1553B data bus and dedicated links where necessary. The flight control bus interfaces to the avionics bus via a dedicated interface.

9.4.4 Utilities Control

Control of the aircraft utilities systems such as fuel, environmental control, brakes and landing gear, secondary power systems, and OBOGS are by means of dedicated controllers connected to a utilities MIL-STD-1553B bus. Also connected to this bus are the full-authority digital engine controllers (FADECs) for the Eurojet 2000 engines and a maintenance data panel. This philosophy in part follows the rationale of an integrated utilities management demonstrated on the EAP described above.

9.4.5 Systems Integration

The aircraft uses a combination of 20 Mbit/s fibre-optic STANAG 3010 and standard 1 Mbit/s MIL-STD-1553 buses to integrate the various avionics subsystems. The STANAG 3910 bus combines high data rate 20 Mbit/s fibre-optic transfers by using wire-based 1553 control protocol as described in Chapter 2. To see how these high-speed buses integrate the Typhoon avionics system, refer to Figure 9.11 which offers a very simplified portrayal; in fact there are a total of two STANAG 3910 and six MIL-STD-1553B in total to integrate all the aircraft avionics subsystem. The aircraft-level data buses may be simply described as follows:

1. *STANAG 3910 avionics buses.* The avionics and attack buses interface with the sensors and displays. There are dedicated interfaces to the defensive aids subsystem (DASS) and flight control system. Two display processors are connected to both the avionics and utilities bus. The stores management system interfaces with the dedicated weapons bus.
2. *MIL-STD-1553B flight control bus.* The flight control system has a dedicated data bus interconnecting sensors, flight control computers and actuator assemblies. There is a dedicated interface connecting the flight control and utilities buses.

3. *MIL-STD-1553B utilities bus.* A dedicated bus interconnects the utility control system (UCS) computers, FADECs and maintenance data panel which facilitates servicing the aircraft.
4. *MIL-STD-1553B weapons bus.* The dedicated 1553/MIL-STD-1760 weapons bus interfaces with the 13 weapons stations as described below.

9.4.6 Survival/Countermeasures

Aircraft survival and countermeasures are provided by an integrated defensive aids suite (DASS) which integrates the following equipment:

1. Wide-band receiver (100 MHz to 10 GHz) providing 360° radar warning receiver (RWR) coverage in azimuth and an active jammer using antennas located on the wing-tip pods and the fuselage.
2. A pulse Doppler missile approach warning (MAW) system is fitted which uses antennas located at the wing roots and near the fin. This system warns of the approach of passive as well as actively guided missiles. Improvements are expected to enhance this system using either IR or UV detectors.
3. Laser warning receiver (Royal Air Force only).
4. Towed radar decoy (Royal Air Force only). This is a derivative of a system already deployed by the RAF on Tornado and other aircraft.
5. Chaff and flare dispenser.

9.4.7 Weapons

The Typhoon is able to carry a wide range of weapons and stores to satisfy the operational needs of the four participating nations and export customers. The Typhoon has a total of 13 weapons stations, four under each wing and five under the fuselage. The full complement of weapons that may be carried is shown in Tables 9.2 and 9.3. Figure 9.12 illustrates several of these weapon fit options.

Figure 9.12 Typhoon weapon carriage options. (BAE SYSTEMS)

Table 9.2 Typhoon weapons stations and stores configurations – air to air

Store	Left wing				Centre – fuselage			Right wing			
	STA 1	STA 2	STA 3	STA 4	STA 5 + STA 6	STA 7	STA 8 + STA 9	STA 10	STA 11	STA 12	STA 13
AMRAAM		1		1	2		2	1		1	
BVRAAM		1		1	2		2	1		1	
Sky Flash					1		1				
Aspide					1		1				
AIM-9	1	2								2	1
ASRAAM	1	2								2	1
IRIS-T	1	2								2	1

Certain variants also carry an internal 27 mm gun.

Table 9.3 Typhoon weapons stations and stores configurations – air to ground

Store	Left wing				Centre – fuselage			Right wing			
	STA 1	STA 2	STA 3	STA 4	STA 5 + STA 6	STA 7	STA 8 + STA 9	STA 10	STA 11	STA 12	STA 13
LGB		1	1	2		1	•[a]	2	1	1	
Storm Shadow			1	1		1		1	1		
Taurus			1	1		1		1	1		
JDAM		1	1	2				2		1	
ALARM		1	1	1				1	1	1	
HARM		1	1	1				1	1	1	
Brimstone		1	1	1			1	1	1		
BL-755		1	1	2				2	1		
DWS-39		1	1	1				1	1	1	
Harpoon		1	1	1				1	1	1	
Penguin		1	1	1				1	1	1	
Fuel			1000 l/ 264 USG			1500 l/ 396 USG			1000 l/ 264 USG		

[a] • Targeting and designator pod

9.5 F/A-22 Raptor

9.5.1 Introduction

The F/A-22 was the outcome of a US Air Force DemVal fly-off between two competing designs for the advanced tactical fighter (ATF): the Lockheed YF-22A and the Northrop YF-23A. The aircraft was selected as the winner in December 1990 and progressed to the engineering manufacturing and development (EMD) phase. After a protracted development phase the aircraft is now entering service with the US Air Force with an original planned purchase of 648 aircraft, subsequently progressively reduced to 232 aircraft owing to rising costs. The aircraft designator was recently changed to F/A-22 from F-22 to reflect an increased emphasis upon the attack role.

The JIAWG architecture for the F/A-22 has already been described in Chapter 2, and many of the development and technology obsolescence issues have already been outlined. The F/A-22 is the only survivor of the triad of programmes for which the JIAWG architecture was intended: the US Navy A-12 [formerly the advanced tactical aircraft (ATA)] was cancelled in 1991, and the US Army RAH-66 Comanche (formerly the LHX) followed in 2004. Chapter 2 also outlined the integrated RF aperture sharing architecture adopted in an initial form by the F/A-22 and to be carried forward to the Lockheed Martin F-35.

The common integrated processor (CIP) integrated rack/LRM technology did find another application in the modular mission computer (MMC) on the F-16E/F, albeit with a 30-module rack rather than the 2×66 module racks of the CIPs. The CIP is reported to contain 33 signal processors and 43 data processors of seven different types that are interconnected to provide a fault-tolerant network. The CIPs are liquid cooled using a poly-α-olefin (POA) coolant and a sophisticated environmental control system (ECS) comprising both forward and aft cooling loops. The forward loop comprises the POA liquid coolant loop while the aft loop is more conventional using an air cycle machine (ACM) and air/fuel heat exchangers to dump rejected heat into the fuel. This system enables the modules to be maintained in a friendly environment of $\sim 68°F$ or $20°C$, allowing LRM mean time between failure (MTBF) figures of $\sim 25\,000$ h per module to be attained.

Given that the F/A-22 architecture and the associated technology have already been thoroughly described in Chapter 2, this portrayal will be confined to the major subsystems that comprise the avionics system (Figure 9.13):

- The APG-77 AESA radar;
- Electronic warfare (EW) and electronic support measures (ESM);
- Communications, navigation and identification (CNI);
- Displays;
- Vehicle management system (VMS);
- Weapons.

9.5.2 AESA Radar

The APG-77 radar is arguably one of the most advanced in service. The concept and benefits of the AESA are described in Chapter 4. The APG-77 has a 1500-element array with a claimed detection range of 125 nm against a $1\,m^2$ radar cross-section target (Figure 4.29). Perhaps more important is the flexibility that an active electronically scanned array confers

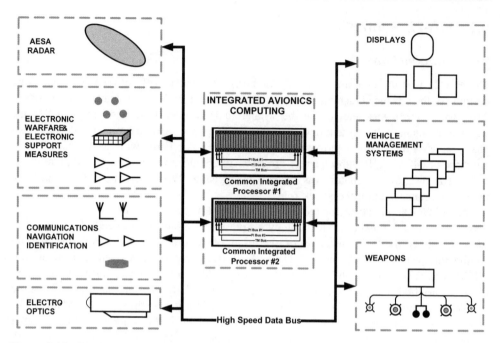

Figure 9.13   F/A-22 – avionics system top-level architecture.

where the beam or beams may be switched virtually instantaneously across the entire scan pattern. Added to this is the fact that the radar can effectively operate simultaneously in several modes, providing the utmost in operational flexibility.

The use of frequency agility to switch the operating frequency of the radar makes it difficult for a potential foe to categorise and classify the radar emissions. The use of sophisticated pulse burst modulation techniques enables information to be extracted that would not be possible using conventional techniques. The availability of a complex array of signal processors enables all the necessary target characteristics to be analysed and the characteristics of the radar to be modified to confuse or nullify the opponent's radars (Fulghum, 2000). The F/A-22 and the APG-77 radar are depicted in Figure 9.14.

9.5.3   *Electronic Warfare and Electronic Support Measures*

The F/A-22 combines its high levels of stealth with a sophisticated array of EW and ESM sensors. The aircraft is able to jam enemy radars using on-board jammers operating in the 2–18 GHz frequency band, but this would not be the preferred mode of operation for a stealthy aircraft since those transmissions may be detected. The trump card is that the aircraft can operate undetected – or at least with a very low probability of detection – while gathering vital data about the enemy radars and the electronic order of battle (EOB). The aircraft carries a range of integral EW/ESM sensors and apertures conformally, that is, within the envelope of the aircraft and flush with the surface (Figures 2.32 and 2.34 and Table 2.1). The aircraft passive EW/ESM antennas include:

- Six radar warning receiver (RWR) spiral antennas: four located port and starboard, forward and aft, one each top and bottom operating over the frequency range 2–18 GHz;

Figure 9.14   F/A-22 as Air Force photo and APG-77 radar. (Northerp German)

- Twelve situational awareness (SA) spiral antennas: six port and six starboard operating over the frequency range 2–18 GHz;
- Six SA spiral antennas: three port and three starboard operating over the frequency range 0.5–2 GHz.

In certain modes of operation, RWR and SA inputs may be combined to provide azimuth (AZ) and elevation (El) direction finding on the emitting source. This facility is called precision direction finding and allows accurate angular measurements of emitters to be correlated with radar returns or stored in a threat library.

The antennas associated with active ECM are:

1. six active ECM log periodic antennas transmitting in the 2–6 GHz frequency band. These antennas are located on the waterline and so are effective in countering threats both above and below the aircraft.
2. six active ECM log periodic antennas transmitting in the 6–18 GHz frequency band. These are located on the waterline.
3. two ECM spiral antennas located top and bottom and receiving signals over the frequency range 2–18 GHz.

While precise details of the performance and capabilities of the EW/ESM system have not been revealed for obvious reasons, the sophistication of the antennas suite is clearly indicative of a highly significant capability.

9.5.4 CNI

With one or two exceptions, the CNI system is relatively conventional, comprising the following equipment and functions:

- VHF communications;
- Secure voice communications;
- UHF communications;
- Inter/intraflight data link (IFDL) – a cooperative data link that allows all the F/A-22s in a flight to share data automatically about the status of the aircraft and the targets being engaged; the antennas are narrow-beam steerable arrays;
- Dual Litton LRG-100 inertial reference set (IRS)/GPS;
- TACAN;
- JTIDS/Link 11;
- ILS: glideslope, localiser and marker receivers;
- Microwave landing system (MLS) – growth/space provision;
- SATCOM;
- IFF interrogator – this incorporates an electronically scanned linear array;
- IFF transponder.

All CNI antennas are conformal to preserve the aircraft stealth characteristics;

9.5.5 Displays/Cockpit

The display suite comprises a HUD and colour AMLCD head-down displays and an integrated control panel as follows (Figure 9.15):

1. The wide-angle HUD has a 30° (H) × 25° (V) FOV.
2. The integrated control panel (ICP) provides the primary means of inputting data into the avionics system.
3. Two 3 in × 4 in up-front displays (UFDs) portray CNI settings, warnings and cautions and standby flight instrumentation and fuel indications.
4. The primary multifunction display (PMFD) is the primary pilot's display providing navigational data such as waypoints and flight route as well as SA information about the threat scenario.
5. Three secondary MFDs (SMFDs) are used to display tactical information as well as aircraft management data: checklists, system status, engine thrust, stores management, etc.
6. HOTAS controls using the side-stick controller and both throttles. Up to 60 functions may be controlled via HOTAS.
7. NVG compatible lighting.

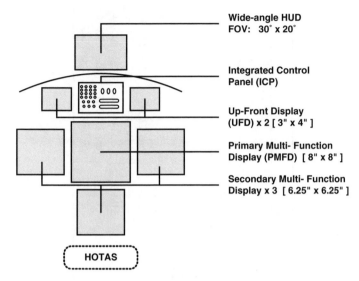

Wide-angle HUD
FOV: 30° x 20°

Integrated Control
Panel (ICP)

Up-Front Display
(UFD) x 2 [3" x 4"]

Primary Multi- Function
Display (PMFD) [8" x 8"]

Secondary Multi- Function
Display x 3 [6.25" x 6.25"]

HOTAS

Figure 9.15   F/A-22 cockpit schematic.

9.5.6 Vehicle Management System

The vehicle management system (VMS) integrates the flight control, engine control and aircraft utilities control into one major subsystem. The integrated vehicle subsystem controller (IVSC) integrates all the utilities subsystems:

- Environmental control system;
- Fire protection;
- Auxiliary power generation system (APGS);
- Landing gear;
- Fuel system;
- Electrical system;
- Hydraulics;
- Arresting system.

9.5.7 Weapons

The weapons on the F/A-22 may be carried in three ways (Table 9.4):

- Inside the centre weapons bay;
- Inside the left and right side bays;
- External carriage of fuel tanks and missiles for ferry purposes.

9.6 Nimrod MRA4

The Nimrod MRA4 is a total redesign of the original MR1/2 aircraft with totally new systems and remanufactured wings and empennage. The MRA4 first flew on 27 August 2004

Table 9.4 F-22 Raptor weapons stations and stores configuration

	F/A-22 weapons options						
			Stealthy Weapons Carriage				
Role	Left wing		Left side bay	Centre bay	Right side bay	Right wing	
Air superiority			AIM-9	6 × AIM-120C or 2 × AIM-120C + 2 × JDAM (1000 lb)	AIM-9		
Interdiction			AIM-9	As above	AIM-9		
External combat configuration	2 × AIM-9	Fuel 2300 l/ 600USG	AIM-9		AIM-9	Fuel 2300 l/ 600USG	2 × AIM-9
Ferry configuration	2 × AIM-9 + fuel 2300 l/ 600USG	2 × AIM-9 + fuel 2300 l/ 600USG	AIM-9		AIM-9	2 × AIM-9 + fuel 2300 l/ 600USG	2 × AIM-9 + fuel 2300 l/ 600USG

Note: The above figures include the M61AA1 20 mm gun with 480 rounds.

and is due to enter service in 2008. The Nimrod MRA4 has a federated architecture as shown in Figure 9.16. The key MIL-STD-1553B data buses are:

- Air vehicle bus integrating NAV/FMS, FDDS/displays and utilities control functions;
- Communications bus integrating the communications systems;
- Sensor bus interfacing the maritime search radar, ESM, magnetic anomaly detector (MAD), acoustics and electrooptic sensors;
- Mission bus integrating mission control, station control and the DASS;
- Defensive aids subsystem (DASS) interfacing to the EW/ESM suite.

Besides this main framework there are many other buses including Ethernet 10BaseT and fast switched ethernet (FDX) 100BaseT COTS buses as most of the mission suite is COTS based. Most of the data buses interface directly with the two main computers.

The main functional areas are:

- Navigation and displays;
- Utilities control;
- Communications;
- Mission system and sensors;
- DASS.

9.6.1 Navigation and Displays

The navigation and displays are based upon civil flight deck architectures using ARINC 429 data buses for intercommunication within the subsystems as well as interfacing to the dual-redundant MIL-STD-1553B air vehicle bus.

The flight management system (FMS) is a dual-computer, dual multifunction control and display unit (MCDU) system based upon that flying in commercial Boeing 737s, albeit with additional modes of operation. The civilian FMS is described in detail in the companion volume (Moir and Seabridge, 2003), Chapter 8 – Navigation. The dual-FMS computers and captain's and first officer's navigation displays and MPCDs are connected as shown in Figure 9.17.

The flight deck display system (FDDS) is likewise based upon an Airbus flight deck display implementation. Two display computers with in-built redundancy supply six colour multifunction displays that are usually configured as follows (see Moir and Seabridge (2003), Chapter 7 – Displays):

- Captain: primary flight display (PFD) and navigation display (ND);
- First officer: PFD and ND;
- Central displays: associated with alerts and warnings, system synoptic displays and status data.

This is typical of civil aircraft platforms that are adapted for military use and allows a cost-effective way of adapting proven civil systems and integrating with the specific needs of the military aircraft mission system.

The air data, attitude and inertial/GPS navigation data are provided by units connected to the air vehicle MIL-STD-1553B bus.

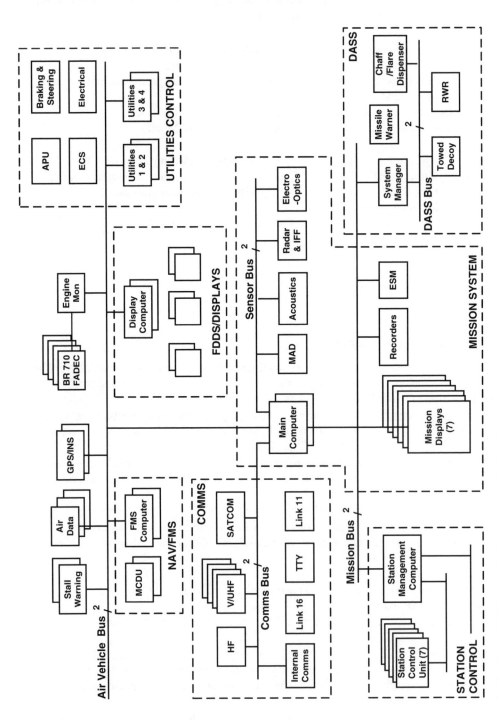

Figure 9.16 Nimrod MRA4 – avionics system top-level architecture.

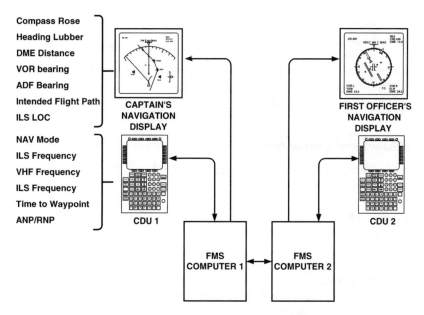

Figure 9.17   Nimrod MR4 FMS based upon the Boeing 737 system.

9.6.2   Utilities Control

Control of the utilities systems is accomplished by the following:

1. An engine monitor unit connecting to the four BR 710 full-authority digital engine control (FADEC) units – one for each engine. The baseline FADEC was designed with ARINC 429 data bus interfaces, and this unit connects the engine controllers to the aircraft systems.
2. Dedicated controllers for the following subsystems:

   - Environmental control system (ECS);
   - Auxiliary power unit (APU);
   - Braking and steering;
   - Electrical system.

3. Four multisystem utilities controllers based upon the integration principles proven on the experimental aircraft programme (EAP) – the technology demonstrator forerunner to Eurofighter/Typhoon.

9.6.3   Communications

The communications system has its own dedicated 1553B bus connecting the following:

- Several V/UHF radio sets;
- HF communications;
- SATCOM;
- Link 11 and link 16 data links;

- Secure communications;
- Internal communications.

With a mission crew complement of up to 11 operators, it is possible, indeed likely, that several will be communicating at any one time using a combination of these communications assets either in secure or non-secure modes.

9.6.4 Mission and Sensor Systems

The main computers connect to the sensor and mission buses. The sensor bus connects the following equipment:

1. Searchwater 2000 MR radar which provides the primary mission sensor. The radar has the following modes:

   - Anti-submarine warfare (ASW) and anti-surface warfare (ASuW) capability in open seas or littoral waters;
   - Synthetic aperture radar (SAR) with swath capability;
   - Inverse SAR (ISAR) mode;
   - Pulsed Doppler mode for air-to-air operation;
   - IFF interrogator.

2. Magnetic anomaly detector.
3. An extensive acoustics/sonics suite.
4. Electrooptics. The aircraft is fitted with a retractable Northrop Grumman Nighthunter II EO turret with a combination of sensors:

   - FLIR using MWIR and LWIR detectors;
   - Laser range finder;
   - Colour TV using CCD devices.

This system embodies Litening pod and F-35 electrooptic surveillance and detection system (EOSDS) technology to provide leading-edge sensing capabilities.

The mission bus connects the following:

1. Electronics support measures (ESM) suite.
2. Interface with the DASS bus.
3. Connection of the seven mission workstations and the station management system which in turn controls the seven station control units.

9.6.5 DASS

The DASS bus connects the following:

- Radar warning receiver;
- Missile warner;
- Towed decoy array;
- Chaff/flare dispenser;

- Future additions are likely to include: laser warning, directional IR countermeasures (DIRCM) and jammers.

9.6.6 Weapons and Stores

The weapons and stores load includes the following located in the bomb bay and various underwing stations:

- Spearfish;
- Air-to-surface/anti-ship missiles (Harpoon);
- Mines;
- Search and rescue stores;
- Flares and smoke markers.

9.7 F-35 Joint Strike Fighter

The evolution of the JIAWG (F/A-22) and JAST (F-35) architectures is described in Chapter 2. Two key features that distinguish these architectures are:

- The centralisation of the avionics computing function into multiprocessor signal and data processing resources in integrated avionics racks with extensive use of high-bandwidth fibre-optic buses for interconnection;
- The rationalisation of RF systems into a common integrated sensor system (ISS) utilising shared apertures and frequency conversion modules.

The F-35 avionics architecture is depicted in Figure 9.18 and comprises the following major subsystems interconnected by fibre channel (FC):

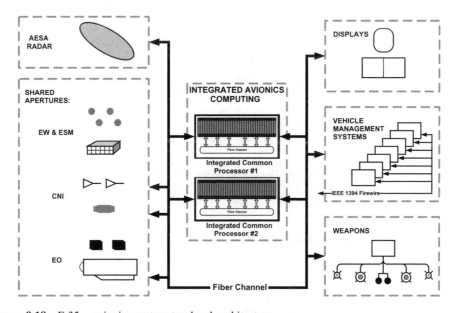

Figure 9.18 F-35 – avionics system top-level architecture.

- Central integrated computing resource comprising two integrated common processors (ICPs);
- AESA radar;
- Integrated EW/CNI and EO systems;
- Display suite;
- Vehicle management system (VMS);
- Weapons.

9.7.1   Integrated Common Processors

The F-35 common integrated processors (ICPs) comprise the backbone computing resource for the aircraft. This function is packaged into two racks with 23 and 8 slots respectively and consolidates all the signal and data processing tasks formerly managed by a range of dedicated processors. The ICPs use COTS components to enable future upgrades to be readily incorporated. The present baseline uses Motorola G4 Power PC microprocessors. At initial operational capability the ICP will be capable of performing calculations at the following rates:

- Data processing: 40.8 billion operations/s (40.8 GOPS);
- Signal processing: 75.6 billion operations/s (75.6 GOPS);
- Image processing: 225.6 billion multiply/accumulate operations/s.

The design incorporates 22 modules of different generic types:

- Four general-purpose (GP) processing modules;
- Two general purpose I/O (GPIO) modules;
- Five special I/O modules;
- Two image processor modules;
- Two switch modules (these are 32 port FC switches that interconnect the FC elements);
- Five power supply modules.

Future growth allows for an additional power supply and eight more digital modules. The ICP uses Green Hills Software Integrity commercial RTOS for data processing and Mercury Systems' commercial multicomputing operating system (MCOS) for signal processing. The CNI and display computers also use the Integrity RTOS to provide an upgrade path to allow for future developments.

9.7.2   AESA Radar

The AN/APG-81 AESA radar has all the multimode operation and benefits of the active radar as described in Chapter 4. The 1200 array AESA radar is said to have a detection range of 95 nm against a $1\,\text{m}^2$ cruise missile target.

9.7.3   Integrated EW/CNI and EO Systems

The EW subsystem comprises two subsystems:

- The ESM subsystem from Northrop Grumman;
- The radar warning and electronic countermeasures system from BAE SYSTEMS.

The integrated CNI subsystem integrates the following elements:

- Aircraft communications;
- Data links;
- Navigation;
- Radar altimeter;
- Identification and interrogation.

These systems use an integrated antenna suite comprising one S-band (2–4 GHz), two UHF (300–1000 MHz), two radar altimeters (~4 GHz) and three L-bands (1–2 GHz) per aircraft.

The EO systems include two major subsystems as already described in Chapter 5:

- Electrooptic targeting system (EOTS) providing an EO targeting system (Figure 5.35);
- Distributed aperture system (DAS) providing 360° coverage (Figure 5.36).

9.7.4   Displays Suite

The F-35 display suite extends the F/A-22 display philosophy to a single panoramic 8 in × 20 in viewing area. Two 8 in × 10 in screens provide the multifunction display system (MFDS) projection displays, each with a 1280 × 1024 pixel display resolution. The system is effectively split into two such that, if one half fails, the system can continue operating with the remaining 'good' half.

A helmet-mounted display system (HMDS) replaces the conventional HUD, resulting in significant cost and weight savings. The HMDS displays flight-critical, threat and safety information on the pilot's visor. The system is also able to display imagery derived from the distributed aperture system (DAS) or via a helmet-mounted day/night camera.

Pilot's commands are initiated using the HOTAS system.

9.7.5   Vehicle Management System

The vehicle management system (VMS) comprises three vehicle management computers which perform calculations for the aircraft flight systems: flight control, fuel systems, electrical system and hydraulics, among many others. The vehicle management computers are interfaced via IEEE 1394b data buses and continuously compare computations across all three computers to ensure integrity. In the event of a disagreement between the units, arbitration techniques permit the 'voting out' of a defective processor or sensor input.

The system also has 10 remote interface units (RIUs) that collect data for interfacing to the VMS computers. These are distributed at suitable points throughout the airframe to act as collection or distribution agencies for all the VMS I/O signals.

9.7.6   Weapons

The F-35 will be cleared for the following range of weapons and stores:

1. Internal carriage:

   - Joint direct attack munition (JDAM);
   - CBU-105 wind-corrected munitions dispenser (WCMD);
   - Joint stand-off weapon (JSOW);

- Paveway II guided bombs;
- AIM-120C AMRAAM.

2. External carriage:

- Joint air-to-surface stand-off missile (JASSM);
- AIM-9X Sidewinder;
- Storm Shadow cruise missile;
- External fuel tanks.

9.7.7 *Gun*

The gun has yet to be selected – the US Air Force (F-35A) will have an integral gun whereas the marine (F-35B) and carrier (F-35C) versions will have an external gun pod.

9.8 MIL-STD-1760 Standard Stores Interface

The military standard, MIL-STD-1760, was developed to ensure a standard interface between weapons or stores and the carriage aircraft electrical and avionics systems. This greatly facilitates the carriage of a particular weapon type across a wide range of platforms, reducing development effort and maximising operational flexibility. MIL-STD-1760D was released on 1 August 2003.

The aim of MIL-STD-1760 is to provide a common interface between aircraft and stores including:

- The electrical and fibre-optic interfaces at aircraft stores stations, the interface on mission stores the interface on carriage stores, and the characteristics of umbilical cables;
- Interrelationships between aircraft and stores interfaces;
- Interrelationships between stores interfaces at different stores stations on the aircraft.

The electrical interfaces covered by the standard are:

- Aircraft station interface (ASI);
- Carriage store interface (CSI);
- Carriage store station interface (CSSI);
- Mission store interface (MSI).

Refer to Figure 9.19 which is equivalent to Figure 1 in MIL-STD-1760D.

There are many possible combinations given the large number of weapons stations on modern fighter aircraft and the wide variety weapon types. Figure 9.20 portrays two examples:

1. Figure 9.20a shows a very simple stores interface.
2. Figure 9.20b depicts a complex arrangement where two intelligent stores are communicating via the aircraft communications network. This might be typical of a EO pod handing off target coordinates to a smart weapon elsewhere on the aircraft.

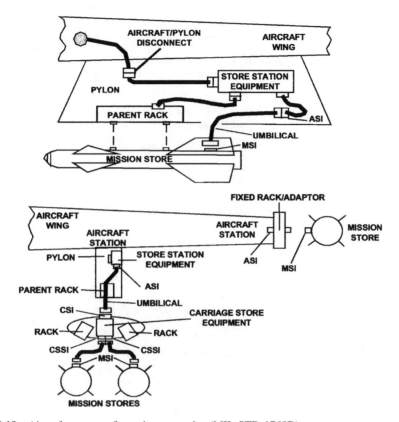

Figure 9.19 Aircraft store configuration examples (MIL-STD-1760D).

The MIL-STD-1760 signal set is listed in Table 9.5. These are further classified for particular applications:

- Class I – basic interface set;
- Class IA – basic interface set plus the auxiliary power lines;

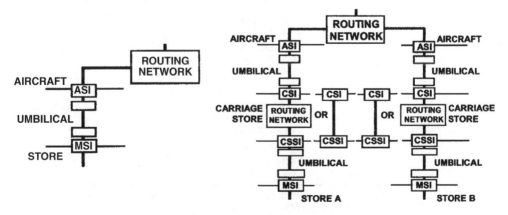

Figure 9.20 Examples of stores interfaces (MIL-STD-1760D)

Table 9.5   MIL-STD-1760 signal set.

Aircraft		Store
	Signal lines	
<<<<	High bandwidth 1	>>>>
<<<<	High bandwidth 2	>>>>
<<<<	High bandwidth 3	>>>>
<<<<	High bandwidth 4	>>>>
<<<<	MUX A	>>>>
<<<<	MUX B	>>>>
<<<<	Low bandwidth	>>>>
<<<<	Fibre optic 1	>>>>
<<<<	Fibre optic 2	>>>>
	Discrete lines	
	Release consent	>>>>
	Interlock	>>>>
<<<<	Interlock return	
	Address BIT 4	>>>>
	Address BIT 3	>>>>
	Address BIT 2	>>>>
	Address BIT 1	>>>>
	Address BIT 0	>>>>
	Address BIT parity	>>>>
<<<<	Address return	
<<<<	Structure ground	
	Power lines	
	28 V DC power 1	>>>>
<<<<	Power 1 return	
	28 V DC power 2	>>>>
<<<<	Power 2 return	
	115 V AC phase A	>>>>
	115 V AC phase B	>>>>
	115 V AC phase C	>>>>
<<<<	115 V AC neutral	
	270 V DC power	>>>>
<<<<	270 V DC return	
	Auxiliary signals	
	Auxiliary signals provide an additional set of 28 V DC power, 115 V AC power, 270 V DC power and interlock and structure ground signals	

Note: <<<< denotes a signal transiting left; >>>> denotes a signal transiting right.
A combination of the two indicates a bidirectional signal.

- Class II – basic interface set excluding high-bandwidth 2 and 4 signals and fibre optics 1 and 2;
- Class IIA – as for class II plus the auxiliary power lines.

This classification scheme therefore allows the stores interface to be standardised but takes account of the fact that a smart weapon employing fibre-optic communication may not be used at every station. The auxiliary power lines allow for the fact that certain stores such as EO pods may have much higher electrical power requirements than simple stores.

High-bandwidth signals may be one of two types: type A from 20 Hz to 20 MHz and type B from 20 Mhz to 1.6 GHz. Low bandwidth signals are those between 300 Hz and 3.4 kHz.

The specific provisions of the D standard which has been recently introduced are:

- New provisions for 270 V DC power;
- Additional data time tagging criteria;
- Characterisation of the GPS RF signals.

9.9 Air-to-Air Missiles

Some of the common air-to-air missiles are briefly described below. These are:

- AIM-9 Sidewinder;
- AIM-120 advanced medium-range air-to-air missile (AMRAAM);
- AIM-132 advanced short-range air-to-air missile (ASRAAM).

These missiles are in service today and are also intended to arm aircraft in development such as the F/A-22 and F-35.

9.9.1 AIM-9 Sidewinder

The AIM-9 Sidewinder is a short-range IR guided missile fitted on the aircraft of many air forces today. The missile has a long pedigree, being developed during the 1950s with the first successful firing in September 1953. The first missiles to be launched in anger were fired by the Chinese Nationalist F-86s in 1958 against Mig-17s of Communist China. The missile has since undergone many upgrades, with the latest version in operational service being the AIM-9M which has the capability of all-aspect engagements. In the early days the electronic technology utilised was the vacuum tube whereas it is solid-state today. Kopp (1994) gives an excellent insight into the development and capabilities of the Sidewinder.

Under normal launch conditions the missile generates an audio tone in the pilot's earphones when the IR seeker senses a target. The pilot then releases the missile which tracks the IR energy being emitted by the target and homes on to the target using a blast fragmentation warhead to kill the target. Another method is to use the Sidewinder expanded acquisition mode (SEAM) where the missile seeker head is slaved to the aircraft radar. As the radar tracks the target the missile seeker head is slaved to the radar boresight, being pointed in the direction of the target.

Finally, for advanced systems using a helmet-mounted display (HMD) it is possible to use slaving cues from the pilot's helmet-mounted sight where the missile takes its guidance from the direction the pilot is looking. This provides the pilot with a 'first look–first shot' capability.

Figure 9.21   AIM-9 Sidewinder.

The AIM-9 complements the longer-range AMRAAM to give a total capability to the aircraft self-defence capability (Figure 9.21 and Table 9.6).

9.9.2   AIM-120 AMRAAM

The AIM-120 advanced medium-range air-to-air missile (AMRAAM) was developed as the follow-on missile to the semi-active AIM-7 Sparrow and UK Skyflash, both of which had semi-active radar guidance. Semi-active guidance suffers from the disadvantage that the launch aircraft has to illuminate the target aircraft throughout the engagement as the missile semi-active guidance head is only able to track an illuminated target. This reduces the ability of the launch aircraft to engage multiple targets and also makes the launch aircraft vulnerable during the flight time of the missile. The AMRAAM overcomes this problem by having an active radar seeker guidance capability. As well as the guidance improvements, the AMRAAM is smaller, faster and lighter than its predecessor and has improved capabilities against low-altitude targets.

The development of AMRAAM grew out of a joint agreement between the United States and several NATO countries that also included the ASRAAM with a view to establishing a common technology baseline and joint production. In the event, the joint agreement lapsed and the United States continued with development of the missile alone but as a joint US Air Force/US Navy programme. The missile exists in a number of versions:

Table 9.6   ASRAAM characteristics

Function	Short-range air-to-air missile
Guidance	IR guidance single sensor for early versions, 12×128 FPA for advanced versions
Speed	Supersonic Mach 3+
Range	10–18 miles depending upon altitude
Weight	85.5 kg (190 lb)
Length	2.87 m (9 ft 6 in)
Diameter	13 cm (5 in)
Warhead	Blast annular fragmentation

Figure 9.22 AIM-132 ASRAAM. (BAE SYSTEMS © 2005)

1. AIM-120A, the original production version with deliveries commencing in 1988. This variant requires hardware modification to reprogramme the missile.
2. The AIM-120B and AIM-120C versions presently in production. These variants feature smaller control surfaces to facilitate internal carriage in the F/A-22. They are also software programmable which enhances operational flexibility.
3. A preprogrammed product improvement (P^3I) version. This features improved software reprogrammability, advanced countermeasures and improvements in the propulsion system.

9.9.3 AIM-132 ASRAAM

The advanced short-range air-to-air missile is a European development to replace the AIM-9 Sidewinder. It was initiated in the 1980s by Germany and the United Kingdom but the two nations were unable to agree upon the details of the joint venture; Germany left the project in 1995 and initiated its own version of the improved Sidewinder – IRIS-T. The United Kingdom continued with the development of ASRAAM and began to equip its aircraft with the missile in 1998. The Australian Air Force purchased the missile in 1998 for use on the F/A-18 (Figure 9.22 and Table 9.7).

9.10 Air-to-Ground Ordnance

The earliest air-to-ground missiles were used during the Vietnam War, examples such as BullPup and Helldog were typical examples. As the war developed it was realised that the release envelopes for close air support weapons were very hazardous for the pilot when used

Table 9.7 ASRAAM characteristics

Function	Short-range air-to-air missile
Guidance	IR guidance plus strapdown IN
Speed	Supersonic Mach 3+
Range	300 m to 15 km
Weight	100 kg (220 lb)
Length	2.73 m (8 ft 11 in)
Diameter	16.8 cm (6.6 in)
Warhead	Blast fragmentation

against concentrated ground fire; small arms fire could down an aircraft just as easily as anti-aircraft artillery (AAA) or surface-to-air missiles (SAMs).

The effectiveness of air power used against strategic targets was boosted with the advent of laser-guided bombs (LGBs) used against bridges and bunkers and anti-radar missiles (ARMs) to engage air defence and missile fire control radars. Later generations of this technique, augmented by inertial navigation (IN) and GPS guidance, have further improved accuracy and effectiveness.

The air-to-ground missiles and ordnance described in the section include:

- Wind-corrected munition dispenser (WCMD);
- Joint direct attack munition (JDAM);
- AGM-88 high-speed anti-radiation missile (HARM);
- Air-launched anti-radar missile (ALARM);
- Storm Shadow/SCALP EN.

9.10.1 Wind-corrected Munition Dispenser

The wind-corrected munition dispenser (WCMD) programme was developed to provide improved guidance for submunition dispensers. These weapons were developed for deployment at low level against heavily defended targets and are intended to produce an area denial effect on targets such as airfields. During Desert Storm, low-level tactics were not the preferred weapon of choice, preference being given to precision weapons launched from medium level. Nevertheless, such weapons can have merit in use against certain types of target and the WCMD is intended to upgrade existing dispensers and provide greater accuracy by countering the effects of wind.

The WCMD includes a tail kit to add to existing ordnance to provide greater accuracy. The guidance is purely by inertial means with no GPS guidance added.

9.10.2 Joint Direct Attack Munition

The joint direct attack munition (JDAM) is a tail kit to be added to existing 'dumb' bombs to allow their use in a precision mode. The JDAM kit consists of a guidance package that fits on to the tail section of the existing bomb and strakes fitted along the side of the bomb to enhance aerodynamic effects.

The guidance used is a tightly coupled INS/GPS system that can operate in both GPS-aided INS and INS-only modes of operation. The specified accuracies of circular error probabilities (CEPs) are 13 m and 30 m respectively, although it is understood that accuracies better than this are regularly achieved. JDAM kits are being fitted to a range of bombs including 250, 500, 1000 and 2000 lb existing ordnance, some with a hard target penetration capability.

As well as improving accuracy, the aerodynamic characteristics of a JDAM-fitted bomb allow a lateral footprint to be accommodated. This means that a bomb may be deployed within reason against targets that do not lie immediately on the track of the launch aircraft. This offers significant operational flexibility so that an aircraft can fly over a dense target environment, engaging targets lying to the left and right of the aircraft track and allowing multiple targets to be engaged in a single pass. The combination of this operational capability combined with the accuracy of the JDAM guidance package can deliver devastating effects, as has recently been demonstrated.

9.10.3 AGM-88 HARM

The AGM-88 HARM missile was developed as a successor to early ARMs such as the AGM-45 Shrike that was used during the Vietnam War. The missile was fitted to the US Air Force F-4G Wild Weasel aircraft before they were withdrawn from service in the early 1990s. Certain versions of the F-16 now discharge this role, and Navy F/A-18s are also capable of deploying the missile.

The launch aircraft provides the missile with data relating to the radar of the target to be engaged. The missile seeker head acquires the target and, after launch, uses the RF energy to home in on the radar, destroying it with a high-explosive (HE) direct fragmentation warhead.

The target radar may counter an ARM by using emission control (EMCON) procedures, namely by switching off the radar when missile launch is detected. In certain circumstances these may achieve the desired operational effect if the launch aircraft is screening a multiaircraft raid. In other situations it may just result in a lost missile.

9.10.4 ALARM

The air-launched anti-radiation missile (ALARM) is a UK missile developed to fulfil the same mission as HARM. ALARM entered service in the early 1990s and was successfully used during Desert Storm. The missile has the capability of ascending to 40 000 ft and loitering if the enemy radar is switched off. This is accomplished by deploying a parachute which allows the missile to descend slowly but still in an active mode. If the target radar recommences radiating, the missile will detect the target, release the parachute and fall to the target using gravity. The missile also has a memory so that, if the target fails to reradiate, the missile will attack the last known position.

9.10.5 Storm Shadow/SCALP EN

Storm Shadow is the outcome of a joint UK/French development of the Apache air-breathing missile; the two governments have evolved a common technical solution, albeit the precise implementations are slightly different. Storm Shadow is a stealthy cruise missile of ~1300 kg (2860 lb), capable of delivering a conventional warhead at ranges of over 250 km (156 nm) against range of high-value targets such as C3 facilities, airfields, ports, ammunition/storage depots and bridges.

The guidance and warhead attributes of Storm Shadow are:

- Guidance based upon a terrain profile matching (TERPROM) system with integrated GPS;
- Terminal guidance using an IR sensor and an autonomous target recognition feature;
- A highly lethal bomb Royal Ordnance augmented charge (BROACH) warhead.

Storm Shadow will be deployed on the Tornado GR4/4A, Harrier GR7 and Eurofighter/Typhoon.

Resources

www.eurofighter-starsteak.net
Flight International F-16E/F cut-away drawing.
Flight International AH-64C/D Longbow Apache cut-away drawing.
Flight International Typhoon cut-away drawing.
Flight International F/A-22 cut-away drawing.

References

Fulghum, D.A. (2000) New F-22 radar unveils future. *Aviation Week and Space Technology*, 7 February.
Kopp, C. (1994) The Sidewinder story – the evolution of the AIM-9 missile. *Australian Aviation*, April.
MIL-STD-1760D (2003) Interface standard for aircraft/store electrical interfaces, 1 August.
Moir, I. and Seabridge, A.G. (2003) *Civil Avionics Systems*. Professional Engineering Publishing/AIAA.

10 Vehicle Management Systems

10.1 Introduction

The utility systems are a collection of fluid, air, mechanical and electrical systems associated with the provision of sources of power or energy to perform the general or utility functions of the basic air vehicle. This control is usually obtained by the performance of some functional activity resulting in the appropriate control of that energy to impart flow or motion. These systems are also known as:

- Flight systems;
- General systems;
- Aircraft systems;
- Power and mechanical systems;
- Vehicle systems.

Figure 10.1 shows the aircraft considered as a set of systems to illustrate the position of utility systems relative to the airframe, the avionics and the mission systems. These systems can and do operate autonomously, but there are important system interactions that need to be considered in the design of each system. Examples are:

1. Fuel is often used to maintain the aircraft centre of gravity within certain limits. This is especially relevant to an unstable fighter aircraft where cg has an impact on manoeuvrability and is dependent on fuel status and external weapons carriage.
2. Fuel is also used as a cooling medium, for example in fuel-cooled oil coolers, and its temperature needs to be carefully monitored.
3. Undercarriage oleo switches are used as an indication of weight on wheels, and these signals are used to inhibit some functions.

Military Avionics Systems   I. Moir and A. Seabridge
© 2006 John Wiley & Sons, Ltd

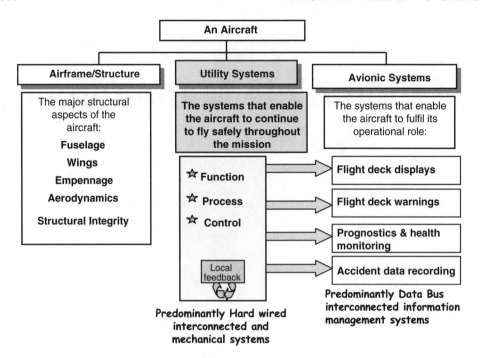

Figure 10.1   The aircraft as a set of systems.

The utility systems of an aircraft have a particular set of characteristics that make them challenging to interface and control, for example:

1. They are predominantly mechanical or electrical.
2. Their operation usually involves a transfer of energy.
3. They have a diverse range of input and output characteristics.
4. In earlier aircraft they were controlled by relay logic, hydromechanical devices or individual controllers and indicators/warnings.
5. In current aircraft there is a need to connect them to data bus interconnected avionic systems, especially to gain access to the human–machine interface with controls and electronic displays.
6. In high-performance, highly integrated aircraft systems there is a need to improve performance and increase knowledge of system operation and failure modes.

In the modern aircraft the utility systems must coexist with the avionic systems. There is an interdependency – the avionic systems need electrical power and cooling air, both of which are generated by the set of utility systems, and the utility systems need to be connected to the human–machine interface in the cockpit, which is largely generated by the avionic systems. In addition there are data generated within each group of systems that must be exchanged. Examples of such data include:

- Air data – altitude, airspeed, Mach number, attitude;
- Weight on wheels;
- Stores configuration;

- Navigation steering commands;
- Fuel centre of gravity.

The vehicle systems exist to provide the airframe with the capability to perform as an air vehicle and to provide suitable conditions for the carriage of the crew and the avionic or mission systems. This they do by providing controlled sources of energy to move the vehicle, and to control its direction, providing electrical energy and cooling, as well as providing for the health and survival of the crew. There is a need for interconnection or integration with the mission avionics systems to improve the effectiveness of the air vehicle. In a modern aircraft the vehicle management system has the ability to facilitate this interconnection by its connections to the aircraft data bus systems, and by the fact that it stores information about the performance of the vehicle systems in a format that is compatible with the mission system computing architecture. Thus, the predominantly mechanical and electrical utility systems can be interconnected with the predominantly data- and information-driven avionic systems. This chapter will provide a brief description of the utility systems and their key interfaces with avionics.

10.2 Historical Development of Control of Utility Systems

The characteristics of utility systems resulted in a gradual development from predominantly mechanical and hydromechanical systems towards a system where electrical and electronic control of functions became feasible and attractive. Although this progress removed the dependence on mechanical items such as cams, levers, control rods or wires and pulleys, it did result in a large number of individual solutions. The consequence of this was a large number of separate control units of differing technologies and shapes, each with its own installation issues and setting and repair tasks.

A determined effort to reduce this diversity was attempted in the mid-1980s which resulted in microprocessor-based architectures for control. Figure 10.2 (Moir and Seabridge, 1986) shows the reductions in the number of control units achieved as a result of an experimental aircraft project in the United Kingdom.

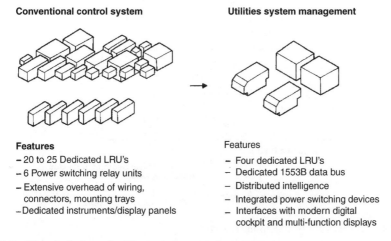

Conventional control system   Utilities system management

Features
– 20 to 25 Dedicated LRU's
– 6 Power switching relay units
– Extensive overhead of wiring,
  connectors, mounting trays
– Dedicated instruments/display panels

Features
– Four dedicated LRU's
– Dedicated 1553B data bus
– Distributed intelligence
– Integrated power switching devices
– Interfaces with modern digital
  cockpit and multi-function displays

Figure 10.2 Historical view of utility systems control equipment.

SYSTEM	Number of Signals	Number of Power Drives
Engine & Associated Systems	141	33
Fuel Gauging & Fuel Management	206	36
Hydraulic Systems including Gear & Brakes	146	13
ECS/Cabin Temperature	59	14
Secondary Power Systems	46	7
Miscellaneous Systems – LOX contents; Electrical system monitoring; Probe heating etc	34	7
Total	632	110

Figure 10.3   Early indication of utility system signal numbers.

Figure 10.3 (Moir and Seabridge, 1986) gives an indication of the number of individual signals connected to control units in a conventional aircraft in the 1980s. This situation also needed to be resolved because each signal generally required more than one wire and the resulting proliferation of wires led to large and heavy cable harnesses with the attendant installation issues and the need for large multipin connectors. It was estimated that a considerable mass and maintenance penalty could be reduced by the use of a utility systems specific data bus.

The resulting solution was based on a small number of processing units widely spaced in the airframe. Interfacing was local to those units so that lengthy wiring runs were reduced to the distance between utility system components and the processing units. Any signals required by the cockpit or by other systems were transferred by a standard serial data bus.

What emerged from this activity was a generic architecture that allowed utility systems to become integrated with the major avionics systems and the various types of data bus. The structure illustrated in Figure 10.4 shows the key interfaces that need to be respected in the

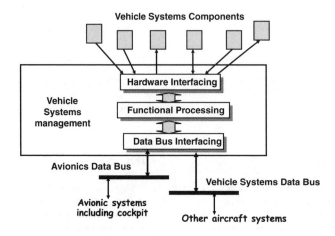

Figure 10.4   Principal utility system interfaces.

modern aircraft. The agreement of standards and conventions for designing these interfaces early in the product life cycle enables the designers of the numerous systems of the aircraft to work independently, secure in the knowledge that their systems will work in harmony.

Thus, the main objectives of the processing units became those of:

- Interfacing with hardware components and converting analogue and discrete signals into data format, as well as converting digital commands into a power command;
- Performing whatever control function was required;
- Receiving signals from, and passing signals in an appropriate fashion to, a data bus.

10.3 Summary of Utility Systems

The utility systems provide the basic functions that enable the airframe to fly and allow the safe carriage of crew, passengers and stores. Not all of these systems will be on every aircraft type, and the varying range of ages of aircraft in service means that the technology will be different. The utility systems may be subdivided into a set of subsystems.

10.3.1 Mechanical Systems

Those systems associated with control of the flight path of the aircraft and with landing, steering and braking:

- Primary flight controls;
- Secondary flight controls;
- Landing gear;
- Wheels, brakes and tyres;
- Arrestor hook/brake parachute;
- Actuation mechanisms.

10.3.2 Crew Systems

Systems associated with the comfort of the crew in normal flight, the safe escape of crew and passengers under emergency conditions and the continued well-being in the presence of chemical and biological threats:

- Crew/passenger escape;
- Aircrew clothing;
- Life support;
- Oxygen/on-board oxygen generation system;
- Canopy jettison/fragmentation.

10.3.3 Power Systems

Those systems associated with providing a source of energy in the form of thrust, rotational or air off-take, electrical or hydraulic energy:

- Propulsion;
- Secondary power;

- Emergency power;
- Electrical power generation and distribution;
- Hydraulic power generation and distribution.

10.3.4   Fuel System

The system that stores fuel and controls the transfer of fuel from the storage tanks to the main propulsion units, providing information to the crew on the status of fuel quantity and location within the aircraft tanks.

- Fuel system architecture;
- Fuel gauging;
- Fuel feed/venting;
- Fuel management.

10.3.5   Air Systems

Systems associated with the provision of clean air at appropriate temperature and humidity to provide a safe and comfortable environment for crew and passengers, to enable systems equipment to operate throughout a wide range of ambient temperature conditions and to provide air for operation of actuators, demisting or de-icing systems:

- Cooling system;
- Cabin cooling air distribution;
- Equipment cooling air distribution;
- De/anti-icing;
- Canopy demist.

10.3.6   Electrical Power Distribution Systems

These systems are associated with the distribution and protection of electric power throughout the aircraft:

- Primary power distribution and protection;
- Secondary power distribution and protection;
- Power switching;
- Load shedding and restoration;
- Internal lighting;
- External lighting;
- Probe heating;
- Ice detection and protection;
- Windscreen de-icing.

10.3.7   Vehicle Management System (VMS)

An integrated computing system to perform data acquisition, functional control and energy control for the utility systems:

- System architecture;
- System interfaces;
- Functional requirements.

In some more recent systems the VMS also integrates engine control with flight control to form an integrated flight and propulsion control (IFPC) system. In others the VMS will comprise the overall integration of flight control, propulsion control and utilities control within the same entity.

This system is essentially the integrating mechanism for utility or vehicle systems – the control and monitoring functions of the systems are brought together in a single computing architecture. This is very close to being 'avionics' in nature – it is an open architecture system with appropriate input/output and data bus interfacing. Although implemented as a separate system today, it is most likely that future aircraft will incorporate this function into a single avionic and mission computing structure.

10.3.8 Prognostics and Health Management

A system that acquires data from the aircraft systems in order to diagnose deteriorating performance of systems and to provide a prediction of failure mode and time in order to provide a preventive maintenance approach with significant improvement in systems availability and operating costs:

- Data acquisition;
- Data processing;
- Data recording.

10.4 Control of Utility Systems

Many of the subsystems described above require some functions to be performed in order to ensure that the subsystems operate satisfactorily. Although many of the systems are predominantly mechanical in nature, the functions are increasingly being performed by software in a computing system. There are many advantages to this, since modifications to functions can more easily be achieved by a software change rather than by redesigning and manufacturing mechanical components.

Figure 10.5 shows a general arrangement of a utility management processing system that allows the following key aspects of interfacing and control to be performed:

1. It provides a means of interfacing a number of different types of vehicle system input component to the processing system characterised by a diversity of type, range, source impedance and slewing rate. These interfaces include:

   - relay or switch   discrete 28 V or 0 V;
   - fuel gauge probe   AC capacitance or fibre optic;
   - gearbox speed   pulse probe;
   - actuator position   potentiometer or variable differential transformer;
   - temperature   thermistor or platinum resistance;
   - demand   position sensor, e.g. throttle lever.

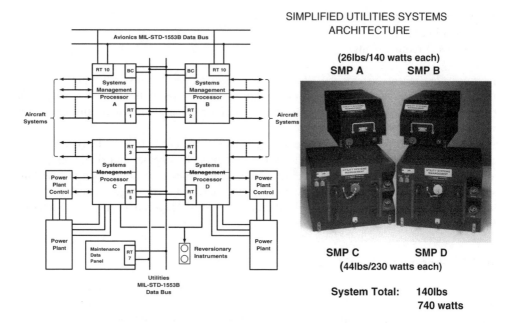

Figure 10.5 World's first integrated utility management system.

2. Provide a means of interfacing a number of different types of effector (a device that reacts to a low-energy input signal to perform a high-energy transfer task, e.g. an actuator or pump) to the processing system characterised by diversity of type, range, load impedance/resistance and reactance. These interfaces include:

- valve   28 V or 0 V discrete power;
- DC motor   DC power drive 28 V DC or 270 V DC;
- actuator drive low-voltage analogue;
- Torque motor   low-current servo drive;
- Fuel pump   high-current AC motor drive;
- Filament lamp   Lamp Load (high in-rush current).

3. It provides a means of connecting to the aircraft data bus structure. These buses include:

- MIL-STD-1553 (Def Stand 00-18, Stanag 3838);
- ARINC 429;
- ARINC 629;
- IEEE 1394 (firewire);
- Fibre-optic buses.

4. It provides a means of performing subsystem control functions in software.
5. It provides a means of performing control functions in hardware such as application-specific integrated circuits (ASICs), programmable logic devices or hardwired means.

A typical subsystem may have a combination of software and hardware control implementations depending upon the nature of the system operation and the level of integrity required.

Figure 10.5 depicts the key features of the first integrated system of its type to fly in the world: the British Aerospace (BAe) – now BAE SYSTEMS – experimental aircraft programme (EAP). This was a single-aircraft flight demonstrator that proved key technologies for Eurofighter Typhoon and that first flew in August 1986.

Each of the processing units shown in this system provided a facility for interfacing with the components of the utility systems. For economic reasons it would be ideal if these units were identical in construction, with only the software load providing each unit with its own characteristics. There are many practical reasons why this rarely happens. Figure 10.6 shows a generic form of such a processing unit.

As the application of utility management has advanced through a number of military and commercial aircraft types, so technology has advanced in areas of semiconductors, memory and data transmission systems. The nomenclature has now changed to vehicle management systems to reflect broader applications than solely aircraft.

The vehicle management system can be designed to observe the same architectural principles as the avionics and mission systems, for example data bus type, data bus protocols, processor and memory type, software language, redundancy requirements and physical

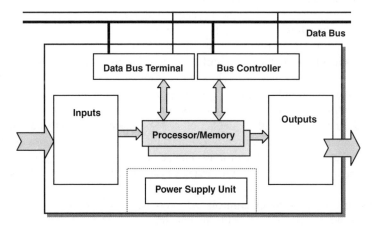

Figure 10.6 Generic arrangement of a utility management systems processor.

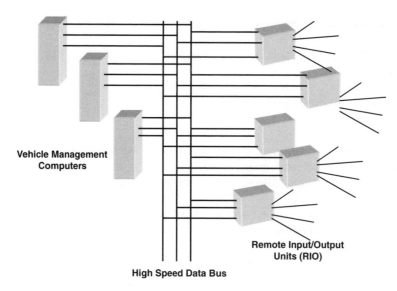

Vehicle Management
Computers

Remote Input/Output
Units (RIO)

High Speed Data Bus

Figure 10.7   Modern VMS architecture.

enclosure. Although contemporary aircraft often implement vehicle systems control in a separate processing structure to avionics, the two are likely to merge in the future to meet demands for more efficient on-board processing and reduced equipment mass.

The modern vehicle systems management architecture has developed to take account of technology improvements, as well as a radical rethink of the installation aspects of the system. It makes considerable sense to install high-throughput, high-dissipation processors and high-capacity memory in a comfortable, cooled environment to achieve the highest availability possible, while changing the method of interfacing with the system components.

Therefore, units associated with input and output interfacing functions, commonly known as remote input/output (RIO) units, can be designed as simple, rugged items packaged so that they can be installed in remote areas of the airframe that may be harsh in terms of temperature and vibration. Local connections to components will be short, and high-speed data links connect the RIOs to the main data bus or processing system. A typical architecture with a combination of VMS computers and RIOs is shown in Figure 10.7.

A characteristic of the majority of utility systems is that they do not generally lend themselves to multiple-redundant architectures. Obvious exceptions are flight control (generally triple or quadruple redundancy) and propulsion control (generally duo-duplex). Otherwise the utility systems are a collection of singular control problems with single components to control. As a result, a number of strategies have emerged that are still practised in the software-dominant systems currently available for controlling the systems and for dealing with failures. A key determining factor in the choice of method is the integrity requirement or safety requirement for the system. These strategies can be summarised as follows:

1. Some systems such as flight control and propulsion control have their own control systems, but may make use of utility management to gain access to data in the utility systems or to access the cockpit displays and other avionic systems.

2. Some systems such as fuel systems are wholly controlled and monitored by utility management with redundancy or with graceful degradation achieved by distributing control among several utility management processors.
3. Some systems are mainly controlled by utility management but may have alternative hardwired means of obtaining manual override or reversionary control, albeit with some degradation in performance.
4. Some systems have no connection to utility management at all. Usually these are systems that are concerned with the safety of the aircraft or the crew and may be required to operate when the majority of other on-board systems have failed, for example the ejection seat.

10.5 Subsystem Descriptions

The utility systems will be described in more detail below. More detailed descriptions can be found in Moir and Seabridge (2001 and 2002), Jukes (2004) and Pallett (1992). Figure 10.8 shows a generic block diagram which can be used to represent each of the following subsystems. This shows a variety of inputs that determine the status of the system:

- Demand from the pilot operating a device in the cockpit – the human–machine interface;
- Sensor inputs from components in the system;
- Other system inputs which may be relevant data or commands;
- Energy to provide power for the processor;
- Feedback from the system to close the loop on any control processes.

Also shown is a variety of outputs:

- Waste products in the form of heat, noise or energy which must be dissipated;
- Signals and data to the cockpit to be presented on the displays or warning system;

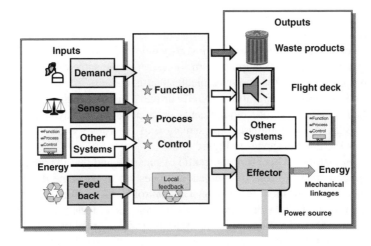

Figure 10.8 Generic utility system block diagram.

- Signals to other systems to demand a response or to provide information;
- Commands for energy to enable an event to take place.

10.5.1 Mechanical Systems

10.5.1.1 Primary Flight Controls

In its most basic form, a flight control system is a set of rate demand control loops that allows the pilot to manoeuvre and control the aircraft attitude in pitch, control and yaw axes. In the modern aircraft this function is provided by a system architecture with appropriate safety and integrity implemented by bus interconnections, computing and software, control law design, actuation systems and pilot's inceptors. The flight control system is always considered to be flight safety critical, particularly in unstable air vehicles, since loss of the system may result in loss of the aircraft and severe injury or death of the crew or the overflown population. Multiple, similar redundancy or multiple redundancy with dissimilar back-up is used as a design solution. In most modern solutions the flight control function is performed in a dedicated set of computers, and not integrated with vehicle systems management. However, improved component reliability and the emergence of robust methods of software design coupled with high-speed data buses may lead to a change in this position.

Systems such as flight controls are flight safety critical since a catastrophic failure can lead to loss of the aircraft. For this reason the system must be designed so that no single failure, and very often no multiple failures, will lead to loss of control. Common mode failures must be avoided, and hence the power sources and supporting sensors must be carefully selected. For these reasons, flight control systems have tended to be self-contained, with minimal interaction with other systems. Hence, in most modern aircraft the flight control system computing, sensors and actuators are virtually independent of all other systems other than providing sources of attitude and air data to other systems and relying on hydraulics and electrical power for sources of energy.

In modern aircraft two types of flight control may be used. All those aircraft derived from civil transport aircraft and some fighter aircraft platforms exhibit a stable form of flight control where the aircraft centre of pressure is located aft of the aircraft centre of gravity (CG) position. This means that the aircraft is naturally stable and will resist natural perturbations during flight. This natural stability is reflected in the aircraft handling characteristics and therefore in the level of integrity required of the system. The disadvantage of this configuration is that the tailplane trim forces act in the same direction as the aircraft weight vector and the design is aerodynamically inefficient.

In the second category the aircraft is naturally unstable, usually in the pitch axis; either for reasons of high manoeuvrability requirements or because of aerodynamic/stealth drivers (F-117A and B-2). A diagrammatic representation of the forces acting upon an unstable aircraft such as the EAP is shown in Figure 10.9.

It can be seen that the CG is aft of the centre of pressure and that the aircraft is unstable. The saving grace is that the trim or manoeuvre forces act in the same direction as the lift vector, making the aircraft more responsive. The control surface configuration of the EAP is depicted in Figure 10.10. This diagram also gives some idea of the primary and secondary flight control surfaces, and a very similar configuration was adopted on Eurofighter Typhoon owing to the successful handling characteristics demonstrated by this aircraft. The fact that the aircraft is highly unstable results in a quadruplex flight control system architecture for

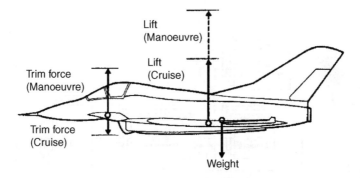

Figure 10.9 Longitudinal control forces on an unstable platform.

sensors, computing and actuators since the total loss of flight control would result in loss of the aircraft. Similar quadruplex flight control architectures are common on many modern high-performance aircraft.

The outline digital quadruplex architecture to control the EAP airframe is shown in Figure 10.11. The key elements are:

- Flight control computer (FCC);
- Aircraft motion sensing unit (AMSU);
- Pilot input sensors;
- Airsteam direction sensing vanes;
- Air data;
- Actuator drive unit.

Modern vehicle management systems with high-speed data buses and high-reliability systems may change this philosophy and lead to more functional integration. In some cases

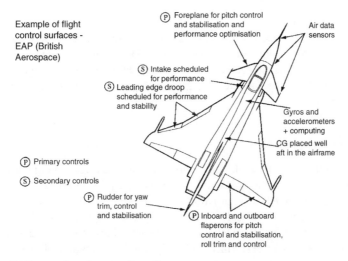

Figure 10.10 EAP control surface configuration.

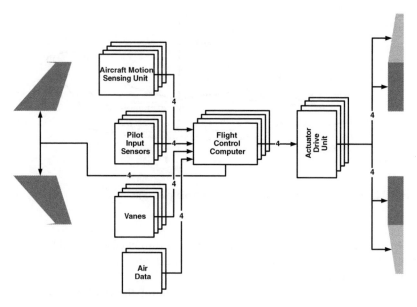

Figure 10.11   EAP quadruplex flight control system architecture.

this type of functional partitioning represents a sound systems approach. However, where distributed systems are required to meet high-integrity requirements, it must be remembered that the interconnecting data buses need to meet the same requirements. This may constrain the application of many standard data buses, particularly those of a COTS origin; in those cases the COTS data bus chips may have to be redesigned and revalidated if the high-integrity requirement is to be supported. Some subsystem examples are described below.

10.5.1.2   Secondary Flight Controls

Secondary flight controls include the provision of secondary control surface mechanisms (flaps, slats, air brakes, speed brakes, pilot interfaces). Some of these systems are relatively simple logical controls with some airspeed or Mach number limitations. There is an increasing tendency to include secondary controls in with the primary flight control system, as shown in the EAP example above. The main utility interface is to collect information for the cockpit displays.

10.5.1.3   Landing Gear

Landing gear systems include the provision of undercarriage configuration and loads, sequencing, doors, locks, indications and warnings, and pilot interface. Position-indicating devices such as microswitches or proximity detectors are used to indicate the position of doors and the gear legs and wheels so that the correct sequence is followed. The utility system task is to perform logical checks on the switching sequence and to ensure that extensive built-in testing is performed to verify correct operation of the doors and gear sequence. In addition it is essential to understand when the undercarriage oleos are

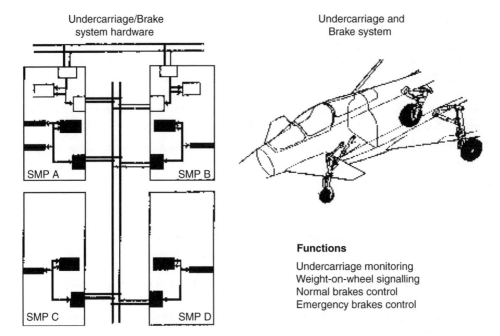

Figure 10.12   Landing gear system example.

compressed – weight on wheels is used to signal to a number of systems that it is safe to operate. Examples of this is the use of weight on wheels to inhibit the arming and release of a weapon when the aircraft is on the ground, and the actuation of thrust reverse only when the aircraft is firmly on the ground. A typical system is shown in Figure 10.12.

10.5.1.4   Wheels, Brakes and Tyres

Wheels, brakes and tyres provide the mechanism for meeting the loads and performance required to match the aircraft role for landing, take-off and ground handling under normal and emergency conditions – for nose wheel steering, brakes/anti-skid, park brake, tyres, runway loads, etc. Braking and anti-skid are safety-critical, high-speed, closed-loop systems and may be duplicated to provide fail-safe operation. The utility control system reacts to demands from the pilot's pedals to activate the braking system, and differential speed sensors in the wheel detect the onset of a skid and adjust the brake demand accordingly.

10.5.1.5   Arrestor Hook/Brake Parachute

These subsystems are to provide suitable methods of arresting the aircraft to supplement the braking system. Aircraft operating from aircraft carriers will routinely be fitted with an arrestor hook. Many fast jets also carry a hook to enable them to engage with arrestor gear on an airfield should the need arise – in the event of failed brakes, reverse thrust or brake parachute, for example. These are simple logical functions which are usually separate from utility system control.

10.5.1.6 Actuation Mechanisms

Throughout there will be a need to provide suitable actuator types and mechanisms for operating surfaces. Specifying these mechanisms requires a knowledge of surface travel, mechanical advantage, flight loads, failure cases, etc. There may be many doors, flaps or vents that need to be operated in response to manual or automatic demands. Hydraulic or air power may be used for this purpose. These functions may be simple logic but may have some dependencies on air speed, weight on wheels or other system modes that can be provided by utility systems control.

10.5.2 Crew Systems

10.5.2.1 Crew Escape

Crew escape systems provide safe aircrew escape under certain conditions of flight, including ejection seat performance, ejection clearances, escape doors, parachute requirements, emergency location devices and survival equipment. Crew escape is a last resort and may be needed when all other functions have been lost. For this reason it is independent of all other systems. Increasingly sophisticated functions such as autoejection and automatic seat trajectory adjustments will be performed by integral controllers. An example of a crew escape system is shown in Figure 10.13.

10.5.2.2 Aircrew Clothing

Military aircrew are provided with crew clothing for anti-*g*, immersion, restraint, comfort and survival in the arduous conditions of a military jet environment. Clothing varies according to the type of aircraft – fighter and fast-jet aircrew need to wear clothing to protect them against high-*g* conditions and the ejection case; larger aircraft with multiple

Figure 10.13 Modern military ejection set – F-35 JSF Seat. (Martin Baker)

crew members tend to operate in a 'shirtsleeve' or flying overall environment. There are some simple functions to be performed such as inflation of garments to anticipate the rapid onset of *g*, but these functions can be independent of utilities control.

There is an important avionic interface with crew systems at the oxygen mask and the helmet. The supply of oxygen under pressure to the mask and the action of pilot breathing in close proximity to the communications microphone can distort speech and contribute to noise in the pilot's headset.

The interface at the helmet is also significant, since the additional mass of sensors and displays can have an impact on the specification of the escape system and may affect pilot health and safety as a result of load on the neck under high-*g* conditions.

10.5.2.3 Life Support

The crew are provided with protection against biological and chemical threats, and with emergency air/oxygen. This protection may be provided by means of the air-breathing system in the case of fast jets, or by portable respirators or personal biological/chemical packs. This is usually an independent function.

10.5.2.4 Oxygen/OBOGS

The aircraft is equipped with systems to provide gaseous oxygen or on-board generated oxygen/enriched air and the mechanism for providing oxygen to the crew. Most fast jets provide oxygen derived by liquid oxygen (or LOX) or in the form of oxygen-enriched air provided by OBOGS. Aircraft with a large cabin are provided with filtered air by the aircraft environmental control system. Gaseous oxygen is carried for emergencies, such as depressurisation, where oxygen masks are deployed to individual crew members. Although the majority of these functions are independent of utilities, there is a need for control of the sequencing of the catalyst beds and monitoring of oxygen concentration in the on-board oxygen generation system which is becoming increasingly common to provide a measure of independence from supplies of liquid or gaseous oxygen.

A typical OBOG system is shown in Figure 10.14 in this case for a two-seat aircraft, although the architecture is the same for a single-seat aircraft, but with only one regulator. The following description is from a Honeywell Aerospace Yeovil paper (Yeoell and Kneebone, 2003).

Engine bleed air enters the preconditioning system element where the temperature is reduced, ideally to less than 70°C, and water is removed as far as possible. In addition it is normal at this stage to use a combined particulate and coalescing filter to remove potential contaminants, including freewater, that may still be contained in the inlet air.

The OBOGS contains a pressure-reducing valve to reduce the inlet air pressure of the air supply to that required by the OBOG generator, typically 35 psig.

The next system element is the oxygen generator or, more correctly, the OBOGS concentrator which uses multiple zeolite beds to produce the oxygen-rich product gas.

The switching between the zeolite beds is achieved using solenoid-actuated pneumatic diaphragm valves controlled by the system monitor/controller which may be located in the vehicle management system. These valves are 'wear free' and allow the concentrator to be a 'fit and forget' system that requires no scheduled maintenance and exhibits high reliability.

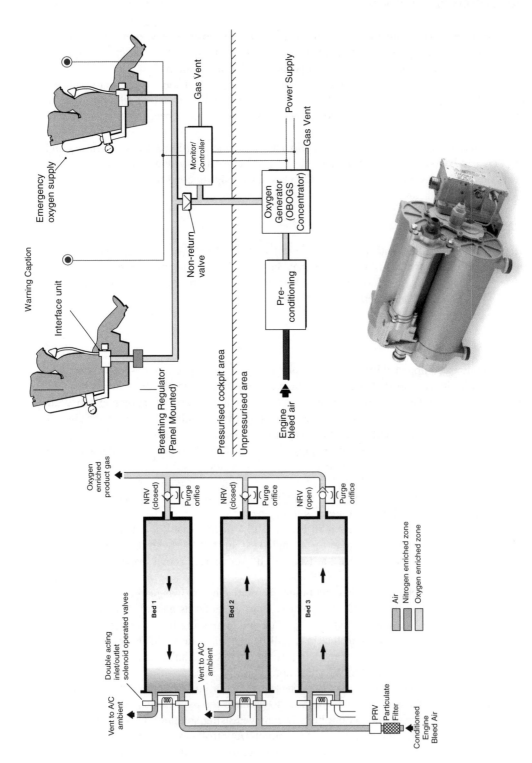

Figure 10.14 On-board oxygen generation system (Honeywell Aerospace Yeovil).

The system monitor/controller is a solid-state electronic device that monitors the PPO_2 level of the OBOGS concentrator product gas and adjusts the cycling of the beds to produce the desired level of oxygen concentration for cockpit altitudes below 15 000 ft. This process is known as concentration control and means that no air mix, or dilution, of the product gas is required at the regulator, hence preventing the ingress of any smoke or fumes from the cabin into the pilot's breathing gas supply.

The breathing gas then passes to the pilot's breathing regulator, in this case a panel-mounted unit is shown, but ejection seat and pilot mounted devices can also be used.

10.5.2.5 Canopy Jettison

Emergency escape of the crew is enhanced by the provision of a canopy jettison or canopy fracture mechanisms to match the requirements for escape. Fast-jet canopies may be jettisoned by explosive or rocket motor to remove them rapidly from the aircraft prior to ejection. Alternatively, the transparency material may be fractured by an explosive device known as miniature detonating cord (MDC) bonded to the canopy material – the pilot then ejects through the fractured material. As this system is associated with crew escape, it is independent of all other systems.

10.5.3 Power Systems

10.5.3.1 Propulsion System

There is an increasing tendency for the propulsion unit or power plant to be developed by engine manufacturers as a self-contained item with integral starting, full-authority digital engine control units and power supplies. Propulsion control has also remained independent of the majority of airframe systems. This separation has been strengthened by a solution in which the engine manufacturer has traditionally been responsible for the design and integration of the total propulsion system. The propulsion system has evolved from early implementations where all the electronic control units were mounted on the airframe in a conditioned environment. This solution needed large cable harnesses to connect the units to the engine across the airframe/engine interface – a considerable installation and maintenance penalty. Improved electronic component reliability and build standards that are able to tolerate harsh temperature and vibration environments have resulted in engine-mounted control systems. This has led to almost total independence of the full-authority digital engine control system with self-contained cooling, vibration isolation and power supplies, the only connections to the airframe being for the pilot's demand (throttle box) and cockpit displays.

10.5.3.2 Secondary Power

Provision of power to start engines and to maintain electrical and hydraulic power during ground operations. This is performed by an auxiliary power unit (APU) which is a small gas turbine started by battery and used to drive a generator and hydraulic pump, as well as to provide air to starter turbines on the engine(s). In some cases auxiliary power for maintenance and engine start is provided by ground power units. The APU controller is usually self-contained with demands from the engine start sequence provided by utility

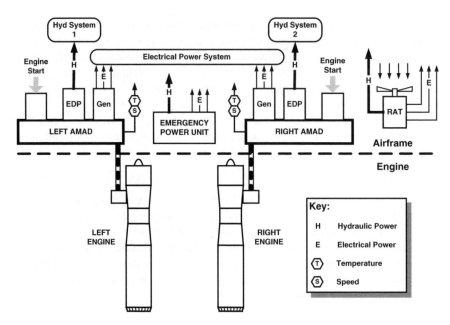

Figure 10.15   A secondary power/emergency power system.

management. In some instances the APU may be started and used in flight to provide an assisted start to an engine that has flamed out.

10.5.3.3 Emergency Power

Provision of power during engine failure conditions to ensure safe recovery of the aircraft or safe crew escape – includes the emergency power unit (EPU), ram air turbine (RAT), fuel cells, etc. There are instances where the aircraft main engine(s) may cease to provide thrust or motive power. In this case an emergency power unit is used to allow the aircraft to be flown to a safe flight condition so that the engine can be relit. A ram air turbine (RAT) is the usual method used, in which a turbine is dropped into the airstream and used to power electrical or hydraulic power-generating devices. This system will be commanded to operate by utility management when loss of rotation of both engines is detected. A secondary power system together with emergency power provision is shown in Figure 10.15.

This figure shows how an airframe-mounted accessory gearbox (AMAD) derives shaft power from the engine and drives the 3000 psi (F-15), 4000 psi (Tornado) or 5000 psi (part of F/A-18E/F and V-22) hydraulic system by means of an engine-driven pump (EDP). In the event that both engines are lost, emergency power is provided by an emergency power unit or by a ram air turbine (RAT) until such time that the engines may be relit. If a single engine is lost it is possible in some cases for the remaining engine to cross-drive both AMAD gear boxes and maintain full electrical and hydraulic services. Electrical power – typically 115 V AC, three-phase or 270 V DC – is provided by generators which supply the power to the electrical power distribution system. Emergency electrical power may also be derived from the emergency power unit or the RAT. The AMAD is also used to start the engine using a start motor which uses air or electrical power as appropriate. The system is subject to extensive monitoring, with temperatures, speed, pressures, status, etc., being signalled to the various controlling LRUs. Alternatively, these control and monitoring functions may be embedded in an integrated VMS.

10.5.4 Electrical Power Generation and Distribution

The electrical power generation and distribution system provides AC and DC power at avionic equipment terminals with voltage and frequency characteristics defined by power system standards. The power system is generally designed to eliminate single-point failures and to maintain electrical supplies under defined failure conditions. The distribution system includes power circuit protection devices to isolate equipment faults and to protect the aircraft wiring.

Under certain primary power source conditions, e.g. loss of engine or generator, it may be necessary to remove loads selectively from the bus bars, a technique known as load shedding. The distribution system design enables groups of electrical loads to be disconnected to allow a progressive degradation in mission performance, while retaining critical avionic functions such as displays and controls, navigation and communications.

Electrical power generation and distribution include the provision of electrical power derived from main engines and secondary or emergency power sources to meet the predicted electrical loads throughout the mission, and the provision of a suitable distribution and protection system. The system design also includes the provision of bonding and earthing, electrical hazard protection and compatibility with external power sources. Generator control circuits, which contain a means of sensing over- and undervoltage as well as under- and overfrequency, may be self-contained or may be incorporated into utility management. This control is safety critical because malfunction may lead to disconnection of the generators and subsequent loss of all electrical power. A key feature of utility management is the detection and announcement of system failures and the determination of corrective action which may include automatic load shedding.

Most military aircraft power systems are conventional and use integrated drive generators (IDGs) to produce two or more channels of 115 V AC, three-phase, 400 Hz electrical power. Two slightly unusual examples of military aircraft power are described below: those for electrical power systems of the F/A-18E/F and the F-22.

10.5.4.1 F/A-18E/F Super Hornet

The F/A-18E/F uses a variable-speed constant-frequency (VSCF) cycloconverter power switching and commutation technique to provide 115 V AC and 28 V DC power to several aircraft buses. This type of system is widely used in the United States, being the primary means of electrical power on the F/A-18C/D, F-117A, U-2 and V-22, and has proved to be very reliable in service. This system is shown in Figure 10.16.

10.5.4.2 F/A-22 Raptor

The F/A-22 uses a 270 V DC electrical system that has been favoured by the US Air Force for the last 10 years or so. This system will also be used on the F-35 joint strike aircraft (JSF). Primary electrical power is generated at 270 V DC, but, as many legacy subsystems and components require 115 V AC and 28 V DC, converters are added to convert power to

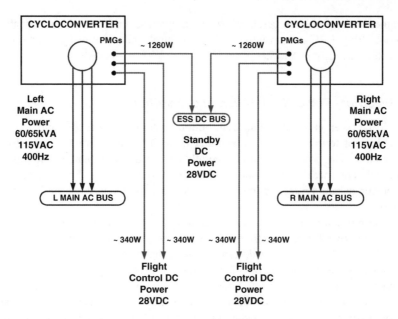

Figure 10.16   VSCF cycloconverter as used on the F-18E/F.

cater for these loads. The electrical power distribution is provided by eight power distribution centres (PDCs): four distributing 115 V AC power and four distributing 270/ 28 V DC power, as shown in Figure 10.17.

   Other systems associated with the utilities system generation of power are:

1. Hydraulic power generation and distribution – provision of hydraulic power derived from the main engines and secondary or emergency power sources to meet the predicted hydraulic loads throughout the mission, and provision of suitable system components to maintain power, provision of pipes, couplings and compatibility with external power sources. A key feature of utility management is the detection and announcement of system failures and the determination of corrective actions (see Figure 10.20).
2. Fire detection and suppression – provision of methods for detecting fire or overheat, warning the crew and providing a means of suppressing/extinguishing the source of heat such as a hot gas leak or naked flame. Fire protection systems are usually independent of all other systems.

10.5.5 *Hydraulic Power Generation and Distribution*

The hydraulic system provides power mainly for flight control surface movement and landing gear operation. There are hydraulic circuits that are used to open doors and hatches, or to deploy sensors by providing motive power. Examples are camera bay doors, radar antenna rotation motors, electrooptic turret deployment and retraction and bomb-bay door operation. Hydraulic power is generated by engine-driven pumps at flowrates and pressure to meet the specification of the flight control system demands. The distribution system – pipes,

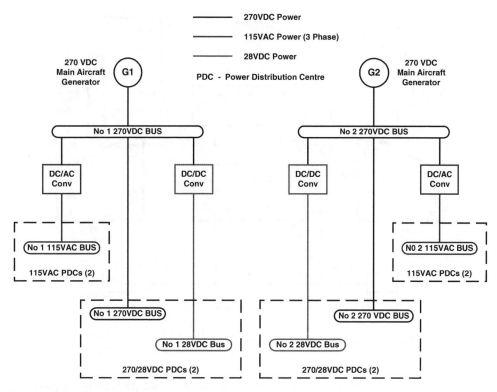

Figure 10.17 F-22 power generation and distribution system.

reservoirs, accumulators and valves – ensures that the flight-critical services are isolated from non-flight-critical services to allow isolation in emergency cases.

10.5.6 Fuel Systems

The primary purpose of the fuel system is to provide a source of energy for the propulsion system. It does, however, have two subsidiary functions: one is to act as a form of movable ballast to maintain the aircraft cg within specified limits, and the other is to act as a heat sink for cooling system loads – heat is dumped from gearbox and avionic loads into a fuel-cooled heat exchanger. Key functional areas include:

1. Fuel system architecture – provision of a suitable shape and location of the internal and external fuel tanks to accommodate the fuel mass required to meet mission requirements, and provision of a gauging and transfer system to meet the tank layout. Requirements for the fuel system form the basis of the functional requirement and the determination of the distribution of interfaces and functions to utility management, implemented in software. An example system is shown in Figure 10.18 in the normal engine feed mode. Such a system has several different modes of operation including:

   • Engine feed from the forward and rear fuel groups to the left and right engine respectively;

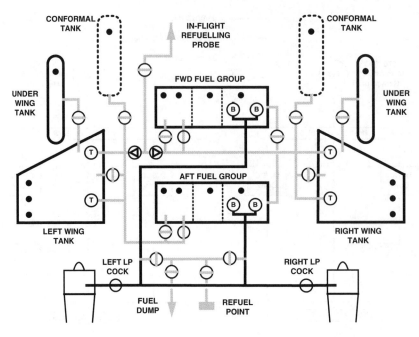

Figure 10.18 Typical fighter aircraft fuel system – simplified.

- Fuel transfer from wing, underwing or conformal tanks in a preordained sequence;
- Refuel/defuel;
- In-flight refuelling;
- Fuel dump.

2. Fuel gauging and level sensing provides a means of determining the quantity and location of fuel in the aircraft. Modelling tank shapes allows designers to define the location of gauging, level and density sensors to obtain accurate measurement of usable fuel under all flight attitude and manoeuvre conditions. Interface definition with the vehicle management system and pilot requirements for control and indications lead to a functional requirement and a human-machine interface definition. The number of fuel probes and the tank shapes dictate the accuracy of the fuel quantity measurement system. The variation in fuel properties and fuel density have to be carefully measured and taken into account. Depending upon the location and type of fuel, the density can vary by almost as much as 10% between the tropics versus the Arctic region. Fuel properties vary greatly at different locations in the world which can impact civil operations but have a huge impact upon military aircraft operating in remote locations in third-world countries.

3. Fuel feed/venting – provision of tank interconnections, couplings, bonding and earthing, refuel and defuel, and provision of suitable venting for jettison, expansion, etc.

4. Fuel management – provision of an appropriate mechanism for transferring fuel from tanks to engine to meet normal fuel demand, cg management and emergency/leak/battle damage conditions. Detection of component failures, leak detection and automatic reconfiguration and provision of system status to the cockpit displays. In unstable aircraft it may be necessary to ensure that the aircraft centre of gravity is maintained within strict

limits so that the pilot is able to make manoeuvre demands to meet his operational demands without causing damage to the aircraft – the flight control system will ensure this. To do so safely, the flight control system requires information about the position of aircraft components. There are consumables on the aircraft that make a significant contribution to the cg, the main items being fuel and weapons and stores. As fuel is used, it is transferred from tank to tank and, although the range of variation in fuel cg is limited by the vehicle management system, its variation is sufficient to affect safe aircraft carefree handling. For this reason, in some aircraft types, the location of fuel is declared to the flight control system together with the initial store load and the use of stores during a mission. The flight control system is able to modify its control laws and limitations to protect the aircraft from damage.

The simplified military aircraft fuel system as shown in Figure 10.18 comprises a number of fuel tanks:

1. The fuselage tanks are usually separated into forward and rear fuel groups.
2. Wing tanks are on the left and right. Total internal: 3986 l/1053 US gallon.
3. Underwing tanks, to extend operational range, are on the left and right. Many aircraft can also carry another external tank on the fuselage centre-line. Total external: 5680 l/1500 US gallons.
4. Conformal tanks, for ferry purposes or to extend operational range, are on the left and right. Total conformal: 1700 l/450 + US gallons.

The figures given relate to the F-16E/F block 60 configuration.

The components that define the system configuration and fluid logic are transfer and refuel valves operating under the control of the fuel management function. Even on this simplified diagram a total of 20 transfer/refuel valves are shown. Fuel transfer is usually accomplished with the help of pressurised air in the case of the external tanks. Electrically driven fuel pumps assist the transfer of fuel from the wing tanks to the forward and rear fuel groups, while booster pumps assist the flow of fuel from these groups to the engine. On the relatively simple EAP demonstrator, which had no external tanks fitted and no in-flight refuelling capability, there were a total of 206 signals and 36 power drives associated with the fuel management and quantity gauging system.

10.5.7 Air Systems

A considerable number of utilities functions are powered by bleed air from the main engines or the auxiliary power unit (APU). The functions powered from bled air taken from the engine compressors include the following:

1. Cooling system – provision of an air refrigeration and decontamination system to condition air taken at high temperature and pressure from the engine connection.
2. Cockpit/cabin cooling air distribution – provision of air at an appropriate temperature and humidity into the crew compartment to maintain crew comfort with minimum cabin noise. Controls are provided to enable adjustment of flow and temperature.
3. Equipment cooling air distribution – provision of air at an appropriate temperature and humidity into the equipment bays/compartments to maintain continued equipment operation. The environmental control system (ECS) provides clean air at appropriate

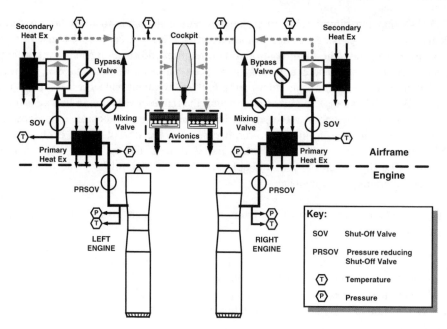

Figure 10.19   Simplified air/environmental control system.

temperature and humidity levels to ventilate equipment bays and to cool avionic
equipment mounted in the bays. The supply of cooling air may be provided as ventilation
by circulation in the bay, by washing air up the side walls of equipment or by providing
air to a plenum chamber in the equipment mounting tray. In certain cases, air cooling is
not sufficient and a liquid cooled (cryogenic) mechanism may be provided for such high-
dissipation devices as radar transmitters or electrooptic sensor packages.
4. De/anti-icing – provision of suitable methods for de-icing or anti-icing of the leading
   edges of appropriate surfaces – wings, empennage, intakes by hot air internal bleed or
   rubber boot inflation.
5. Canopy demist – provision of hot air to demist internal canopy surfaces.

A typical simplified layout for a military fighter aircraft is shown in Figure 10.19. This
diagram depicts a bleed air supply taken from the engine compressor and fed through the
ECS systems; the system effectively comprises two independent channels, each extracting
air from the respective engine. The main attributes are:

- Pressure-reducing shut-off valves (PRSOVs) to moderate the pressure of the engine bleed
  air to a constant pressure irrespective of throttle setting;
- Primary and secondary heat exchangers to reject heat overboard;
- Shut-off valves (SOVs) to provide an independent shut-off capability in addition to
  PRSOVs;
- Closed-loop, variable-mixing and turbine bypass valves to control the flow of warm and
  cold airstreams before entering the mixing plenum chamber;
- Distribution system to allocate air of the necessary temperature to the cockpit and
  avionics bays;

- A number or temperature and pressure monitoring points (as well as the very basic monitoring set shown in Figure 10.19, there will be a significant number of sensors to monitor system performance and status and to provide display data and warnings for display to the pilot when required).

The foregoing system example is a greatly simplified portrayal. As well as this system there will be a number of other dedicated cooling subsystems – some of which will be using liquid cooling for high-power devices – that will also be interfacing with and rejecting heat into the system. On a relatively simple aircraft like the EAP demonstrator, which had a minimal mission fit, the number of signals associated with the ECS was ~59 with 14 power drives.

10.5.8 Electrical Utilisation Systems

The miscellaneous electrical utilisation systems listed below require simple logical control functions and may be independently controlled. However, in some cases there are dependencies such as weight on wheel interlocks, or automation such as automatic dimming of lighting. Typical electrical utilisation systems include:

1. Internal lighting – designed to provide a balanced lighting solution across the cockpit in a variety of lighting conditions with the ability to dim panels, instruments and equipment lighting. The system provides all power sources for all types of flight deck, cockpit and cabin lighting including integral lighting, floods, wander lamps, anti-flash and maintenance bay lighting.
2. External lighting – provision of lighting solutions for external lights, e.g. navigation, anti-collision beacons, high-intensity strobes, formation lights, refuel probe lights and land/ taxi lamps.
3. Probe heating – provision of heating for external air data, temperature and icing probes.
4. Ice detection and protection – provision of methods for detecting icing conditions and ice accretion and icing protection/de-icing systems. These systems may be used to provide a warning of icing conditions, or to initiate de-icing methods automatically.
5. Windscreen de-icing – provision of anti/de-icing methods for canopy, windscreens or direct vision (DV) windows.

10.5.9 Prognostics and Health Management

The majority of aircraft in service today include a mechanism for capturing and recording failures that are detected during a flight. Most systems contain some form of in-built testing to monitor their continued satisfactory operation. Built-in test (BIT) is initiated automatically at system power-up (PBIT), or may be initiated by the ground crew or even by the aircrew (IBIT), or may take place continuously throughout system operation (CBIT). All of these forms of test will produce a result when a failure is detected, requiring some form of corrective action at the next convenient opportunity.

This discrete form of recording failures has disadvantages:

1. The next landing site may not be the best place to perform maintenance, especially if the aircraft is on deployment or in transit through a site where spares and skilled mechanics may not be available.

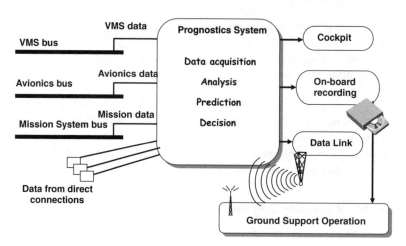

Figure 10.20 Prognostics system block diagram.

2. It is better to be warned of an impending failure so that maintenance can be planned.
3. With multiple-redundant systems it may be possible, even desirable, to despatch with a known failure, particularly in times of conflict.

For these reasons, prognostics systems are developing that enable decisions to be made on the basis of known deteriorating performance and a statistical prediction of time remaining before action is required. All systems will be monitored for detected faults resulting from BIT, and system parameters will be measured to determine changes in performance (e.g. times, rates of changes, pressures, stress, etc.) and to measure structural limitations and exceedences.

Connections are made to systems to measure such signals as flow, pressure, rates of change, numbers of operations and duration of operation. This enables decisions to be made by comparing measured values with datum values. An informed condition report can be generated by the system which will inform the ground crew when a maintenance action should be performed. Connection of the prognostics system to data link allows information to be transmitted to the destination airfield so that spares and skilled ground crew are available to repair the aircraft. An example system block diagram is shown in Figure 10.20:

- System architecture – provision of a computing architecture to meet the requirements for gathering of information, processing of information and interfaces with the customer's support environment;
- System interfaces – provision of interface structures or remote interfacing units to match the vehicle system components requirements: I/O type, impedance, slewing rates, load, etc.;
- Software – provision of functional requirements, algorithms and software.

10.6 Design Considerations

10.6.1 General

Given the safety-critical and safety-involved nature of the utility systems, great care must be taken in the design of the utilities control system. This starts with developing a clear

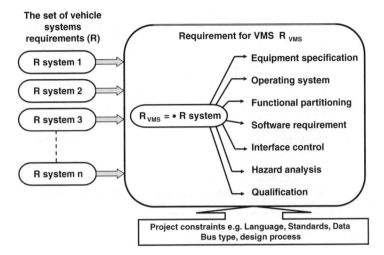

Figure 10.21 Vehicle systems requirements.

understanding of the requirements of the individual utility systems. This is not a trivial task. The total utility vehicle systems requirement is the sum of a large number of individual requirements of differing degrees of complexity. Figure 10.21 shows how the requirements flow from the individual systems, are then consolidated into a single vehicle system requirement and than used to influence the architecture and design principles. Functions are then isolated and allocated to individual processing units on the basis of integrity, separation, processor loading, etc. This leads on to development, testing and qualification of the total system.

10.6.2 Processor and Memory

An analysis of the functional requirements of the utility systems and the partitioning of those functions into individual processors provides an indication of the throughput requirement for the processor and the memory capacity. The designer must ensure that there is sufficient growth potential in the selection to account for the following eventualities:

- Growth in the requirement during the early stages of detailed definition and design;
- Changes resulting from errors in the design and encountered during testing;
- Changes resulting from flight testing;
- Growth potential to allow for system capability enhancement during use.

The processor and memory configuration is shown in Figure 10.22.

10.6.3 Interfacing

The early implementation of utility systems management integrated the input and output interfacing with the processor and memory in a small number of processing units of a standard case size. The reason for this was to minimise the number of processing units and the number of suppliers in order to achieve a level of standardisation or commonality in the architecture. There is now a trend with modern vehicle management systems to segregate

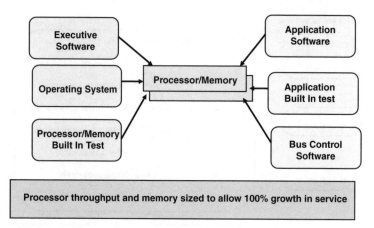

Figure 10.22 Processor and memory configuration.

the interfacing circuitry into remote input/output (RIO) units that can be distributed to the extremities of the aircraft. This reduces the majority of interface wiring to the locality of the RIO. The information is encoded in the RIO into a data format that is compatible with the processing unit.

The interfacing circuits must take full account of the electrical characteristics of the component. The RIO should ideally perform a 'degree of goodness' check, confirming that the input signal is within reasonable bounds and ensuring that it is a valid value; in other words, that the data the sensor is providing lie within a window of upper and lower values that gives confidence that they are valid.

10.6.4 Software

The choice of language will probably be driven by the need to have a single common language for a project, and may be driven by a customer need to have a language common with other assets in their inventory.

Whatever the language, there is a need for a robust design process that is consistent with the level of safety defined for the individual utility systems. This is usually safety critical or safety involved.

10.6.5 Obsolescence

The long life of military products – there are many examples of aircraft still in active service 40 or more years after their initial design – means that component obsolescence is a certainty rather than a risk. An obsolescence plan is required to detail all components that are likely to become difficult to replace after a period of time. There is sufficient experience available to determine those components most likely to become obsolete as a result of technological advances, and there are existing mitigation plans that can be used as a model. However, obsolescence can also be driven by health, safety and environmental legislation, where materials once considered safe to use are no longer acceptable because they are dangerous to health in use, or their disposal poses a health or environmental hazard. This situation is less predictable and requires a continuous monitoring of legislation.

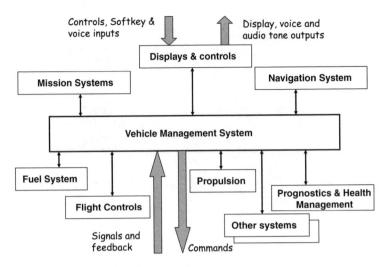

Figure 10.23   Human–Machine Interface.

The generally lower performance – in terms of throughput, clock rate, memory and executed instructions – of many of the processing devices used within VMS applications may offer alleviation to the obsolescence problem rather than protection from it. The problem of COTS is a major problem in the high-performance avionics systems, as already discussed in Chapter 2.

10.6.6   Human–Machine Interface

The human–machine interface in the modern aircraft cockpit is largely based on multi-function displays combined with multifunction or 'soft' keys, as described in Chapter 11. There will be some independent, hardwired controls, but these will be minimised to those essential instinctive controls. The main interface between the individual utility systems and the pilot will be provided by the vehicle management system. The VMS will interpret messages from the cockpit and use them in the control functions to place appropriate demands on a system to modify a control function or directly to command an action. Data from the systems are used in control functions and transferred to the cockpit displays or to other avionic systems via the appropriate data buses. In this way the pilot can make demands and observe the behaviour of the utility systems, making maximum use of the carefully designed cockpit layout.

Two items of aircraft equipment straddle the boundary between utility systems and avionics. These items are the throttle lever handles and the stick top, associated with propulsion and flight controls respectively. Both items have been furnished with switches and controls to perform a number of avionic functions, over and above their basic use as a comfortable grip for the throttle and flight control demand levers. This allows the pilot to perform a number of control functions without removing hands from the two demand levers. This has become know as hands-on throttle and stick, or HOTAS. The positioning and actuation loads of the switches installed in the handles has to be designed with great care to meet ergonomic requires for instinctive finger movements. Figure 10.24 shows an example HOTAS arrangement.

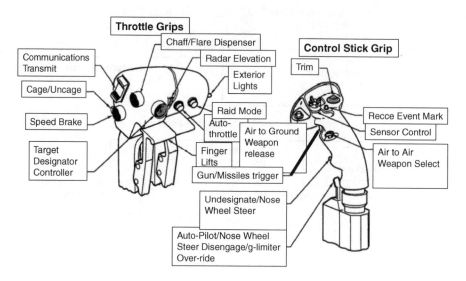

Figure 10.24   HOTAS concept in the F/A-18 Hornet.

References

Jukes, M. (2003) *Aircraft Display Systems*, Professional Engineering Publishing.
Moir, I. and Seabridge, A.G. (1986) Utility systems management. *Royal Aeronautical Society Journal – Aerospace*, **13**(7), September.
Moir, I. and Seabridge, A.G. (2001) *Aircraft Systems*, Professional Engineering Publishing.
Moir, I. and Seabridge, A.G. (2002) *Civil Avionics*, Professional Engineering Publishing.
Pallett, E.H.J. (1992) *Aircraft Instruments and Integrated Systems*, Longmans Group.
Yeoell, L. and Kneebone, R. (2003) On-board oxygen generation systems (OBOGS) For in-service military aircraft – the benefits and challenges of retrofitting. Aero India 2003 – International Seminar on *Aerospace Technologies: Developments and Strategies – Flight Testing and Man Machine Interface*, 8 February 2003.

Further Reading

Bryson Jr, R.E. (1994) *Control of Spacecraft and Aircraft*, Princeton University Press.
Conway, H.G. (1957) *Landing Gear Design*, Chapman and Hall.
Currey, N.S. (1984) *Landing Gear Design Handbook*, Lockheed Martin.
Hunt, T. and Vaughan, N. (1996) *Hydraulic Handbook*, 9th edn, Elsevier.
Lloyd, E. and Tye, W. (1982) *Systematic Safety*, Taylor Young.
Pallett, E. H. J. (1987) *Aircraft Electrical Systems*, Longmans Group.
Pratt, R. (2000) *Flight Control Systems: Practical Issues in Design and Implementation*, IEE Publishing.
Principles of Avionics Data Buses (1995) Avionics Communications Inc.
Raymond, E.T. and Chenoweth, C.C. (1993) *Aircraft Flight Control Actuation System Design*, Society of Automotive Engineers.

11 Displays

11.1 Introduction

The history of aircraft displays can broadly be divided into three technology eras, the mechanical era, the electro-mechanical (EM) era and the electro-optical (EO) era.

Although the design boundaries are clear, the time boundaries are vague. The catalyst for the electro-optical era was the dramatic increase in performance and capability of digital electronics in the late 1960s that led to an impetus to change the means to display information on the flight deck. Cathode Ray Tube (CRT) technology was the first multifunction display medium, latterly superseded by Active Matrix Liquid Crystal Display (AMLCD) technology. Both provide a more flexible means for the display of information than had hitherto been possible. Multifunction displays (MFDs) can show many formats on the same display surface and portray the same piece of information in a variety of different ways.

This chapter provides examples of 'glass' electro-optical military fighter cockpits, or crew stations as they are now known, tracing the evolution from the electro-mechanical gyro gunsight and the radar 'scope' to current electro-optical head-down, head-up and Helmet-Mounted Displays for advanced tactical and global situational awareness in today's digital battlespace.

The Head-Up Display (HUD) has been applied predominantly to military fast jet fighter aircraft. The optical principles of the Head-Up Display will be described, along with its principles of operation for air-to-air and air-to-ground weapons aiming.

The rotorcraft community were the first to apply Helmet-Mounted Display (HMD) technology. The operating principles and the role of the Helmet-Mounted Display will be described. The Head-Up Display is placed on the head of pilots, greatly enhancing their ability to acquire and designate targets and to release weapons off-boresight. Indeed, as currently planned for the joint strike fighter, it is entirely likely that the Helmet-Mounted Display will become the primary flight instrument in future fast jet fighters, and the head-up Display will no longer be fitted.

This chapter will also discuss the principles of operation and device features of the shadow-mask CRT and the active matrix LCD head-down displays presenting primary flight,

Military Avionics Systems   I. Moir and A. Seabridge
© 2006 John Wiley & Sons, Ltd

navigation, topographical and tactical map information, systems and sensor (radar and FLIR) images and weapons information to the crew.

Emerging and potential future display technologies of rear projection reflective LCD, transmissive LCD (liquid crystal on silicon – LCoS) and digital micromirror devices (DMD) will be reviewed, as well as the means to apply these technologies to large-area megapixel head-down displays.

Finally, the crew-station ambient lighting conditions in day, dusk/dawn and night will be described, together with the principal accepted industry visual performance metrics and optical test methods to achieve display viewability. The requirements and means to achieve compatibility with night-vision imaging system (NVIS) devices will be discussed.

A more in-depth discussion of the topic areas in this chapter can be found in Jukes (2004).

11.2   Crew Station

Undeniably, the first use of electro-optical (EO) devices in the cockpit was during World War II with the display of airborne intercept radar contacts on the Cathode Ray Tube. Soon the CRT was used in the gyro gunsight to produce a collimated (focused at infinity) aiming reticle. Very quickly the CRT gyro gunsight developed into a sophisticated projection device through which the operator could correlate the position or vector of the aircraft or weapon with the outside world. The Head-Up Display (HUD) had been invented.

As CRT technology improved, it became possible to augment the role of the radar 'scope' also to provide flight information. The multifunction head-down display (HDD) had been invented. However, CRT brightness and contrast technology limitations meant that early CRT multifunction displays were monochrome (green), and information presentation was limited to character/symbolic images. In these early 'glass' cockpits the moving map display was produced by optical rear projection of a 35 mm filmstrip topographical map. However, it was not long before further advances in colour CRT technology made it possible to present a full colour topographical map image in the severe lighting conditions and environment of a military fighter cockpit.

11.2.1   Hawker Siddley (BAe) Harrier GR.Mk1 and GR.Mk3 (RAF) and AV-8A (USMC)

The first UK aircraft to enter service with a 'designed-in' Head-Up Display was the vertical take-off Hawker Siddley (BAe) Harrier. These aircraft equipped four operational squadrons, one in the United Kingdom and four in Germany. The US Marine Corps (USMC) took delivery of 110 aircraft, designated the AV-8A.

The cockpit of the Harrier GR.Mk3 is shown in Figure 11.1. The flight instruments are conventional electro-mechanical counter pointer type. In the centre of the instrument panel is the moving map. This large and complex instrument projects the image of a filmstrip topographical map on to a rear projection screen. The film transport mechanism is driven by aircraft inertial reference coordinates and heading to indicate aircraft current position and course/track with reference to ground topography.

Above the moving map is the Head-Up Display. The HUD was small by today's standards. It was a refractive design with a 4 in exit lens and provided a modest instantaneous field of

Figure 11.1   Harrier GR.Mk3 cockpit (RAeS).

view of about 16°. The HUD itself contained only the CRT and the collimation optics. The CRT high-voltage supplies and the CRT beam deflection electronics were remote from the HUD to minimize space and weight in the cockpit.

The HUD operated in cursive (stroke) mode to provide a daylight-viewable symbolic image. Display modes included:

- Navigation;
- Approach and landing;
- Precise local fix (IN update);
- Air-to-air attack (guns, rockets and missiles);
- Air-to-ground attack (freefall and retarded bombs, CCIP manual and CCRP automatic release).

11.2.2   McDonnell Douglas F/A-18 Hornet

The F/A-18 Hornet is generally considered the true beginning of the electro-optical (EO) era and is pivotal in cockpit display design. The cockpit of the night-attack F/A-18C is shown in Figure 11.2 and Plate 1.

Figure 11.2   McDonnell Douglas F/A-18C cockpit.

The HUD incorporates a raster mode for night-time use, presenting a collimated outside-world image from the forward looking infrared (FLIR) pod overlaid with conventional stroke symbology.

The left and right MFDs present limited colour formats using time-sequential liquid crystal shutter technology. This technology superimposes the additional dimension of red and orange symbology overlays onto the high-resolution, high-brightness green raster weapon imagery.

The projected map display utilises a multipurpose, high-brightness, high-resolution, full-colour stroke and raster shadow-mask multipurpose CRT display (MPCD) with a 5×5 in square format usable screen area. The map image is generated remotely from the display in a digital map computer using CDROM technology.

The cockpit lighting is fully compatible with night-vision goggles and at night the pilot may aid his night vision with the use of 'cat's eye' goggles attached to the helmet.

In the early 1990s, McDonnell Douglas commenced development of the F/A18E & F Super Hornet. The centre CRT-based MPCD is replaced with a larger area (6×6 in square format display) employing AMLCD technology to provide enhanced brightness/contrast and resolution.

The F/A-18 cockpit truly broke new ground, but its introduction represented only the tip of a technological iceberg in terms of the challenge for the cockpit designer to show to the crew the massive amount of data now made available by digital processing without saturating them with data overload. The answer is to present only those data required for the current phase of the mission and to configure the display format accordingly. Initially, this reconfiguration was performed by the operator who decided what to display and when. Unfortunately, excessive operator involvement was found to be counterproductive in terms of reducing workload. Today, progressively more sophisticated decision aids predict the crew information requirements and configure the display formats accordingly (Garland *et al.*, 1994).

As will be seen in subsequent cockpit designs, there is a continuous struggle to reduce the bulk of the display device itself while increasing the display surface area and flexibility of

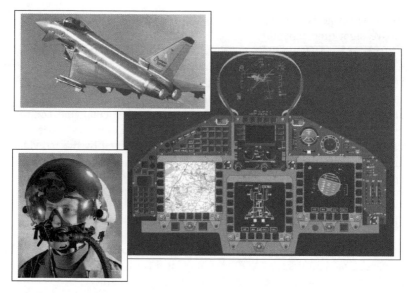

Figure 11.3   Eurofighter Typhoon cockpit (BAE SYSTEMS).

the information content. The ultimate aim, possibly to be achieved in the joint strike fighter, is to provide the operator with one contiguous, controllable display surface.

11.2.3   Eurofighter Typhoon

The Eurofighter consortium was formed in June 1986 by the three countries that developed the Tornado, namely the United Kingdom, Germany and Italy, and was shortly joined by Spain. The development programme was launched in November 1988 and the first two prototype aircraft undertook their maiden flights on 27 March and 6 April 1994.

The Eurofighter Typhoon cockpit is shown in Figure 11.3 and Plate 2. The main instrument panel comprises three colour multifunction head-down displays (MHDDs). In the prototype aircraft these displays used shadow-mask CRTs to provide daylight-viewable, full-colour, high-brightness, high-resolution images in both cursive (stroke) and hybrid (stroke + raster) modes. In production the CRTs have been superseded with high-resolution 6.25×6.25 in square format Active Matrix Liquid Crystal Displays. The MHDDs incorporate 18 multifunction keys around the bottom, left and right edges of the display. Each key contains a daylight-viewable LED matrix of two rows of four 7.5 characters plus underline.

The HUD uses holographic technology to achieve an ultrawide $30° \times 25°$ field of view (FoV). The HUD provides stroke (cursive) operation for daytime use plus raster for nighttime use with outside-world sensor video. The HUD incorporates a sophisticated up-front control panel with a 4×3 in daylight-viewable LED matrix display. The HUD is the primary flight instrument.

An HMD is planned, configured into two variants. The daytime variant provides symbology for the targeting and release of off-boresight weapons. The night-time variant adds night-vision goggles (NVGs) to the helmet to provide the pilot with enhanced night vision. The NVG image is electrically mixed with the CRT symbology image to provide a comprehensive night-time capability.

To either side of the HUD the left and right glareshield panels provide essential controls and warnings. The right-hand panel incorporates the standby attitude display employing AMLCD technology. The farthermost part of the right-hand glareshield flips open to reveal a set of standby get-u-home instruments in the unlikely event that there is a major power failure.

The Eurofighter Typhoon provides direct voice input (DVI) command control for non-mission-critical functions such as communications equipment. The DVI speech recogniser has a vocabulary of about 100 words. The DVI system is trained by the individual user to function under all operational conditions including high-*g* manoeuvres and low-speed passes with significant wind buffet (Birch, 2001).

11.2.4 Lockheed Martin F-22 Raptor

In April 1991 the USAF announced that it had selected the Lockheed Martin F-22 Raptor to meet its advanced tactical fighter (ATF) requirement for a fighter combining low observability, supersonic cruise, long range and a very high level of agility.

The F-22 advanced avionics system provides the pilot with fused situational awareness in the battlefield environment. The advanced crew-station layout shown in Figure 11.4 includes a number of large-format full-colour AMLCD multifunction head-down displays, numerous general- and special-purpose knobs and switches, a Head-Up Display and a Helmet-Mounted Display. The displays contain integral video processing and graphics generation (Greeley and Schwartz).

A central single 8×8 in primary multifunction display is flanked by three 6×6 in secondary multifunction displays. Two 3×4 in up-front displays are arranged either side

Figure 11.4 F-22 Raptor crew station.

of the HUD. Below the HUD there is an integrated control panel housing some dedicated switches and alphanumeric text read-outs. Bezel option selection buttons (OSBs) surround each of the displays and are used for menu navigation, paging and function select.

The prototype aircraft also included touch-sensitive screens, voice recognition and three-dimensional audio, but these technologies were deemed to be immature and were not included in the production design.

The primary multifunction display (PMFD) is used to display the situation display (SD) or the attitude director indicator (ADI) display. The SD is a tactical format showing the entire track file icons as well as navigation data, with multiple levels of OSB menus for feature control. SD symbology is centred on ownship. The ADI is a back-up to the HUD.

The three secondary multifunction displays (SMFDs) are used to display either of the following:

- Attack display (AD);
- Defence display (DD);
- Expand display (EXD);
- Situation display – secondary (SD-S);
- Stores management display (SMD);
- Fuel display, engine display;
- Mission data edit (MDE) display family;
- Back-up integrated control panel display;
- Flight test display (FTD);
- Surface position display (SPD);
- Electronic checklist (ECL).

The up-front displays (UFDs) are used to display the communication, navigation and identification (CNI) display or the standby flight group (SFG) display. Additionally, the integrated caution, advisory and warning (ICAW) data are shown in the centre column of the CNI display. The SFG display shows a simplified small version of the ADI and HUD attitude data with digital read-outs for altitude, speed and heading.

The Head-Up Display (HUD) provides a $24°$ horizontal $\times 20°$ vertical binocular wide-field-of-view monochrome (green) image. The HUD is the primary flight display (PFD) and provides attitude, flight path, navigation and weapons deployment symbology in general conformance to MIL-STD-1787.

The Helmet-Mounted Display is the US standard joint helmet-mounted cueing system (JHMCS) with a $20°$ instantaneous field of view. The HMD is monochrome (green) monocular (right eye only), and is used to show similar symbology to the HUD with some special symbology for depicting sensor volume limits. Like the HUD, extensive user-editable controls are provided to declutter symbology.

11.2.5 Boeing (McDonnell Douglas/Hughes) AH-64D Longbow Apache

The AH-64D Longbow Apache best illustrates the attack helicopter electro-optical 'glass' crew station. The Apache is a tandem, two-seat helicopter with advanced crew protection systems, avionics and electrooptics plus a weapon control system that includes the nose-mounted target acquisition and designation sight/pilot's night-vision sensor (TADS/PNVS). The copilot/gunner sits in the front seat, with the pilot in the higher rear cockpit. Both use

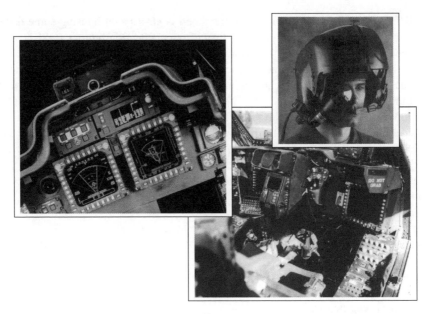

Figure 11.5   AH-64D Longbow Apache Cockpit (RAeS).

sophisticated sensors and systems for the detection and attack of targets, including the integrated helmet and display sight system (IHADSS), which provides a monocular helmet-mounted designator sight. The copilot/gunner has the primary responsibility for firing both the gun and the missiles, but can be overridden by the pilot in the back seat.

The AH-64D crew stations, shown in Figure 11.5, each have two 6 × 6 in full-colour, high-resolution, active matrix liquid crystal (AMLCD) displays replacing the monochrome CRT displays of earlier versions.

11.3   Head-Up Display

The Head-Up Display has been proven over many years to be a means of providing flight navigation, aircraft data and weapon release parameters. The earliest gunsights provided little more functionality than an aiming system for the guns, but, with the introduction of increasingly sophisticated computational capability, complex manual air-to-air and air-to-ground weapon release became possible with enhanced accuracy. Now the HUD forms part of an integrated weapon aiming and release system where automatic modes of release are available for the different stores and release conditions. In recent years it has been possible to introduce sensor video from forward looking infrared (FLIR) on to the HUD such that operation by night can be achieved with much the same capability as daytime operation (Quaranta, 2002).

The HUD is installed in an area of 'prime real estate' in the instrument panel, and therefore the HUD tends to be 'designed in' to the aircraft to optimise operational capability and performance within cockpit geometry and the available space envelope. The configuration of a HUD typically comprises the pilot's display unit (PDU) and a remote display processor. The PDU is rigidly fixed to the airframe and accurately aligned to the aircraft axes by a process known as boresighting or harmonisation.

11.3.1 HUD Principles

The optical principles of the HUD are straightforward, although to integrate these principles into the restricted space confines of a fast jet fighter crew station results in some ingenious light-bending solutions.

The Head-Up Display injects a virtual image of an object (the symbology and/or sensor video) into the pilot's line of sight with the key attribute of being collimated, or focused at infinity, so that to the pilot the image appears to be in the same focal plane and be fixed on features of the outside-world scene. This attribute allows the aircraft systems to place symbology in a manner that is often referred to as 'in contact analogue' with (or conformal to) the real world. Conformal symbology is used to identify the flight path vector, the horizon and sightlines to targets in the real world. In addition, primary flight data (speed, height, heading, etc.) and flight guidance cues allow the pilot to fly the aircraft without having to refocus his eyes to look inside the crew station at his head-down instruments during critical phases of the mission.

There are two means to form virtual collimated images (shown in Figure 11.6):

1. HUDs using the principle of refraction are often known as collimating. The lens system generates a collimated image of display symbology reflected to the eye via a flat combiner. The instantaneous field of view is limited by the size of the collimating lens and the pilot has to move around the eyebox to see the total field of view.
2. HUDs using the principle of reflection are often known as pupil forming. The lens relay system generates an intermediate image of the display symbology which is then collimated by reflection from the curved combiner. The instantaneous field of view is larger, defined by the size of the curved combiner.

The curved combiner in the pupil-forming HUD may use the optical principle of diffraction provided by a hologram rather than reflection to provide optical power, and so this arrangement is sometimes also known as a diffractive HUD (or variously DHUD or DOHUD – diffractive optics HUD). The spectrally selective feature of a hologram is sometimes also used to fabricate the plane combiner/mirror.

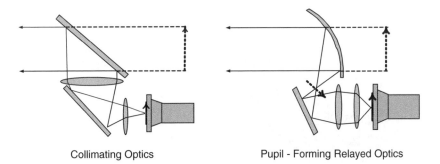

Collimating Optics Pupil - Forming Relayed Optics

Figure 11.6 HUD optical arrangements.

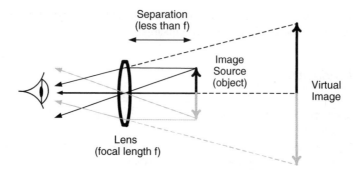

Figure 11.7 Optical refraction.

11.3.2 Collimating (Refractive Optics) Head-Up Display

The optical principle of refraction is shown in Figure 11.7. The image source (object) is placed just inside the focal length, f, of the collimating lens, shown in the figure as a simple convex lens. The observer sees a magnified virtual image of the object. This is the principle of the magnifying glass. If the object is placed in the focal plane of the lens, then the rays of light refracted by the lens from each point on the object will be parallel, i.e. collimated.

To form a practical HUD, as shown in Figure 11.8, the image source must be placed out of the line of sight. The collimating lens, together with field-flattening lenses, focuses the CRT image at infinity and the image is introduced into the pilot's line of sight by a semi-reflective plane mirror, known as the combiner, so called because the pilot sees the HUD image combined with the normal forward view of the outside world. A second mirror or prism is usually introduced between the image source and the collimating lens to produce a compact design for convenience of installation and, more importantly, to ensure there is no obstruction to the pilot should the need arise to eject from the aircraft.

In practice the simple convex lens comprises a number of elements to minimise aberrations together with a number of elements in the image plane to 'flatten' the field.

The HUD CRT generally has a faceplate with an active area typically 40–50 mm in diameter.

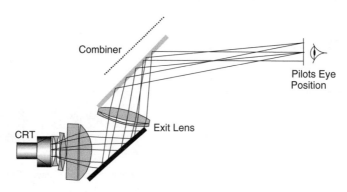

Figure 11.8 HUD using refractive optics.

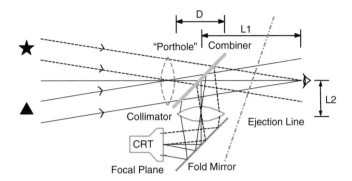

Figure 11.9   Refractive HUD instantaneous field of view.

The combiner is mostly transmissive (typically better than 90%) to the wide-band daylight-visible spectrum so as not to degrade the pilot's view of the outside world in low-light conditions. The CRT phosphor is chosen to have a narrow waveband emission in the green part of the visual spectrum. The combiner is optically tuned to reflect the CRT spectral wavelength with high efficiency (Jukes, 2004 – Chapter 6).

11.3.3   Field of View

The total field of view (TFoV) of any HUD is the total angle subtended by the display symbology seen with head movement from any location.

The instantaneous field of view (IFoV) of a refractive HUD is the display field of view seen from one head position using one eye and can be simply derived from the size of the collimator exit lens as seen by the observer, reflected in the combiner. A typical arrangement shown in Figure 11.9 for an HUD with a 150 mm (6 in) exit lens viewed 450 mm (18 in) from the combiner yields an instantaneous field of view (IFoV) of about 15°.

If the pilot's head moves forwards towards the combiner, the instantaneous field of view will increase. Similarly, if the pilot's head moves laterally, angles further off-axis will be seen. In practice, of course, the pilot sees two portholes from one head position, one with each eye. This describes the binocular field of view (BFoV).

It is possible to increase the instantaneous field of view in the vertical axis by adding a second combiner parallel to and vertically above the first. This has the effect of allowing the observer to view the CRT image through a second porthole vertically above and super-imposed over the first. This arrangement is shown in Figure 11.10.

11.3.4   Collimating (Refractive) HUD – Examples

11.3.4.1   British Aerospace Harrier GR.Mk1 and GR.Mk3

The Head-Up Display of the Harrier GR.Mk1, shown in Figure 11.11, and contained only those optical and mechanical supporting structures described above together with the CRT and its electron beam deflection yoke and focus magnet. All other electronics was contained remotely. This architecture minimized the size and weight of the pilot's display unit (PDU) to facilitate its installation into the cockpit.

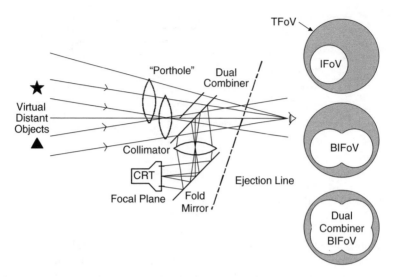

Figure 11.10   Dual combiner (including FOV).

The PDU had a 4 in collimating lens and provided an IFoV of about 16° and a TFoV of 25°. The single combiner was servo-ed to travel in the fore–aft direction to maintain the aiming vector within the pilot's normal field of view under all normal manoeuvres.

11.3.4.2 McDonnell Douglas/British Aerospace Night-Attack Harrier II (GR-7 and AV-8B)

The night-attack Harrier II entered service in 1989 and incorporates a high-performance wide-field-of-view refractive HUD capable of day/night operation. The HUD shown in

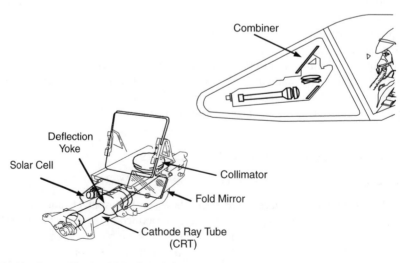

Figure 11.11   Jaguar/Harrier HUD installation (Smiths).

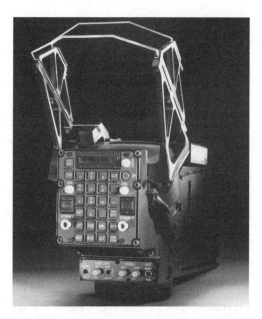

Figure 11.12   AV8B/Harrier GR7 night-attack HUD (with FOV) (Smiths).

Figure 11.12 incorporates the CRT, the deflection amplifier, the high-voltage power supplies and all the services to make the HUD a stand-alone line replaceable unit.

The HUD employs a dual combiner. The collimating exit lens is truncated in the fore and aft axes to allow the optical axis of an effectively larger-diameter exit lens to be installed nearer the pilot without encroaching upon the ejection envelope. In the Harrier II, the binocular IFoV is 20° azimuth × 15° elevation. The TFoV is 22°.

The HUD presents daytime images in cursive (stroke) mode and night-time images from the FLIR sensor as a raster video image with stroke symbology overlay during the raster field retrace period.

11.3.5   Pupil-forming (Reflective/Diffractive) Head-up Displays

It is highly desirable to offer a larger field of view than that obtainable from a collimating HUD to support more aggressive manoeuvring and to give pilots the sensation of flying under visual (VMC) conditions at night using a projected FLIR image, enabling them to use familiar visual cues to judge speed and terrain clearances and to navigate and identify targets.

Figure 11.13 illustrates the significant operational improvement that would be achieved if the HUD IFoV were to be increased from the 20° × 15° practical limit of a refractive collimating HUD to an IFoV of 30° × 20°.

Using the optical principle of reflection (shown in Figure 11.14), it is possible to place the collimator closer to the pilot and therefore achieve a larger IFoV without infringing the ejection line.

If an object is placed within the focal length of a concave mirror, the rays of light are reflected by the mirror towards the observer, who sees a magnified virtual image of the

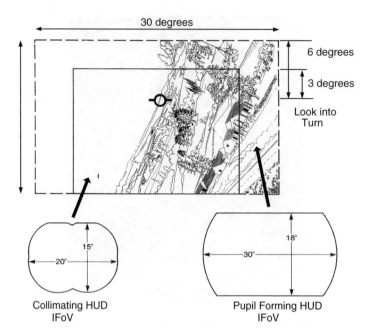

Figure 11.13 Operational improvement of increased IFoV (BAE SYSTEMS).

object. This is the principle of the shaving mirror. By placing the object in the focal plane of the mirror, the rays of light from each point on the object will be reflected by the mirror to emerge in parallel, i.e. collimated.

As with refraction, it is necessary to design an optical path that allows the CRT image to be introduced into the pilot's outside-world line of sight without interfering with it. This is more complex than with the refractive arrangement. Some of the possible optical configurations are shown in Figure 11.15 (Fisher). On the right is a class of diffraction optics designated 'off-axis', which represents the most elegant solution but is the most optically complex. The curvature of the combiner provides the principal collimating function but is too great to allow the use of a planar doublet to sandwich the reflective coating. The

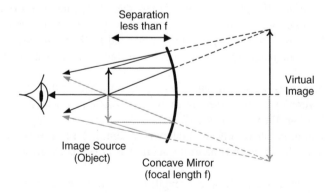

Figure 11.14 Optical reflection.

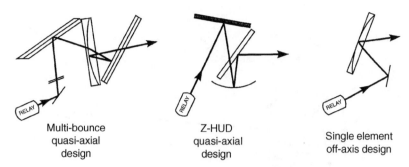

Figure 11.15   Some possible reflective optic HUD configurations.

necessary protection has to be provided by two pieces of curved glass. This reduces the thickness and weight of the element but is more complex to manufacture than a plane combiner, and care must be taken to ensure that the apparent distortion of the real world seen through it is minimal. The large off-axis angle causes significant optical aberrations, which must be corrected by introducing compensating aberrations in the reflective combiner itself and in a complex relay lens.

By comparison with the off-axis systems, the other class, termed quasi-axial, have a much-reduced critical angle of incidence for reflection. Aberrations are minimal and do not require compensating aberrations in the reflective element itself: the limited corrections necessary can be implemented in a simple relay lens. The Z-HUD (so called because the optical path makes the letter Z) has the further advantage that the reflecting function is separated from the combining function. While this is obviously attractive, it probably places the collimating mirror over the ejection line in a real installation.

It is theoretically possible to realise these optical systems using conventional refractive/reflective optical elements, but the optical efficiencies would be poor and so in practice they have been fabricated using holographic diffractive techniques (Jukes, 2004 – Chapter 6).

11.3.6   Pupil-forming (Reflective/Diffractive) HUD – Examples

11.3.6.1   F16 LANTIRN HUD – Multibounce Quasi-axial Configuration

The first application of the multibounce quasi-axial pupil-forming diffractive HUD config-uration was in the F-16 as part of the USAF low-altitude navigation and targeting infrared for night (LANTIRN) system (Hussey, 1981).

The optical arrangement is shown in Figure 11.16. Obviously, all glass elements above the glareshield must be transparent to allow the pilot an uninterrupted view of the real world. They must also sometimes be transparent to CRT light and sometimes reflect it strongly to allow the HUD image to reach the pilot. This apparent paradox is resolved by the use of highly angularly selective holographic optical coatings that will only reflect a narrow band of green wavelengths emitted by the CRT phosphor when incident at a particular critical angle. The effect of this is to make the real world seem to the pilot to have a slightly pinkish tinge.

The complete HUD is shown in Figure 11.17 and Plate 3, together with its installation in the F-16 LANTIRN crew station.

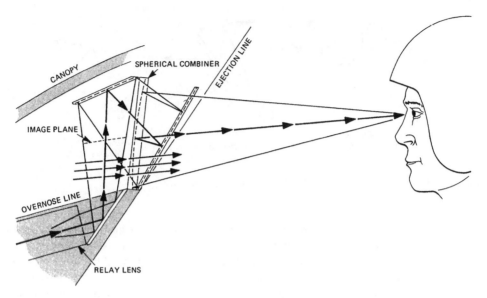

Figure 11.16   F-16 optical configuration (BAE SYSTEMS).

The HUD provides an instantaneous field of view (IFoV) of 30° azimuth × 18° elevation and operates in both raster and stroke modes to provide day/night operation. It is fully compatible with night-vision goggles.

11.3.6.2   Eurofighter Typhoon HUD – Single-element Off-axis Configuration

The much more aesthetically elegant solution is the single-element off-axis configuration shown in Figure 11.18. This configuration achieves a wide-field-of-view HUD with no

Figure 11.17   F-16 LANTIRN HUD (BAE SYSTEMS).

Plate 1   F/A 18 C Cockpit (McDonnell Douglas).

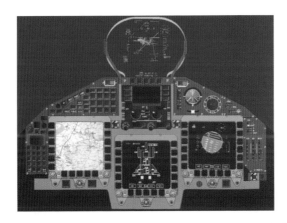

Plate 2   Eurofighter Typhoon Instrument Panel Layout (BAE SYSTEMS).

Plate 3 F16 Cockpit (Lockheed Martin).

Plate 4 Eurofighter Typhoon (BAE SYSTEMS).

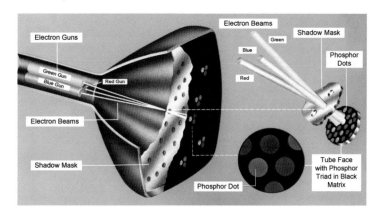

Plate 5   Colour Shadow Mask CRT.

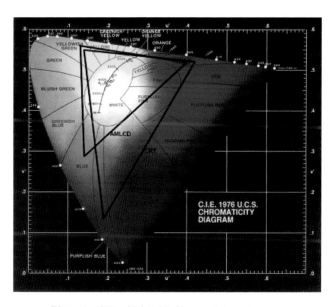

Plate 6   CIE 1976 UCS Chromaticity Diagram.

Plate 7   Eurofighter Typhoon MHDD (Smiths).

Plate 8   JSF 'Big Picture' Projection Head-Down Display (Rockwell Collins).

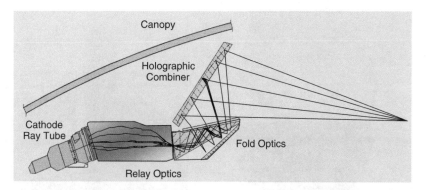

Figure 11.18 Eurofighter pupil-forming optical configuration (BAE SYSTEMS).

incursion into the ejection envelope and with no upper mirror to obstruct the upper field of view. It comprises a single optical element between the pilot and the outside world; the semi-transmissive curved collimating mirror/combiner.

The elegant simplicity of this configuration belies its optical complexity, which arises because, by its very nature, the intermediate image subtends a significant 'off-axis' angle to the collimating mirror. This means significant optical correction needs to be applied to correct for distortions.

The collimator must emulate a complex aspheric surface in order to ensure all rays of light from the reflected image emerge in parallel (i.e. collimated). It is only possible to fabricate this element using holographic techniques in which the hologram itself is computer generated.

The relay lens is complex also. It contains several aspheric elements to provide compensation for the image distortions produced by the off-axis collimating combiner. Finally, complex geometric distortions have to be applied to the CRT image. These are produced by correspondingly distorting the electron beam deflection current drive waveforms.

Notwithstanding the above complexities, the clean lines and low forward obscuration make this optical configuration the configuration of choice for high-capability, high-performance and wide-field-of-view applications. It is now introduced on to production prestige fast jet fighters such as Eurofighter Typhoon, Rafale, F-15 and Gripen.

The Eurofighter HUD is shown in Figure 11.19 and Plate 4. It employs advanced computer-generated holographic optics to provide a $30° \times 25°$ total field of view (TFoV). The instantaneous field of view is identical to the total field. It provides stroke (cursive), raster and hybrid modes of operation with outstanding display luminance of 2700 ft.L ($9200 \, cd/m^2$) in stroke (daytime) mode and 1000 ft.L ($3500 \, cd/m^2$) in raster (night-time) mode, this latter being viewable in daytime under cloud and haze. The outside-world transmission is 80%.

The HUD also provides a comprehensive up-front control panel with a large-area daylight-viewable LED matrix display and programmable keys. The HUD is a high-integrity design and is used as the primary flight display in Eurofighter.

11.3.7 Head-Up Display Functional Description

Unlike the Harrier HUD discussed earlier, most modern HUDs contain all the electronic services to support the HUD functions, providing a low-level signal interface to the aircraft

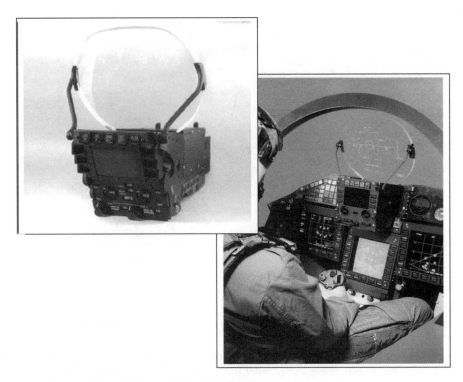

Figure 11.19   Eurofighter Typhoon HUD (BAE SYSTEMS).

computer systems. A typical HUD block diagram is provided in Figure 11.20. The HUD comprises the following functional elements:

1. *Optics assembly.* A set of optical elements comprising the final collimating lens for the collimating HUD or the relay lens for the pupil-forming HUD, image field flattener lenses and the fold mirror with coatings and filters to reduce sunlight reflections.
2. *Combiner.* A pair of optically flat parallel glass plates for the collimating HUD or a curved collimator for the pupil-forming HUD, with semi-reflective coatings tuned to the peak spectral emission of the CRT phosphor.
3. *Cathode ray tube.* A high-brightness, high-resolution CRT used to produce bright, precision fine-line stroke-written graphics symbology and raster sensor video imagery.
4. *X and Y deflection amplifiers.* These are high-precision power amplifiers that source current into the CRT X and Y magnetic deflection yokes to cause the CRT electron beam to trace out the stroke (cursive) symbology and also the raster sensor video scan waveforms.
5. *Video amplifier.* Controls the CRT beam current by adjusting the CRT cathode bias with respect to the grid electrode to turn the beam on or off in stroke mode and modulate the beam with the sensor video image.
6. *Ramp generator.* Strips synchronization pulses from the sensor video and generates the raster scan waveforms to align/harmonise the sensor video with the outside world.
7. *High-voltage power supply.* Provides the final anode potential (usually around 18 kV), the A1 (focus) potential (usually around 2–4 kV) and the grid potential (usually around 200 V).

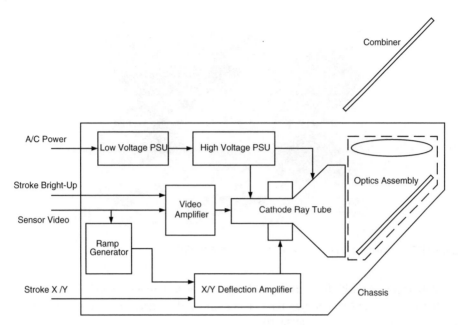

Figure 11.20 HUD functional block diagram.

8. *Low-voltage power supply.* Provides all the low-voltage rails for the electronic circuits, typically $+5\,V$, $\pm 15\,V$, $\pm 20\,V$, etc.
9. *Chassis.* Usually a complex precision casting, the chassis provides environmental protection to the HUD functional elements and affords a means of boresighting the HUD to the aircraft structure.

The physical realization of these components into a fully operational HUD tends to be aircraft specific to match the operational capability requirements within the available cockpit envelope. The HUD electronic assemblies are packaged to make the best possible use of the available space envelope. The subassemblies of a typical HUD are shown in Figure 11.21.

11.3.8 Image Generation

To achieve maximum display brightness and contrast, the HUD day image (symbology) is written in stroke mode. The information content is in the X and Y beam deflection signals. The line width of the graphic elements is the CRT electron beam width. For best accuracy and finest spot size, the CRT is usually electromagnetically deflected and electrostatically focused. To avoid flicker, the entire image is redrawn or refreshed at a frequency greater than 50 Hz (20 ms). The absolute brightness range is typically four orders of magnitude (10 000 : 1) to span the ambient illumination range from bright sunlight to night-time conditions.

Day/night Head-Up Displays provide the capability to present raster video images of forward looking infrared (FLIR) sensors on the HUD, requiring the HUD image to be

Figure 11.21   Typical HUD subassemblies (BAE SYSTEMS).

accurately registered with the outside world. It is necessary therefore to be able to adjust the size, the position and sometimes the orientation of the raster scan waveforms to match precisely the field of view of the sensor.

A day/night HUD provides hybrid operation in which stroke-written symbology overlays the raster sensor image. A typical HUD FLIR hybrid (raster video plus stroke overlay) is shown in Figure 11.22.

11.3.9 HUD Symbology and Principles of Use

HUD symbology has historically been designed specific to the aircraft type, although some standardization is now emerging. Principal areas of symbol generation include the following:

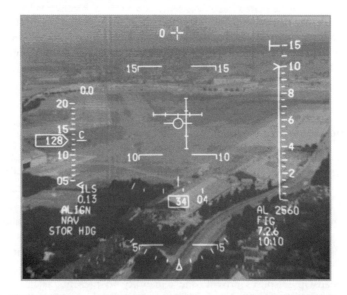

Figure 11.22   HUD FLIR image plus stroke symbology overlay (BAE SYSTEMS).

- Primary flight data;
- Navigation symbology;
- Air-to-air weapon-aiming symbology;
- Air-to-air weapon-aiming symbology.

11.3.9.1   Primary Flight Data

Primary flight data displayed on the HUD usually encompass:

- Flight-path vector/marker (also known as velocity vector);
- Attitude (pitch and roll);
- Speed;
- Altitude;
- Heading;
- Vertical speed (rate of climb/dive);
- Angle of attack.

A typical example of HUD primary flight symbology is shown in Figure 11.23.

The flight-path marker presents the flight-path vector in 'contact analogue' form. This symbol indicates the instantaneous velocity vector, or direction, in which the aircraft is flying (and will continue to fly if the current manoeuvre is maintained).

Attitude (pitch and roll) is also indicated in 'contact analogue' form as a horizon line in register with the real horizon, if visible. Additional lines (the pitch ladder) parallel to the horizon line indicate positive and negative pitch angles, usually at 5° intervals.

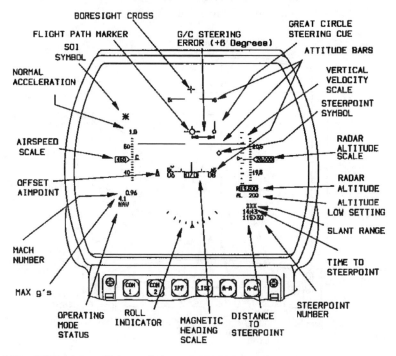

Figure 11.23   HUD primary flight symbology (BAE SYSTEMS).

Speed and height are presented as numerical values and a linear tape scale. Heading is usually presented as a numerical read-out on a horizontal scale, emulating a compass card viewed edge-on. Other numeric read-outs may also be included if they are generally helpful to the pilot to enable him to prosecute his mission effectively. These include instantaneous *g*, barometric set, waypoint lat/long, time/distance to go to next waypoint, etc.

11.3.9.2 Navigation Symbology

Navigation symbology is added to the primary flight information to provide en-route navigation to/from the mission target area. This information typically includes:

- Lateral and vertical guidance to maintain the preselected flight plan;
- Next waypoint position (latitude and longitude);
- Bearing and range to next waypoint;
- Time to go to next waypoint.

On approaching the next waypoint, a conformal fix cross indicates the computed sightline to the waypoint on the basis of the best estimate of present aircraft position. The fix cross should precisely overlay the visually sighted waypoint. Any error represents an error in estimation of present position and can be corrected manually by the pilot.

11.3.9.3 Air-to-Surface Weapon Aiming

On entering the target area, the pilot selects the weapon to be deployed, and the HUD symbology changes accordingly. Figure 11.24 shows typical symbology for the release of dumb cast-iron, free-fall bombs on to a preplanned target.

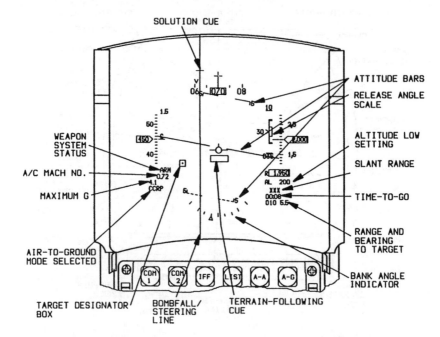

Figure 11.24 Typical HUD air-to-surface bomb symbology (BAE SYSTEMS).

The following description is illustrative only:

1. A target designator box indicates the sightline to the target. The target designator should overlay the visually sighted target. The pilot should manually correct any errors.
2. A continuously computed impact point (CCIP) indicates the sightline to the point on the ground where the selected bomb type will fall if fired now. This point is computed from a knowledge of the aircraft manoeuvre, height above target, bomb ballistics, ejection velocities and an estimation of wind over the target.
3. A bombfall line indicates the locus of impact points on the ground of all future bomb releases and extends forwards from the CCIP marker taking due account of wind. The pilot steers to this line.
4. A safe-pass height marker is often associated with the bombfall line to provide an indication of the fragmentation zone of the bomb. The bomb should not be released if the CCIP marker is close to or infringes the safe pass height marker.

For a successful release the pilot should:

- Fly the aircraft to position the bombfall line over the target ahead of the CCIP marker;
- Continue to fly towards the target, tracking the target with the bombfall line – the target (and target marker) will move down the bombfall line as the distance to the target decreases;
- When the target marker reaches the CCIP marker, the bomb should be released, either manually or automatically;

Specific adjustments are made to the air-to-surface symbology for different weapons classes, different manoeuvres and of course different ballistics.

11.3.9.4 Air-to-Air Weapon Aiming

Should pilots elect to fire guns or missiles, then the appropriate aiming symbology appears to enable them accurately to aim and fire their weapons. Typical symbology is shown in Figure 11.25.

A track line indicates the computed trajectory of the gun or missile on the basis of the current aircraft manoeuvre and weapon ballistics. A ranging circle is positioned on the line at the range of the target aircraft. A target designator box indicates the radar sightline to the target. Other information derived by the air-to-air radar about the target is also displayed. This might include, for instance, target motion relative to own aircraft motion (closure rate), possibly augmented with target slant range and g.

Infrared heat-seeking missiles can in general be commanded to seek targets at angles significantly 'off-boresight'. A missile target marker indicates the missile sightline. Typically, the missile will be slaved to the radar. Once the missile detects an IR source, it will lock to that source and track it independently of the radar. Using the HUD, the pilot can confirm that the missile is locked to the correct source before launch commit.

A firing solution exists when the radar target designator and the ranging circle are coincident and stable. Usually, the ranging circle will have a number of other bugs that indicate the minimum and maximum firing range, together with the probability of success

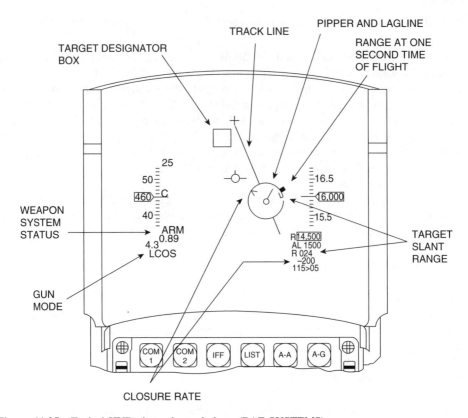

Figure 11.25   Typical HUD air-to-air symbology (BAE SYSTEMS).

based on the manoeuvres executed by both aircraft and on the manoeuvre required by the missile for a successful intercept post-launch.

11.4   Helmet-Mounted Displays

The field of view of the HUD is a very limited field of view when compared with the total hemisphere of regard afforded to the pilot in the bubble canopy of a modern fast jet fighter aircraft, and limited when compared with the field of regard of a modern radar (typically 120°) and the acquisition cone of modern air-to-air missiles (typically 90°), as graphically shown in Figure 11.26.

Helmet-mounted displays overcome this limitation. By the use of miniature display technology producing a display for each eye, combined with accurate head tracking, it is theoretically possible to present a stereoscopic, full-colour image to the user in any direction.

Although display technology and the processing speed of graphics generators have not yet matured to the point where the virtual world image is possible in an aircraft environment, it will undoubtedly materialize in the near future (Garland *et al.*, 1994). Putting the HUD onto the pilot's head, even with the same field of view, but now free to move wherever the pilot is looking, makes it possible to cue, acquire, designate, track and release weapons at targets

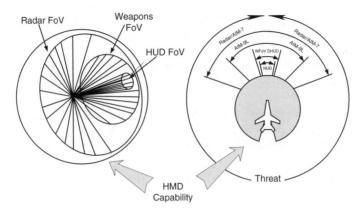

Figure 11.26 Field-of-regard comparisons.

significantly off-boresight without having to manoeuvre the aircraft. The increase in operational effectiveness is a significant 'force multiplier'.

Figure 11.27 shows some of the operational possibilities, which include:

- Off-boresight target cueing by directing the pilot/weapon aimer to look in the direction of potential targets or threats detected by on-board sensors and/or advised by data link from cooperating aircraft and ground stations;
- Designation 'off-boresight' once the target has been recognised – tracking accuracy of the on-board systems and weapons can then be monitored to assure lock is maintained;

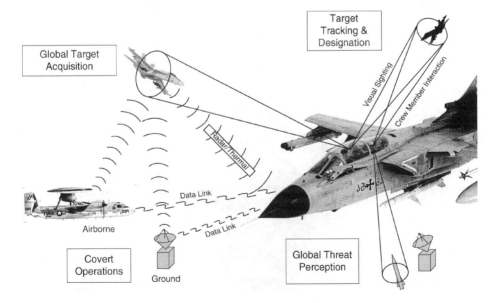

Figure 11.27 Possible uses of a HMD to cue, designate and aim 'off-boresight'.

- Weapon release can be commanded once a launch solution is reached without having to manoeuvre the aircraft to be within the traditional success cone determined by the HUD field of view;
- Handover is facilitated between crew members and between cooperating aircraft flying the same mission.

The first practical HMD recorded was made by Autonetics in the United States during the early 1960s. By 1970 a system had been flown in a US Navy F-4, which proved the key points. That HMD had a sighting reticle and operated with an optical tracker, facilitating the pointing of radar and missiles.

The joint helmet-mounted cueing system (JHMCS) installed on the F/A-18 aircraft and other USAF and US Navy fast jets provides a monocular display for use in daytime operations.

The application of a binocular, day and night capability has taken longer to mature on fast jets. The Eurofighter Typhoon will probably be the first. However, the next generation of fast jet fighters, typified by the JSF, is likely to replace the HUD with an HMD, and to have the HMD as the primary flight instrument.

11.4.1 HMD Physiological and Environmental Aspects

The HMD is considerably more than an HUD on the head; the helmet already provides a number of facilities and life-supporting functions with which the display components must be integrated. These are illustrated in Figure 11.28, including:

- Communications microphone;
- Earphones (possibly with active noise reduction);
- Oxygen mask;
- Retractable sun visor;

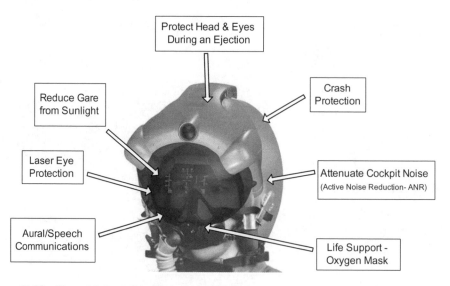

Figure 11.28 Normal helmet functions.

- Possibly NBC protection;
- Head protection during ejection.

The provision of a Helmet-Mounted Display requires the addition of some extra components, namely:

- An image source (e.g. Cathode Ray Tube) or more than one source if a binocular image is required;
- Relay optics to position the HMD image at the focal plane of the collimator;
- Collimating optics to generate a virtual image focused at infinity;
- A partially transparent/partially reflective combiner to introduce the collimated image into the pilot's line of sight.

Adjustments need to be made to centre the optical image axis onto the pilot's eyeball. Typically, these include:

- The vertical position of the eyeballs with respect to the pilot's head;
- The distance between the eyeballs, the interpupilary distance;
- The eye relief between the eyeball and the first optical element.

Furthermore, the additional HMD components must not introduce features that may jeopardise the integrity of the helmet to protect the pilot from hazards such as:

1. Birdstrike. It is potentially possible that a birdstrike may cause the canopy to shatter.
2. Ejection. In the event that pilots have to eject from aircraft, the helmet must protect them from the acceleration forces of the ejection itself and from the windblast (in excess of 600 knot).

The range of human anthropomorphic geometry is vast. Usually it is necessary to have a range of helmet sizes. The helmet must be a firm fit to the pilot's head so that no movement can occur once the optics have been 'boresighted' to aircraft axes, but at the same time be comfortable to wear and keep the head cool. Often this is accomplished using a permanently deformable liner.

The weight of the helmet must be minimised to reduce fatigue. A typical helmet before the addition of an HMD weighs less than 2.0 kg (4.5 lb). Additionally, the CG of the helmet should be in line with the pivotal point of the head on the spine to minimize out-of-balance forces.

With all the HMD optical components added to the helmet, the helmet becomes a high-value item. For this reason, more recent designs allow the HMD optical elements to be removed from the helmet to reduce its intrinsic value.

Achieving all of these requirements represents a significant challenge to the Helmet-Mounted Display designer. The requirements are often conflicting.

11.4.2 Head Tracker

To be effective as a sight, accurate knowledge is required of the head-pointing angle. The head can move vertically, laterally and fore/aft with respect to the body torso and can rotate in all three axes about the pivotal hinge point of the head on the spine.

Figure 11.29   Optical head tracker (Kentron).

11.4.3   Optical Head Tracker

A number of schemes have been used. An early design used a number of sources positioned around the cockpit to generate swathes of scanning beams. Sensors mounted on the helmet detected these beams, providing signals to the head-tracking electronics to decode the interrelationship between the scanning beams and the time of detection and derive the head pointing angles.

A more recent scheme, and that being used on the Eurofighter Typhoon, uses twin-tracking CCD cameras to sense clusters of LEDs on the surface of the helmet (Figure 11.29).

11.4.4   Electromagnetic Head Tracker

Both ac and dc electromagnetic fields have been used. In general, a source transmitter placed just behind the pilot generates an electromagnetic field in the cockpit. A three-axis orthogonal sensor mounted on the helmet detects the local electromagnetic field and provides signals for the head-tracker electronics to decode the amplitude and phase signals and derive the head pointing angles.

The electromagnetic field is significantly distorted by the metal structures within the cockpit, and it is necessary to map the cockpit to characterize it fully. This mapping needs to take account of movable features within the cockpit, for instance, seat height position. Usually it is necessary to map only one aircraft of a type, although the mapping will have to be repeated if there are any subsequent equipment changes which might impact the magnetic properties of the cockpit.

11.4.5   HMD Accuracy and Dynamic Performance

HMD pointing accuracy comprises two elements, the first an optical accuracy similar to that of a HUD system plus a tracker error, which increases rapidly at extended angles from

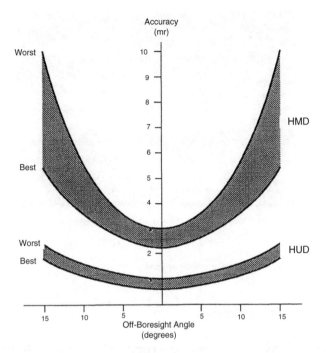

Figure 11.30   Comparison of HUD and HMD accuracy (BAE SYSTEMS).

boresight and at the margins of the head motion box. Typical HUD and HMD system accuracies are shown in Figure 11.30 (Bartlett).

Since the helmet-mounted sight is not fixed to the aircraft structure, it is necessary to provide a means to 'boresight' the HMD system before each mission. Typically, the boresighting procedure comprises sighting a collimated target conveniently located in the cockpit.

Using a helmet-mounted sight to launch a weapon with 'smart' guidance has been shown to be highly effective, especially since the target can be acquired and designated and the weapon launched 'off-boresight'. The HMD accuracy is adequate for this task. However, using a helmet-mounted sight to target a fixed gun or release dumb weapons is generally recognised as being inappropriate.

The dynamic performance requires careful consideration since rapid movements of the pilot's head (30 deg/s) are entirely possible. System bandwidth and delays must be commensurate with the operational role in order to avoid disorientation that might be caused by 'swimming' symbology. For this reason it is usual closely to couple the head-tracking electronics with HMD symbol generation, minimising any data latency problems.

11.4.6   HMD Optical Configurations

The optical configuration should be selected according to operational use. Display performance needs to be balanced with complexity. Bigger is not necessarily better if it weighs more.

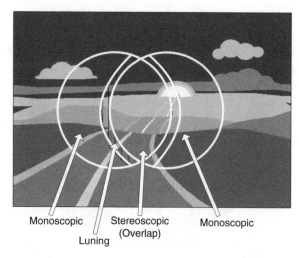

Monoscopic   Stereoscopic   Monoscopic
                  (Overlap)
        Luning

Figure 11.31   Monocular or binocular configuration.

The choices broadly are as follows:

1. *Sight or display.* A sight is used simply as a target designator, possibly with the addition of simple direction-finding cues (such as look-up/down, left/right) and event annunciators (such as weapons lock/fire). A display generally adds other symbology such as primary flight data, weapons data and imagery at night.
2. *Monocular or binocular.* Monocular is suitable for daytime use when the pilot's attention is mostly on the outside world. However, a binocular configuration is preferred at night when vision is augmented by other aids such as FLIR or NVGs.
3. *If binocular, then one image source or two.* Both configurations have been used. If dual sources are used then it is possible to extend the azimuth field of view having a central area in which the image is seen by both eyes with two monocular areas either side, as shown in Figure 11.31.
4. *Off-visor or periscope optical system.* The periscope optical system is easier to design and manufacture and therefore was the first to be used. Off-visor systems are less obstructive, more technically elegant but demand more complex solutions if wide fields of view are required.
5. *Day, night or day/night operation.* It is a self-evident requirement that a daytime display must be viewable in direct sunlight. At night the image may be augmented with sensor video and/or night-vision goggles.

The optical configuration choices for a Helmet-Mounted Display are identical to those used in Head-Up Displays. Typical arrangements are shown in Figure 11.32. The lens diameters are considerably reduced when compared with those of a Head-Up Display since the effective exit lens is close to the pilot's eyeball and thus a 30–40° field-of-view porthole is obtainable with lens diameters of around 1 cm. However, the intermediate optics to relay the image from its source (typically a miniature CRT) to the focal plane of the collimator are

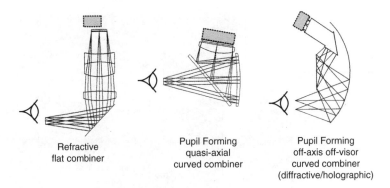

Figure 11.32 Typical optical configurations.

often very complex since they need to bend the light path around the head without being obstructive and at the same time maintain the CG close to the head natural CG.

Figure 11.33 shows a generic HMD optical system. The EO component elements are:

1. *Image source.* Usually a high brightness, high resolution miniature CRT (though early sights used a small LED matrix. Emerging technologies are AMLCD (transmissive and reflective, electro-luminescent (EL), organic LED (OLED) and low-power laser scanning directly into the eye-ball.
2. *Relay and fold mirrors.* The relay is a group of mirrors, lenses and prisms, which translate the image from the source into the correct place to be projected into the pilots' line of sight by the combiner.
3. *Combiner.* The combiner projects the image from the relay lens into the pilots' line of sight. It may be flat or have optical power. It is semi-transparent to the outside world, but reflective to the specific wavelength of light from the image source.

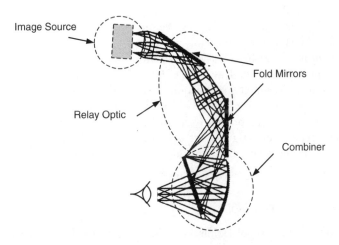

Figure 11.33 Generic HMD optical system.

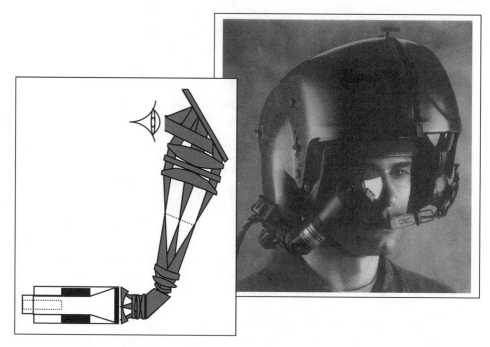

Figure 11.34   IHADS helmet (Honeywell).

11.4.7   Helmet-Mounted Displays – Examples

11.4.7.1   Integrated Helmet and Display Sight System (IHADS)

The first users of HMD technology in aerospace applications were the rotorcraft community. The HMD weight issues are less significant in a rotorcraft, which is not able to pull as much *g* as a fast jet fighter. Neither is ejection or windblast a concern.

Figure 11.34 shows the Apache AH-64 attack helicopter Honeywell-supplied integrated helmet and display sight system (IHADS) helmet worn by the copilot/gunner who has primary responsibility for firing both guns and missiles.

The IHADS provides a $40° \times 30°$ field-of-view, monocular (single-eye), collimated image from a miniature Cathode Ray Tube (CRT). The images constructed on the CRT are derived from the nose-mounted target acquisition and designation sight/pilot night-vision sensor (TADS/PNVS). The TADS/PNVS and the gun are slaved to the copilot/gunner head line of sight. The system has proven itself to be operationally effective; however, there have been some reports of fatigue after extended use at night, possibly caused by the contention created in the brain of a monocular sensor image in one eye and the real world view in the other.

11.4.7.2   Helmet-mounted Sight

The simplest and earliest form of Helmet-Mounted Display used in fast jet applications was the helmet-mounted sight (HMS), used in daytime operations to acquire and designate targets off-boresight. The HMS installed and worn by RAF Jaguar pilots is shown in

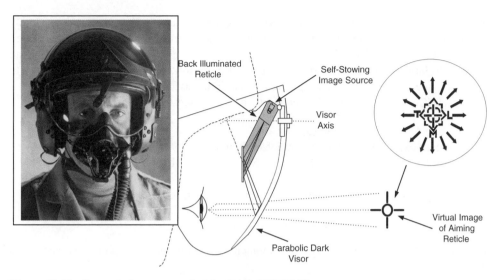

Figure 11.35 Jaguar helmet-mounted sight (BAE SYSTEMS).

Figure 11.35. The Jaguar HMS comprises a simple fixed aiming reticle formed from an array of bright LEDs. Simple optics collimate the image, which is projected off the visor into the pilot's line of sight.

11.4.7.3 Joint Helmet-mounted Cueing System (JHMCS)

The JHMCS programme is the culmination of extensive trial evaluations performed using helmets such as the Kaiser agile eye.

The JHMCS, shown in Figure 11.36, is designed to provide first-shot, high off-boresight weapons engagement capabilities enabling the pilot to direct weapons against enemy aircraft

Figure 11.36 Joint helmet-mounted cueing systems (JHMCS) (Kaiser Electronics).

while performing high-g aircraft manoeuvres. The system can also be employed accurately to cue the pilot to ground targets. Targeting cues and aircraft parameters are displayed directly on the pilot's visor. The JHMCS is designed to have low weight, optimised CG and in-flight replaceable modules to enhance operational performance – including the ability to be reconfigured in-flight to meet night-vision requirements.

The JHMCS Helmet-Mounted Display has the following features:

- A monocular field of view of 20°;
- An 18 mm exit pupil with an eye relief of 50 mm;
- The display module is compatible with US Air Force HGU-55/PU.S and US Navy HGU-68/P helmets;
- The weight is 4 lb (1.82 kg) with mask. MTBF is 1 000 h.

11.4.7.4 Eurofighter Typhoon HMD

The Helmet-Mounted Display in development for Eurofighter Typhoon is shown in Figure 11.37. Features are:

1. Provision of primary aircrew protection.
2. Attachment to inner helmet.
3. Provision of a lightweight but stiff platform for the optical components:

    - CRTs/optics/mirror;
    - Night-vision cameras;
    - Blast/display and glare visors;
    - Head tracker diodes (infrared).

Figure 11.37   Eurofighter Typhoon helmet-mounted display (BAE SYSTEMS).

4. Removable night-vision cameras, autodetach during ejection.
5. Mechanical function:

- Protection features in line with survivability limits;
- Life support up to the limits of human functionality;
- Comfort and stability to support display requirements.

6. Display features:

- $40° \times 30°$ binocular field of view;
- Corresponding $40° \times 30°$ binocular night-vision camera;
- Display of sunlight visible symbology and/or imagery.

The optics arrangement (dual binocular) uses a 1 in high-brightness, high-resolution, monochrome (P53 green) CRT as the image source. A complex relay lens with a brow mirror introduces the relayed image into the focal plane of two spherical diffractive mirrors which are deposited on the visor by holographic techniques. The field of view is $40°$ and the exit pupil is 15–20 mm. The optical arrangement introduces significant geometric distortion, which is corrected electronically.

The Helmet is a two-part arrangement (see Figure 11.38).

The inner helmet fits inside the display outer helmet and can be swapped with a respirator hood version to give NBC protection:

1. It facilitates individual user fitting.
2. It maximises comfort and stability:

- The brow pad is moulded to exact fit;
- Air circulation around the head is assisted.

3. It embraces a wide anthropomorphic range:

- Advanced suspension system;
- Lightweight oxygen mask.

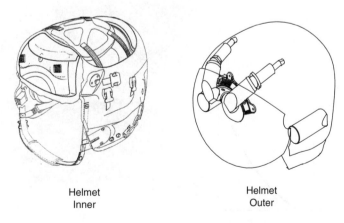

Helmet
Inner

Helmet
Outer

Figure 11.38 Eurofighter Typhoon two-part helmet (BAE SYSTEMS).

4. It provides optimum hose/cable routing.

The display outer helmet attaches to the inner helmet. It provides:

1. Primary aircrew protection.
2. A lightweight but stiff platform for the optical components:

   - CRTs/optics/mirror;
   - NVE cameras;
   - Windblast/display visor (clear) and glare visors.

3. Head tracker diodes.
4. Removable night-vision cameras.

11.4.7.5   *Joint Strike Fighter HMD*

The joint strike fighter (JSF) will have a binocular day/night Helmet-Mounted Display and no Head-Up Display. The HMD will provide all the targeting information and sufficient primary flight information to meet the mission objectives.

Figure 11.39 shows a prototype of the JSF HMD. The helmet provides a $30° \times 50°$ total field of view from two overlapping $30° \times 40°$ monocular images. The optics is a pupil-forming arrangement with the relayed image collimated and introduced into the pilot's sightline by a diffractive holographic semi-transparent curved combiner introduced into the visor. The exit pupil is 18 mm. The image source is a high-resolution (SXGA) transmissive LCD, illuminated with a bright LED backlight.

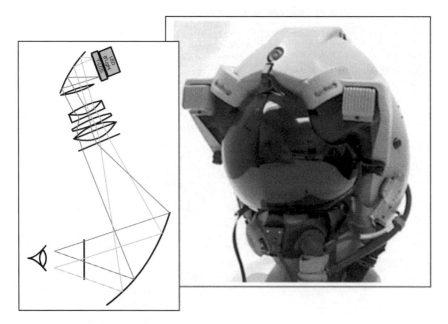

Figure 11.39   JSF helmet mounted display (VSI).

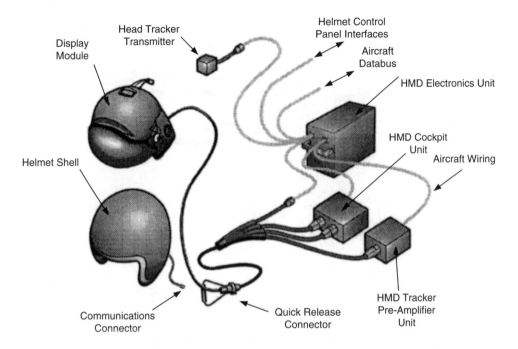

Figure 11.40 Typical HMD system components.

11.4.8 Helmet-Mounted Display Functional Description

A Helmet-Mounted Display is a heavily distributed system. For obvious reasons, only the minimum of components are carried on the head to minimise weight.

A typical HMD system is shown in Figure 11.40 and comprises the following line replaceable units (LRUs):

1. *Basic helmet assembly.* The basic helmet provides a high level of pilot protection and comfort and is fitted with communications equipment, a form-fitting system and display module-mounting points. The basic helmet is form fitted to an individual pilot and becomes part of his personal equipment.
2. *Display module assembly.* This often simply clips on to the basic helmet shell and contains the display components in a single-sized lightweight module. Dependent on the functionality, the components located within the display module are dual-image intensifiers, dual CRT displays, optical assemblies, a dual-visor system, a helmet tracker receiver, autobrilliance sensor, battery pack and umbilical cable.
3. *Helmet electronics unit.* This is located in the avionics bay and contains the main HMD system electronics functions including system interfaces, processing, display drive and helmet tracking.
4. *Helmet tracker transmitter.* This small unit is mounted either on the aircrew seat or the canopy, and is part of the helmet tracker function.
5. *Boresight reticle unit.* This provides accurate optical reference for boresighting the helmet tracker system.

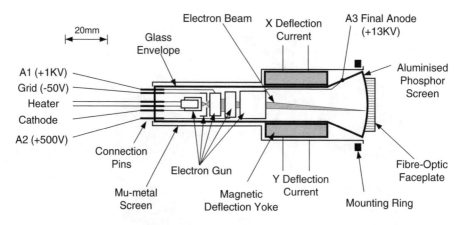

Figure 11.41 Typical HMD miniature CRT.

6. *Quick-release connector cable assembly.* This comprises the cable and connectors running to the HMD quick-release connector.

Most current Helmet-Mounted Displays use a miniature CRT as the image source (Figure 11.41). The CRT has a faceplate image of typically 20 mm (3/4 in) diameter and an overall length of around 100 mm (4 in). The CRT image is generated by an electron gun and focused electrostatically. The final anode potential is typically 13 kV and the electron beam current is around 100 mA. Two orthogonal scan coils electromagnetically deflect the electron beam. To achieve high-brightness daytime viewability, the image is constructed in cursive (stroke) mode. If the HMD is also used for displaying night-time sensor imagery, then the combined sensor plus symbology image is constructed in raster or hybrid (stroke + raster) mode. A fibre-optic faceplate produces a flat image plane and enhances the image contrast.

Next-generation HMDs may employ an LCD as the image source together with a bright LED backlight. This technology would avoid the need for high voltages.

11.4.9 Binocular Day/Night HMD Architectures

The Eurofighter Typhoon and next-generation HMDs are seeking to integrate NVGs with the HMD image generation function. There are two means to achieve this, optically and electronically (Jukes, 2004 – Chapter 8).

11.4.10 HMD Symbology

A typical symbology format is shown in Figure 11.42. In all current HMD installations the HUD is generally the primary flight instrument. Primary flight data on the HMD play a secondary role and are only used subliminally for quick orientation. Pilots defer to the HUD if they become seriously disoriented. In this example, attitude information is presented as a simple line indicating pitch and roll against a fixed circle. The extensions at the end of the

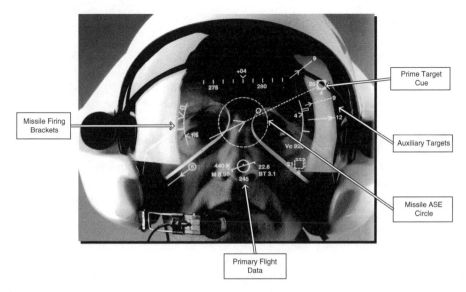

Figure 11.42   Typical HMD symbology format (Thales).

intersecting line indicate normal/inverted flight. Speed, height and heading are presented as simple numerical read-outs.

The primary aspects of the display format relate to targeting data to several priority targets. If the target is not within the HMD field of view, then a line with an arrowhead indicates the direction in which the pilot should look to acquire the target visually. Radar data such as range to target and target *g* level are displayed alongside the target marker. If missiles are selected and have acquired a target, then a marker also shows the line of sight to the missile target. Additional information indicates probability of launch success.

11.4.11   HMD as a Primary Flight Reference

In the future it is probable that HMD systems will replace the HUD as the primary flight reference. Doing so will free up the significant cockpit real estate occupied by the HUD as well as greatly increase the look-through display field of regard available to the pilot for weapon aiming and sensor display. HMDs provide a virtually unlimited field of regard for off-boresight targeting, weapons employment and display of sensor information.

The seeming advantages of this increased area of display space must be approached with caution, however. There is a strong potential for targeting sensor information presented off-boresight to draw the pilot's attention away from the traditional on-boresight flight reference information for longer periods of time. During periods of limited visibility, the increased off-boresight time could result in reduced spatial awareness if some sort of flight information is not presented in the off-boresight display. While the presentation of primary flight information in HUDs is well understood and documented, methods for presenting off-boresight primary flight information are not as mature.

Ongoing work is being undertaken by the USNR Naval Air Warfare Center Aircraft Division Patuxent River and Boeing to design and evaluate operational HMD symbology formats that fulfil the primary flight reference (PFR) requirements (Foote).

11.5   Head-Down Displays

The multifunction head-down display, variously called the MFD, the MHDD, the MPD multipurpose display (MPD) and the multipurpose colour display (MPCD), provides the flight-deck/crew-station/cockpit designer with a flexible display media on which to present data in a variety of ways according to the information needed by the crew for the current phase of the flight or mission.

At the outset of the electro-optical era, display capability was limited primarily to that able to be provided by the monochrome (usually green) CRT. The significant breakthrough came with the application of the shadow-mask CRT to airborne environments. The shadow-mask CRT is perhaps the first example of commercial off-the-shelf (COTS) technology applied to airborne applications. It took significant developments (albeit evolutionary rather than revolutionary) to make the shadow-mask CRT both bright enough and sufficiently robust to be able to be used in the severe environment of airborne applications.

However, CRT technology is bulky, heavy and power hungry and requires extremely high voltages to provide a high-brightness display, posing serious limitations to the crew-station designer. The advent of AMLCD technology offered the promise of eliminating all those problems. In the event, AMLCD technology has proved as difficult if not more difficult to apply to airborne applications. Although small (i.e. flat), low weight and low voltage, the performance of an AMLCD varies with viewing angle (particularly an issue for instrument reading cross-cockpit), and it is temperature sensitive and fragile. Furthermore, as was learnt very painfully, extreme attention to manufacturing process control is required to achieve consistent product quality. This generally only comes with high-volume production. Custom-size low-volume aerospace product is to be avoided if at all possible.

However, these problems are now largely solved and the AMLCD is the technology of choice. Display sizes for new crew-station designs are transitioning from the custom square-format sizes historically used in aircraft to rectangular laptop PC and professional/industrial glass sizes. The focus now is to ruggedise COTS AMLCD devices, not design custom aerospace glass.

11.5.1   CRT Multifunction Head-Down Display

The Cathode Ray Tube was the first fully flexible display device to be used for primary flight displays in the cockpit.

11.5.1.1   Shadow-mask CRT

The shadow-mask CRT shown in Figure 11.43 and Plate 5 comprises an evacuated glass bulb in which an electron gun emits electrons at high velocity to impact on a phosphor screen. The electron gun comprises three cathodes, today usually the in-line configuration, although early applications used the delta configuration.

The shadow-mask CRT is a large heavy device, and not easy to package. Typically, a 6.25 in × 6.25 in CRT is 35 cm (14 in) long and weighs 5 kg (11 lb), including magnetic components and shield.

Electron emission is modulated by three grid electrodes, focused by a multistage electron lens and finally accelerated towards the faceplate through a shadow mask. A series of red,

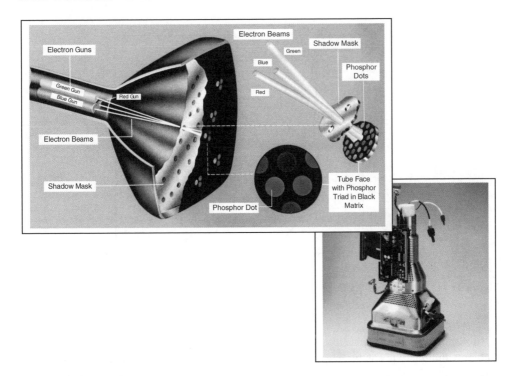

Figure 11.43   Shadow-mask CRT.

green and blue phosphor dots are deposited on the CRT faceplate in a process that uses the shadow mask itself to maintain precise registration. The electron gun, shadow mask and phosphor dot geometry is arranged so that electrons emitted by each cathode only illuminate phosphor dots of its designated colour. The composite beam bundle is deflected in the X and Y axes by a magnetic yoke to scan the CRT display surface. Modulating the individual beams produces a full-colour image comprising three registered primary colour images.

The phosphors deposited on the CRT faceplate emit light in the red, green and blue primary colours. The red and green are narrow-band, short-persistence ($<10\,\mu s$) phosphors; the blue is broadband, longer-persistence (1 ms) phosphor.

The gamut in which colours can be produced by a shadow-mask CRT is best described in CIE u′ v′ colour space. By suitable combination of cathode drive voltages, any colour can be produced within the triangle described by the three primary colours:

	u′	v′	Dominant wavelength
Red	0.42	0.52	610 nm
Green	0.14	0.55	550 nm
Blue	0.19	0.134	460 nm

Figure 11.44 and Plate 6 shows the colour gamut of a shadow-mask CRT. Also shown is the gamut available from an AMLCD for comparison. The CRT has a purer blue, although it can be argued that this is of little operational advantage.

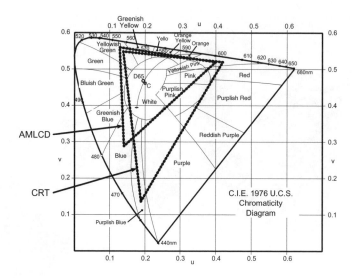

Figure 11.44 CRT and AMLCD colour gamut.

To display a daylight-viewable, full-colour topographical map in raster mode in the high-ambient illumination conditions to be found in a fast jet fighter requires a considerably brighter shadow-mask CRT than used in domestic TV. A specialised custom CRT was developed by Tektronix (later Planar) with funding from industry and the US DoD (Jukes, 2004 – Chapter 5).

11.5.1.2 X/Y Deflection Amplifier

To produce a high-brightness, daylight-viewable, high-quality text and symbolic display, the shadow-mask CRT is usually operated in stroke (cursive) mode, that is, the CRT beam is caused to trace out the desired symbology as a pen plot. The beam current is modulated 'on-off' when raising/lowering the pen, and 'off' when moving (slewing) to the next symbol. In this way the beam can be moved more slowly (typically 10 times slower) than when producing a raster scan and the image is correspondingly brighter. Format content is therefore limited by the deflection writing rate.

When displaying a topographical map or sensor video, the CRT is operated in conventional raster mode, that is, the CRT beam traces the whole display surface area from top to bottom as a series of left-to-right horizontal lines, then retraces back to the top to start again.

The video field retrace time can be used to write stroke symbology over the raster. The slower stroke writing rate usefully emphasises the stroke symbology. This mode of operation is called 'hybrid'. Typical hybrid X, Y and video waveforms are shown in Figure 11.45.

A detailed description of the constituent functional components of a CRT multifunction display can be found in Jukes (2004 – Chapter 5).

11.5.1.3 Shadow-mask CRT Characteristics

The shadow-mask CRT has a number of characteristic features that require careful attention if acceptable performance is to be obtained in aerospace applications:

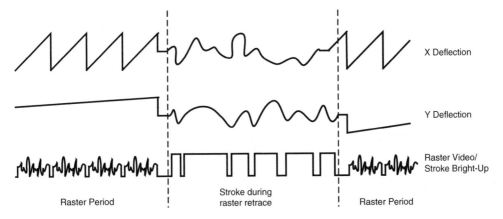

Figure 11.45   'Hybrid' (stroke + raster) operation.

1. *Contrast.* Unilluminated by the electron beam, the phosphor dots are a whitish-grey colour. A contrast enhancement filter has to be fitted to the CRT to reduce the sunlight reflections that otherwise would occur in the adverse lighting conditions of the cockpit.
2. *Colour purity.* Impure colours can be produced by small residual magnetic fields. Changes in the earth's magnetic field are sufficient, hence the need to provide protection afforded by a mu-metal shield.
3. *Convergence.* It is essential to maintain the constituent parts of the image in perfect registration; that is, the three electron beams must converge at all points on the CRT faceplate. Convergence is maintained by static and dynamic electromagnets placed close to the electron gun.

11.5.1.4   CRT MFD: Principles of Operation

The block diagram of a typical colour CRT technology based MFD is shown in Figure 11.46. The functional elements of the CRT MFD comprise:

- Shadow-mask CRT;
- High- and low-voltage power supply;
- *X* and *Y* deflection amplifier;
- Video amplifier;
- Ramp generator;
- Encoder;
- Keypanel;
- Microcontroller.

11.5.1.5   F/A-18 and AV-8B Multipurpose Colour Display (MPCD)

The MPCD shown in Figure 11.47 uses a taut shadow-mask CRT to achieve a full-colour, high-brightness, high-resolution display in stroke, raster and hybrid modes. The cardinal point performance specification is set out in Table 11.1

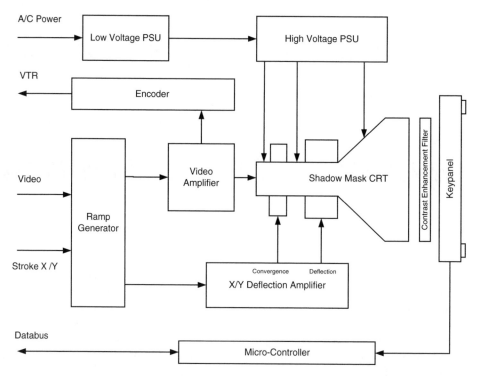

Figure 11.46   Colour shadow-mask CRT MFD block diagram.

Figure 11.47   F/A-18 and AV-8B multipurpose colour display (MPCD) (Smiths).

Table 11.1 Cardinal point performance specification of the MPCD

Usable screen area	5 in × 5 in square format	
Luminance (ft.L):	Stroke	Raster
white	570	260
red	120	45
green	400	200
blue	50	15
Contrast ratio at 10 000 ft.C:		
	Stroke	Raster
white	6.7	3.6
red	2.2	1.4
green	5.0	3.0
blue	1.5	1.1
Stroke writing speed	0.8 mm/μs	
Line width	0.018 in (0.46 mm) to 50 % point	
Resolution	65 line pairs/in	
Dimensions	6.7 in wide × 7.01 in wide × 17.3 in deep	
Weight	24 lb	
Power	115 V, three-phase, 400 Hz, 180 VA	
Dissipation	165 W at maximum brightness	
MTBF	3500 hours	

11.5.2 AMLCD Multifunction Head-Down Display

Active matrix liquid crystal display (AMLCD) is now the accepted technology for all new applications. The principle advantages of the AMLCD (with backlight) are:

- Significantly less depth than a CRT and easier to package (typically <60 mm);
- Significantly less weight than the CRT (typically <1 kg);
- Significantly less power than a CRT (typically <20 W);
- No high voltages;
- No magnetic components and no influence from external magnetic fields;
- Perfect registration (no misconvergence);
- Fixed pixel (spot) size at all display brightness levels.

11.5.2.1 AMLCD Display Head Assembly

In an AMLCD, liquid crystal (LC) material is introduced between two glass plates (known as the active plate and the passive plate). The spacing between the glass plates is critical and is around 4 μm (LC material specific).

The following is a much-simplified explanation of the operation of the cell. Reference should be made to Figure 11.48.

Liquid crystal material is a viscous organic fluid containing long polymer chains, which have the property of rotating the plane of polarisation of light according to their alignment axis. The alignment axis can be changed with electrical bias.

The plane of polarisation of light alters as it passes through the cell. Placing two crossed polarisers either side of the cell (bonded to the glass plates) makes the cell into a light valve.

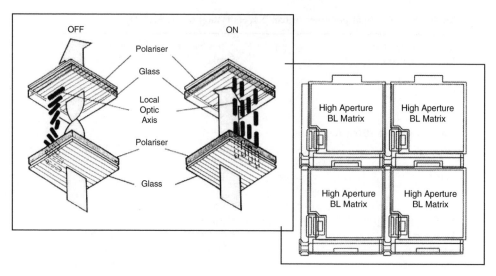

Figure 11.48 AMLCD cell.

With the cell unbiased, light is plane polarised by the first polariser, its plane of polarisation is rotated by 90° as it passes through the LC material and it then passes through the second polariser. In the biased state, light is not rotated by the LC material and does not pass through the second polariser. Greyscale is obtained by applying an intermediate voltage, which partially rotates the molecules so the cell is partially transmissive.

Accurately maintaining the cell gap is vital for correct operation. To achieve this, the LC material is filled with very accurately machined glass beads in suspension that become randomly distributed throughout the cell and are sandwiched between the glass plates under carefully controlled pressure during manufacture.

Row and column address lines form a matrix with an amorphous silicon thin-film transistor (TFT) at each intersection. The processing steps are similar to any silicon foundry process except in this case on to large glass sheets. The row and column drive signals to the TFT control charge applied across the local cell defined by the TFT itself plus an associated pair of transparent electrodes placed opposite one another on the active and passive plate. This charge is the bias for the local light valve.

Red, green and blue colour filters are deposited on the passive plate in exact register with the subpixels to form a colour group or pixel. A number of colour group arrangements have been popular, as shown in Figure 11.49, but now the accepted norm is the R:G:B stripe.

A typical X-video graphics adaptor (XVGA) compatible panel is shown in Figure 11.50. This panel has been ruggedised for the airborne environment. It is 6.25 × 6.25 in square and has 768 × 768 R:G:B pixels; a total of 1.8 million pixels.

The TFT row and column address lines are brought out to the edge of the glass plate. That is a total of around 1700 connections for the XVGA panel described. The lines are driven by row and column driver integrated circuits (ICs). Drive voltages are typically in the range −5 to + 15 V.

The driver ICs are mounted on printed circuit boards which are folded back at 90° to the AMLCD glass panel. The ultrahigh-density interconnect between the drivers and the

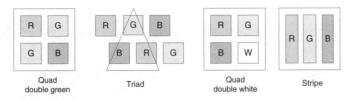

Figure 11.49 Colour group pixel structures.

AMLCD glass panel is made with tape automated bonding (TAB). The TAB process uses copper traces in a flexible polyamide base layer. The driver ICs are mounted on the TAB. Connections are made between the TAB and the glass panel using a pressure-sensitive anisotropic adhesive. The adhesive contains small silver balls that touch each other under pressure, and make contact between circuit traces in the TAB and circuit traces on the glass. The balls do not touch in the axis perpendicular to the pressure axis and isolation is maintained between adjacent traces. Once the connections are correctly made, the adhesive is heated and permanently sets.

11.5.2.2 Backlight

The AMLCD is a light valve but with a transmission of less than 7%. To obtain a usable display requires the addition of a bright backlight (Figure 11.51). This is an important distinction between the CRT and the AMLCD. In the CRT the image and the light source are one and the same, namely the phosphor light emitter. In the AMLCD the image is separated from the light source; the image is made by the LCD; the backlight is the light source.

The fluorescent lamp is the technology of choice. A number of fluorescent lamp backlight configurations have been used, hot or cold cathode, single or multiple lamps. The single,

Figure 11.50 AMLCD panel with drivers (Korry Electronics).

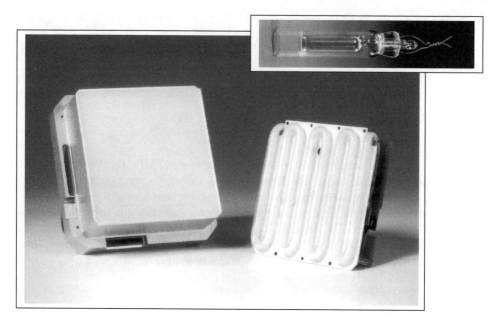

Figure 11.51　Single, cold-cathode, serpentine backlight (Korry Electronics).

serpentine, cold-cathode lamp has become the configuration preferred by most designers (Jukes, 1997). An optical stack is placed between the fluorescent tube and the backlight to provide a uniform diffuse light source to back illuminate the AMLCD.

The luminance of the backlight must be controlled to match the ambient illumination. For normal daytime operation the backlight is operated as a conventional fluorescent tube; that is, the arc is struck with a high-frequency alternating current (typically in the tens of kHz range). To dim the lamp, the hf pulses are gated to provide pulse width modulation of the lamp.

A detailed description of the constituent functional components of an AMLCD multi-function display can be found in Jukes (2004 – Chapter 5).

11.5.2.3　AMLCD Characteristics

There are a number of characteristic features of the AMLCD, different to those of the CRT, that require careful attention if acceptable performance is to be obtained in aerospace applications:

1. *Viewing angle.* The optical performance of an AMLCD is critically dependent on the cell gap. At increasing viewing angles from the normal to the display the optical cell gap effectively increases and the optical performance of the cell degrades. It no longer acts as a perfect light valve, and some light 'leaks' through the cell. This light leakage reduces the display contrast off-axis. For adequate visual performance a contrast ratio of 50:1 is desirable, and 30:1 is acceptable over the normal viewing cone. This represents a considerable challenge for AMLCD technology.

2. *Greyscale.* Greyscale is obtained by partially switching the cell between its 'on' and 'off' states. The greyscale transmission curve is not linear. It is viewing angle dependent and requires temperature compensation since it is sensitive to temperature.
3. *Black level uniformity.* This characteristic has been perhaps the most problematic feature of AMLCD (second only to viewing angle). It is particularly a concern for symbolic images painted on a black background; it is not so significant for video images which typically have little black content. Many factors can result in a blotchy black background, which can be most distracting. This feature, more than any other, has received adverse criticism for the AMLCD. Fortunately, improvements in materials and in process control, driven by the volume laptop PC market, are overcoming this deficiency.
4. *Thermal management.* Many of the parameters associated with the performance of the AMLCD are temperature sensitive and it is necessary actively to manage the thermal environment of the AMLCD. The performance of the fluorescent lamp is temperature dependent also. Active thermal management strategies are discussed in detail in Jukes (2004 – Chapter 5).

11.5.2.4 AMLCD Sourcing

Airborne AMLCD displays have fairly obviously sought to retain the pre-existent electro-mechanical and CRT instrument sizes of 3ATI, 5ATI, ARINC C (6.25 in square) and ARINC D (6.7 in square). These are all unique to the aerospace industry. However, the critical performance parameters of a high-quality AMLCD are heavily dependent on good process control. Nevertheless:

1. Good process control only comes with volume manufacturing experience.
2. Volume manufacture is not compatible with aerospace glass requirements.
3. AMLCD high-volume manufacturers are all in the Far East.
4. Volume glass sizes are aimed at the laptop PC market.

By bundling the whole aerospace industry requirements together, it has been possible to convince a few Far East AMLCD glass foundries to process high-volume batches of bare glass cells in common aerospace sizes and hold them in a benign environment to be custom ruggedised for each specific aerospace application later.

The custom ruggedisation process typically comprises:

- Bonding of a front anti-reflective cover glass;
- Bonding of a rear indium tin oxide (ITO) coated heater glass;
- TAB attachment of row and column driver printed wiring boards;
- Mounting of the end product in a rugged frame;
- Compensation for performance over the intended operating temperature range.

11.5.2.5 AMLCD MFD: Principles of Operation

The block diagram of a typical colour AMLCD technology based MFD is shown in Figure 11.52.

The functional elements of an integrated display unit (IDU) comprise:

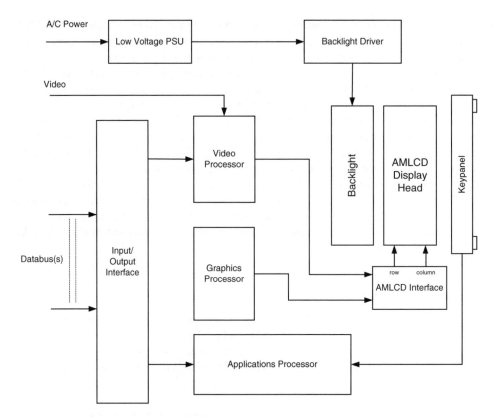

Figure 11.52   Active matrix liquid crystal display MFD block diagram.

- The active matrix liquid crystal (AMLCD) display head assembly (DHA);
- Backlight and associated backlight driver;
- AMLCD interface;
- Video processor;
- Graphics processor;*
- Input/output (I/O) interface;*
- Applications processor;*
- Keypanel;
- Chassis, power supply and interconnect.

Note that a dumb AMLCD MPD without integral graphics processing omits the functional elements marked with an asterisk.

11.5.2.6   Integrated Display Unit

A fully integrated AMLCD multifunction display is shown in Figure 11.53 (with another example given in Plate 7). This unit is intended for dual-use operation in rotary-wing, civil and military applications. Although shown with a 6.25×6.25 in square format AMLCD, the unit is a modular construction and has been designed to accommodate other AMLCD glass sizes

Figure 11.53   Integrated display unit (Smiths).

by replacing the front display-head assembly module with 10.4 in and 15 in rectangular format COTS AMLCDs. The cardinal point performance specification is set out in Table 11.2.

11.6   Emerging Display Technologies

11.6.1   *Microdisplay Technologies*

Microminiature LCDs and DMDs (digital micromirrors) have recently revolutionised digital projector technology. LCD/DMD projectors produced for the consumer market can fit in a fraction of a cubic foot, cost a thousand dollars and can illuminate a large viewing screen in daylight. These technological breakthroughs in the commercial projector market are at the

Table 11.2   Cardinal point performance specification of the AMLCD

Usable screen area	6.25 × 6.25 in
	10.4 in and 15 in options
Resolution	768 × 768 (XVGA) stripe pixels
Greylevels	256 per colour, 16 million colours
Luminance (maximum)	350 ft.L white (maximum)
Luminance (minimum)	0.05 ft.L (dimming range > 10 000:1)
NVG compatibility	NVIS class B
Viewing angle	± 55° horizontal
Interfaces (video)	525 line 30:60 Hz
	625 line 25:50 Hz
Interfaces (digital)	ARINC 429 and MIL-STD-1553B
Processing	Integral anti-aliased COTS graphics processor
	Integral application processor

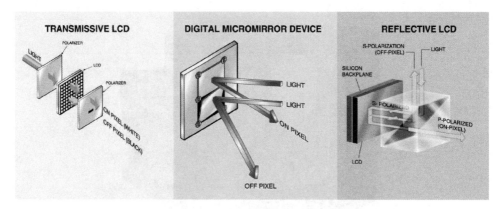

Figure 11.54   Microdisplay technologies.

forefront of candidates for the development of new, cost-effective and robust airborne displays (Tisdale and Billings, 2001).

There are currently three microdisplay technologies employed in display projectors, and these are shown in Figure 11.54. [For a more detailed description of these technologies, see Jukes (2004 – Chapter 9).]

11.6.2   High-intensity Light Sources

What makes the CRT so useful, and therefore so difficult to replace, is that it functions as both an image modulator and a light source. Although the LCD and DMD have proved to be much more reliable image modulators than the CRT, there is still a need to find a viable new light source. Typical projector applications employ high-intensity arc lamps, but there are many obstacles to overcome in adapting arc lamps to avionics applications. Emerging alternative technology light sources are in development, such as light-emitting diodes (LEDs) and high-frequency fluorescent lamps.

11.6.3   Transmissive LCD

Transmissive LCDs are the dominant display device for low-end conference-room projectors. The operating principle of the transmissive LCD microdisplay is very similar to the direct-view AMLCD except that the device is of course much smaller and designed for projection applications. A typical T-LCD is shown in Figure 11.55 and can be thought of as the electronic equivalent of a 35 mm transparency. It is illuminated by a very intense light source.

11.6.4   Reflective LCD

Reflective LCDs, also known as liquid crystal on silicon (LCoS) displays, are rapidly emerging as the technology of choice for medium-end applications. In a reflective display, the light source is positioned so that light reflects from the LCD on to the projection screen.

Figure 11.55 Typical transmissive LCD device.

These devices utilise a backplane of conventional crystalline silicon technology that is aluminised to increase the reflectivity. An LCD material is applied and sealed with a cover glass.

For a monochrome display (as in an HUD or an HMD) only a single device is required, illuminated by a monochromatic light source. For a full-colour display, typically three LCoS devices are employed, with beam-splitting optics to generate three R:G:B images that are recombined and projected to the screen. The spectrum of a bright, white light source is split into its three primary components and illuminates three reflective micro-LCDs. The three images are recombined and, through a projector lens, are focused on to a rear view screen. The white light source is typically a mercury arc lamp. Dimming has to be achieved by mechanical/optical shuttering in the light path. The lamp will dissipate about 60 W, so forced air cooling is required. The functional principles of a reflective LCD projector are shown in Figure 11.56.

Reflective LCD (or LCoS) devices themselves are multisourced and are in widespread use in tabletop projectors. The imaging device, shown in Figure 11.57, is essentially semi-conductor wafer technology using LC principles to polarise reflected light through a simply attached liquid crystal cell placed over the device.

11.6.5 Digital Micromirror Device

The DMD is the primary competitor to the LCD for the image-modulating component of a projector, appearing more widely in high-end applications.

Because of the improved reflection efficiency and high switching speeds of DMD compared with LCoS, it is possible to build a compact projector with a single DMD device, obtaining full-colour rendition using sequential colour processing. This arrangement elim-

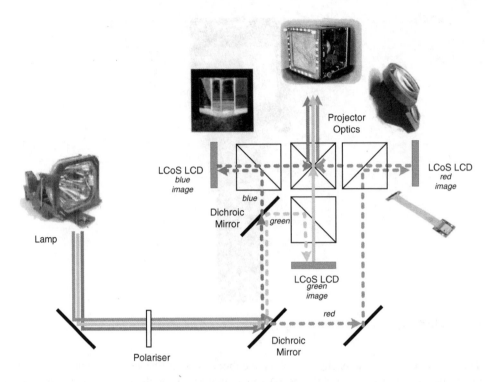

Figure 11.56   Reflective LCD projector (Kaiser Electronics).

inates the need for a beam splitter and the subsequent problems of alignment/reconvergence of the three colour images. The result is a significantly clearer image than LCoS technology, and in smaller volume.

The development of the digital micromirror device (DMD), by Texas Instruments, has yielded a robust, reliable, high-performance device which can now be packaged with other components to produce a range of display solutions. This resultant digital light projector (DLP) brings together the DMD with a light source, optics, colour filters and a projection

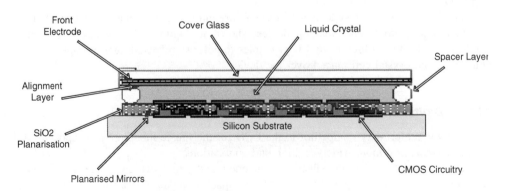

Figure 11.57   Reflective LCD (LCoS) device structure.

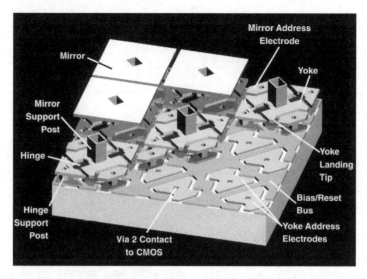

Figure 11.58   DMD array (Texas Instruments).

lens to produce an image on a display surface (i.e. rear projected on to a screen). The arrangement is now widely used for PC and cinema projection systems (Hornbeck).

The DMD is an array of aluminium micromirrors as shown in Figure 11.58. These are monolithically fabricated over an array of CMOS random-access memory cells, each of which corresponds to a micromirror. This allows each mirror to be individually addressed, causing it to tilt by approximately 10°, limited by a mechanical stop.

A bright light source illuminates the DMD array. Depending on the state of each micromirror, light is either reflected towards the projector objective lens or towards a light absorber on the wall of the projector.

Full-colour operation is obtained by interposing a colour wheel comprising red, green and blue colour filters between the light source and the DMD device, as shown in Figure 11.59.

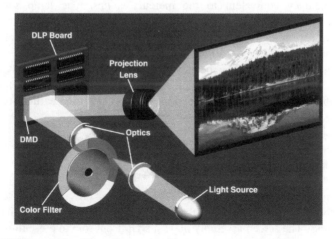

Figure 11.59   DMD projector (Texas Instruments).

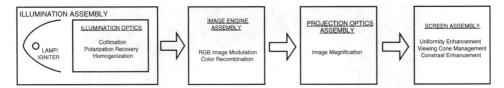

Figure 11.60   Projector optical elements.

Complex algorithms in the associated processor/DLP chip set provide signals to each DMD pixel to produce a full-colour image with at least 256 grey levels per colour by suitable mark-space modulation of each pixel in synchronism with the rotation of the colour wheel.

11.6.6   Rear-projection 'Big Picture' Head-down Display

Rear projection is a possible alternative technology to direct-view AMLCD, able to provide large-area head-down instrument displays. Rear projection has the advantage of being independent of the size vulnerability of using COTS direct-view displays.

The technology is scalable and obsolescence proof. A single optical engine design combined with slightly modified folded projection optics can be used for a variety of display sizes from, say, 4 in square to 32 in diagonal. As the commercial market drives higher-resolution and more efficient display devices (LCoS or DMD), improved image generation components can be incorporated without major redesign. There are no single high-cost components in the product design.

The basic optical elements of a projection system are illustrated in Figure 11.60 and comprise:

1. *Illumination assembly.* Every projection system using a non-emissive image source must have a source of illumination that contains sufficient energy in the red, green and blue wavelengths. Conference-room projection systems most commonly employ high-pressure mercury lamps. Other lamp technologies are in use (e.g. xenon and metal halide) but do not offer efficiency equivalent to the mercury lamps. The lamp should have the characteristics of a point source: that is, the light should emit from as small an area as possible. Collimation and shaping are performed to create a beam of near-parallel rays that will illuminate an area just slightly larger than the display devices.
2. *Image engine assembly.* This is the heart of any projection system. While numerous architectures are possible, the choice of display device technology is the single most important influence in architecture selection. Options identified earlier are:

   - Transmissive LCD;
   - Reflective LCD or LCoS;
   - Digital micromirror (DMD).

3. *Projection optics assembly.* This magnifies the image and provides a focused image at the rear of the screen, free of chromatic aberration, distortion and other optical defects.
4. *Screen assembly.* The screen assembly plays a significant role in presenting the pilot with a high-quality image that is viewable in the luminous environment of the cockpit. For

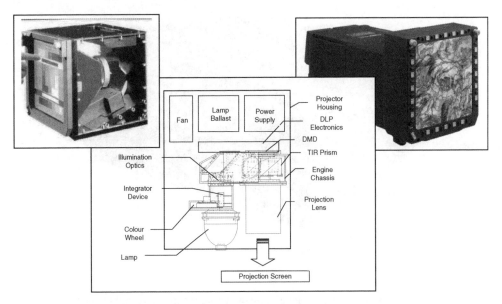

Figure 11.61 Rear-projection head-down display (Kaiser Electronics and Brilliant Technology).

high-performance avionic applications, when compared with commercial applications, the screen must provide:

- Improved uniformity;
- Shape and steer of the viewing cone;
- Enhanced contrast.

Figure 11.61 shows an open-box view of a typical rear-projection head-down display. The lower section of the box contains the image engine and lamp. The projection lens is at the rear of the box.

Multiple projectors can be arranged to fabricate a 'big picture' display encompassing the whole of the instrument panel area. The potential redundancy of this architecture makes it attractive. Overlapping portions of the optical engines achieve the expanded width of the display.

Rear projection is the chosen technology for the joint strike fighter (JSF) crew station, using a multiple-projector arrangement to achieve a 20 in × 8 in 'big picture' display as shown in Figure 11.62 and Plate 8.

11.6.7 Solid-state Helmet-Mounted Display

Current CRT technology HMDs impose large design burdens on the engineer. They are bulky and heavy and require high voltage and expensive quick disconnect interfaces.

The next generation of HMDs is in development using solid-state technology that will provide greater performance and lower cost of ownership owing to high reliability. Most importantly, they reduce helmet size and head-borne weight.

Figure 11.62 Multiple projector 'big picture' HDD (Kaiser Electronics).

Figure 11.63 shows the HMD in development for the joint strike fighter (JSF).

The image sources are two 1280×1024 transmissive AMLCDs. They achieve a contrast ratio of 100:1 with a response time of <9 ms. T-LCDs typically have a transmission of approximately 10%. A small LED backlight has been developed that can produce greater than 26 000 ft.L light output which is sufficient to achieve good readability in a 10 000 ft.C (108 000 lux) day environment. The LED requires around 0.8 mA and dissipates approxi-

Figure 11.63 JSF helmet-mounted display (Kaiser Electronics).

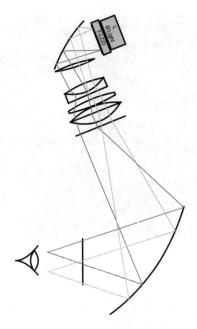

Figure 11.64   JSF HMD optical arrangement (Kaiser Electronics).

mately 1.5 W per T-LCD, a significant achievement. The electrical power requirements in the region of the pilot's head have decreased from approximately 13.5 kV required by the CRTs to only 5 V (Casey, 2002).

The use of T-LCDs allows the use of a smaller optical chain, shown in Figure 11.64, resulting in less complex optics, reduced weight and a more compact design, which has a minimum impact on the helmet profile.

The remote display processor receives incoming video from the sensor suite, digitally processes and overlays the video signals and performs helmet tracking and input/output (I/O) control functions. A real-time image-warping engine performs dynamic scaling and warping of high-resolution video data. Dynamic image warping is required to compensate for the geometric distortion of an off-axis optical design.

11.6.8   Organic Light-emitting Diodes (OLEDs)

The operation of an LED is based upon the junction of p-type and n-type materials. When a voltage is applied, electrons flow into the p-type material, and holes flow into the n-type material. An electron–hole combination is unstable; they recombine and release energy in the form of light. This can be a very efficient process.

Light-emitting diodes, based upon semiconductors such as gallium arsenide, have been around since the late 1950s. These crystalline LEDs are expensive, and it is difficult to integrate them into small high-resolution displays. However, there is a class of organic compounds that have many of the characteristics of semiconductors in which p-type and

Figure 11.65   OLED microdisplay (Cambridge Display Technology).

n-type organic materials can be introduced to make light-emitting diodes. [For a more detailed description, see Jukes (2004 – Chapter 9)].

While the early OLEDs did not have sufficient efficiency or life to be commercially attractive, significant progress is being made to improve these factors.

The advantage of these polymer devices is the ability to spin on the layers and, in some cases, to pattern the films with photolithography. An alternative approach is to employ an active matrix using standard semiconductor techniques. This is eminently suited to making microdisplays, because a small silicon chip can be used as the substrate and the necessary driver circuits can be incorporated into the silicon chip along with the matrix structure (Howard).

A typical OLED microdisplay device is shown in Figure 11.65.

OLED technology is now emerging to be of benefit for both direct-view and microdisplay applications. OLEDs offer higher efficiency and lower weight than liquid crystal displays since they do not require backlights or reflective light sources (Howard).

11.6.9   Virtual Retinal Displays

The virtual retinal display (VRD) offers the potential for a virtual cockpit. The VRD paints an image directly on to the wearer's retina using a modulated, low-power beam of laser lights. Operating at extremely low power, there is no danger to the eye. The wearer sees a large, full-motion image without the need for a screen. [For a more detailed description see Jukes (2004 – Chapter 9)].

To create the image, the VRD uses a photon source (or three for a full-colour display) to generate a coherent beam of light. The only required components are the photon sources, the scanner and the optical projection system. Scanning is accomplished with a small micro-machined electro-mechanical scanner (Figure 11.66). The projection optics are incorporated into the front reflecting surface of a pair of glasses (Collins, 2003).

A prototype helmet-mounted VRD display (Figure 11.67) has been developed for military rotorcraft applications.

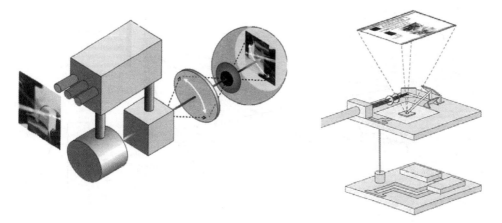

Figure 11.66   Virtual retinal display components (Microvision Inc.).

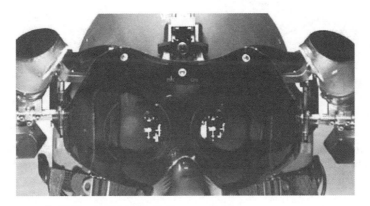

Figure 11.67   Prototype VRD helmet for rotorcraft applications (Microvision Inc.).

11.7   Visibility Requirements

11.7.1   Military Requirements

The requirements for legibility and readability of military combat aircraft displays can be found in the following documents:

JSSG-2010-3	Joint service specification guide – crew systems – cockpit/crew-station/cabin handbook
JSSG-2010-5	Joint service specification guide – crew systems – aircraft lighting handbook
MIL-HDBK-87213	Electronically/optically generated airborne displays
MIL-STD-3009	Lighting, aircraft, night-vision imaging system (NVIS) compatible
EFJ-R-EFA-000-107	Cockpit lighting standard (NATO restricted document)

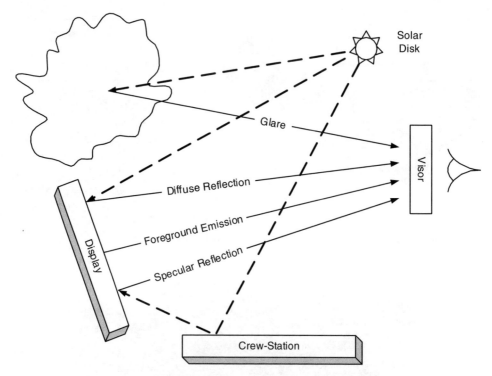

Figure 11.68   HDD high ambient – sun rear scenario.

The real-world natural ambient illumination is both too variable and too complex for every condition to be considered in the design of cockpit displays. However, it is useful to identify a few worst-case scenarios.

11.7.1.1 Head-down Display: High Ambient – Sun Rear

This scenario (Figure 11.68) is the high ambient illumination condition that degrades the visibility of emissive displays by 'washing out' the presented information. The condition is experienced when:

1. The aircraft is flying straight and level at 30 000 ft above 8/8 cloud.
2. The solar disc is low (30° elevation) and to the rear of the aircraft.
3. The display is bathed in direct sunlight – around 100 000 lux.
4. Display specular reflections of the general cockpit area predominate.
5. The forward ambient scene is diffused – around 15 000 lux.
6. The aircrew helmet tinted visor is down.

11.7.1.2 Head-down Display: High Ambient – Sun Forward

This scenario (Figure 11.69) arises when the principle effect of the ambient illumination is to produce 'solar glare' which degrades the pilot's perceptual capability. The condition is experienced when:

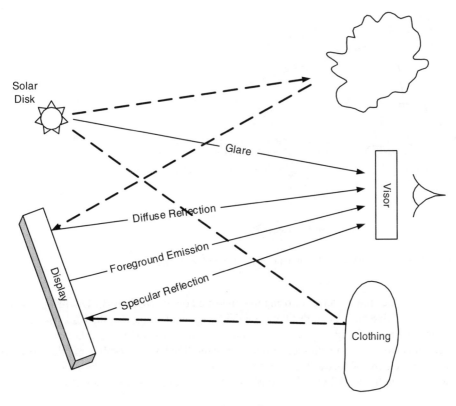

Figure 11.69   HDD high ambient – sun forward scenario.

1. The aircraft is flying straight and level at 30 000 ft above 8/8 cloud.
2. The solar disc to very low (15° elevation) and forward of the aircraft.
3. The display is in shadow and illuminated by diffused skylight – around 8 000 lux.
4. Display specular reflections of the pilot's flying suit predominate.
5. The aircrew helmet visor is down and the forward scene is dominated by the solar disc – around 110 000 lux.

11.7.1.3 *Head-Up Display and Helmet-Mounted Display: High Ambient – Sun Forward*

Practical flight experience indicates that the worst-case ambient illumination condition for the HUD transmissive display visibility is when the display symbology is presented against a cloud face, with a clear high luminance solar disc producing additional glare just outside the display field of view.

    This scenario (shown in Figure 11.70) is as follows:

1. The aircraft is flying straight and level at 30 000 ft above 8/8 cloud.
2. The solar disc is within 15° of the display central vision line.
3. The display field of view is dominated by an illuminated cloud face – around 25 000 cd/m^2.
4. The aircrew helmet visor is down.

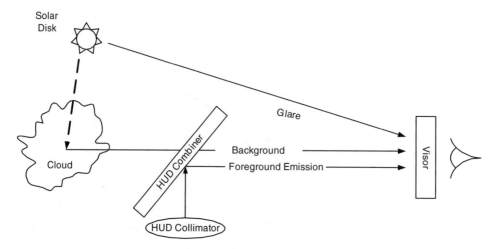

Figure 11.70   HUD/HMD high ambient – sun forward scenario.

The ability of the crew to detect information presented on the HUD or HMD will again depend on the visual difference or contrast between the foreground image and the background. In this case the foreground image is the HUD symbology reflected off the display combiner. The background is forward ambient scene illumination modified by the combiner optics. As before, there is an element of forward ambient veiling glare, which reduces the crew's perceptual capability through the HUD/HMD. All of the above modified by the visor, if down.

11.7.1.4   Low Ambient – Dusk/Dawn Transition

The final scenario is the low dusk/dawn transient illumination condition, when the sun is forward of the aircraft and close to the horizon. The clear air conditions result in the cockpit being in hard shadow while the pilot is subject to solar glare. Reflective displays are most affected, because the limited solar glare is sufficient to degrade the pilot's perception of the display low reflected luminance. This scenario is as follows:

1. The aircraft is flying straight and level at 30 000 ft in clear air.
2. The solar disc is forward of the aircraft close to the horizon line – around 1 500 lux forward ambient.
3. The sky hemisphere is clear and of low luminosity – around 200 lux.
4. The cockpit lighting control 'Night' is selected and set to maximum.
5. The aircrew helmet ND visor is up.

11.7.1.5   Night

In practice, night ambient illumination is not a single condition but the summation of multiple and complex radiation sets based on lunar reflected sunlight, various sky-glow derivatives and individual starlight spectra:

1. Night illumination can be assumed to consist of a diffused illumination over the range 1.0 – 0.0001 lux.
2. The aircrew may be flying night adapted without any vision aids.

3. Alternatively the crew may be using night-vision goggles to aid perception of the external scene.

The ability of the crew to detect information presented on the cockpit head-down displays will depend on the visual difference or contrast between the foreground image and its background. In practice there are three display components, two resulting from the ambient illumination plus a component of veiling glare resulting at the eye, all modified by the visor transmission, if down. These are:

- Display emission;
- Display diffuse and specular reflection;
- Veiling glare.

For further description of these terms, see Jukes (2004 – Chapter 12).

11.7.2   US DoD Definitions and Requirements

The requirements identified in the above referenced US DoD documents embrace the legibility and readability requirements for electronic and electro-optical display contrast and luminance in a single combined environment of both diffuse light [producing 108 000 lux (10 000 ft.C) on the display face] and the specular reflection of a glare source [with a luminance of 6800 cd/m^2 (2000 ft.L)].

There is also a 'minimum luminance difference' requirement, which basically requires displays to have a high luminance in addition to achieving adequate contrast to combat veiling glare.

The combined diffuse and specular environment is intended accurately to simulate the lighting conditions in a fighter crew station in direct sunshine. The 108 000 lux (10 000 ft.C) diffuse requirement represents sunlight of 130 000 lux (12 000 ft.C) to 160 000 lux (15 000 ft.C) (outside sunshine ambient at high altitude) passing through an aircraft canopy (typically 80 – 90% transmission) and striking a display somewhat off-axis. The 6800 cd/m^2 (2000 ft.L) glare source represents a reflection of the sun from interior parts of the crew station or from the pilot's flight suit or helmet.

The visibility requirements, under the prescribed conditions, are shown in Table 11.3.

11.7.3   European (Eurofighter Typhoon) Definitions and Requirements

The requirements identified in the Eurofighter Typhoon document describe the legibility and readability requirements for electronic and electro-optical display contrast and luminance in

Table 11.3 MIL-STD-85762 Table II

Information type	Luminance difference (min)	Contrast (min)
Numeric only		1.5:1
Alphanumeric	100 ft.L (343 cd/m^2)	2.0:1
Graphics plus alphanumerics		3.0:1
Video (high ambient) (six grey shades)	160 ft.L (550 cd/m^2)	4.66:1
Video (dark ambient) (eight grey shades)		10.3:1

a comprehensive but complex manner using the concept of perceived just noticeable differences (PJNDs). The concept is based on the premise that the ability of the crew to detect information presented on a cockpit display will depend on the visual difference between the foreground image and its background.

The PJND values for a particular display device can be computed from three sets of data:

- The display measured performance characteristics;
- The 'worst-case' ambient lighting conditions applicable to that display;
- A set of perception equations that represent a 'standard pilot's eye'.

For further description, see Jukes (2004 – Chapter 12).

Although technically elegant, and a viable analysis of product performance during product formal qualification testing, it is impractical to perform this level of testing on a 100% basis on series-production articles.

11.7.4   Viewability Examples

11.7.4.1   AMLCD Head-down Display

The simplified worst-case scenario is shown in Figure 11.71. The cardinal point performance specification is set out in Table 11.4.

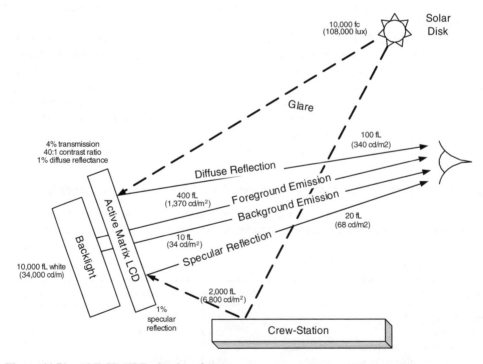

Figure 11.71   AMLCD HDD visual performance.

Table 11.4 Cardinal point performance specification of the AMLCD head-down display

Active area	6 in square
Backlight luminance	10 000 ft.L white (34 000 cd/m²)
Transmission	4%
Contrast ratio	40:1
Diffuse reflectance	1%
Specular reflectance	1%
Ambient illumination	10 000 ft.C diffuse sunlight (108 000 lux)
	2 000 ft.L point source (6 800 cd/m²)

The display background is made up of a diffuse component of sunlight reflected off the AMLCD black matrix, plus a specular component which is reflected off the front surface of the AMLCD, plus the transmission of the 'off' (black) pixels. The display foreground is the AMLCD transmission of the 'on' (red, green and blue) pixels:

$$\text{Background luminance} = (10\,000 \times 1\%) + (2000 \times 1\%) + (10\,000 \times 4\%)/40$$
$$= 130\,\text{ft.L} \ (450\,\text{cd/m}^2)$$
$$\text{Display luminance} = (10\,000 \times 4\%) = 400\,\text{ft.L white}(1370\,\text{cd/m}^2)$$
$$\text{Contrast ratio} = 1 + \text{display foreground/background}$$
$$= 1 + (400/130) = 4.1:1$$

11.7.4.2 Head-Up Display

The simplified worst-case scenario is shown in Figure 11.72. The cardinal point performance specification is set out in Table 11.5.

The display background is the illuminated cloud component attenuated by the combiner transmission. The display foreground is the CRT emission attenuated by the relay optics transmission and the HUD combiner reflectance at the wavelength of the CRT phosphor:

$$\text{Background luminance} = (7200 \times 90\%) = 6480\,\text{ft.L}$$
$$\text{Display luminance} = (10\,000 \times 70\% \times 25\%) = 1750\,\text{ft.L} \ (6000\,\text{cd/m}^2)$$
$$\text{Contrast ratio} = 1 + \text{display foreground/background}$$
$$= 1 + (1750/6480) = 1.27:1$$

11.7.4.3 Night-vision Imaging System Compatibility

Night-vision goggles (NVGs), also called the night-vision imaging system (NVIS), are passive, helmet-mounted, binocular image intensification devices. The NVIS operates by converting photons of the outside night scene into electrons using a gallium arsenide photocathode (Figure 11.73). The photocathode releases one electron for every photon it receives, thereby converting the light energy to electrical energy. The electrons are multiplied by passing through a wafer-thin microchannel plate, which is coated to cause secondary electron emissions that are accelerated by the electric field and finally collide with a phosphor screen. The phosphor screen then converts the electrons back into

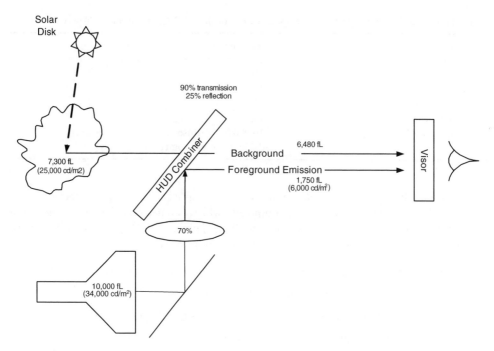

Figure 11.72   HUD visual performance.

photons, displaying the image, which now is amplified around 2000 times. The image is shown in green.

For aviation use, the goggles are mounted on the front portion of the helmet and hang down in front of the pilot's eyes. Figure 11.74 shows one type of NVIS in which the image-intensified image is projected into the pilots' line of sight through a prismatic combiner. The viewing eyepiece sits about 20 mm in front of the eyes, which enables the pilot to view and scan the outside world by looking at the image, and also enables the pilot to see all the cockpit instruments and displays by looking underneath the goggles.

To achieve compatibility of the crew-station lighting with the NVIS, the crew-station lighting should have a spectral radiance with little or no overlap into the spectral response of the NVIS image intensifier tubes. Figure 11.75 illustrates the requirements of NVIS-compatible crew-station lighting.

The rationale and criteria for achieving compatibility of crew-station instruments, displays and lighting with night-vision goggles is encapsulated in the US DoD document MIL-STD-3009 which is derived from the earlier MIL-L-85762A.

Table 11.5   Cardinal point performance specification of the HUD

Bare CRT stroke luminance	10 000 ft.L (34 000 cd/m^2)
Relay optics efficiency	70%
Combiner reflectance	25% tuned to CRT phosphor wavelength
Combiner daylight transmission	90%
Ambient illumination	7 300 ft.L (25 000 cd/m^2) illuminated cloud face

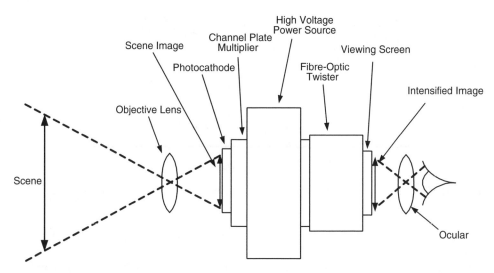

Figure 11.73 Diagram of an image intensifier.

To achieve compatibility, both the NVIS and the displays are fitted with complementary filters. Three filter classes are generally accepted for the NVIS:

1. The class A filter maximises the NVIS sensitivity but only allows blue, green and yellow lights to be used in the crew station. Red cannot be used in the crew station because the

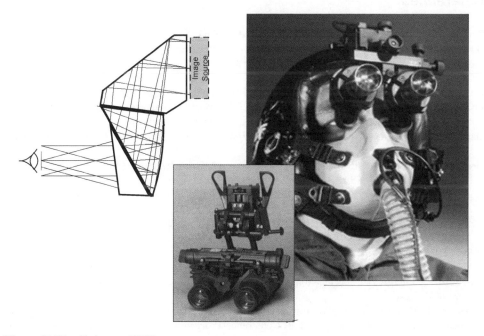

Figure 11.74 Cat's eyes NVIS.

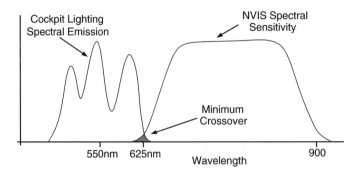

Figure 11.75   Conceptual diagram of the spectral distribution of NVIS-compatible lighting.

class A NVIS is extremely sensitive to colours with radiance at wavelengths longer than 600 nm.

2. The class B filter was developed primarily to allow three-colour CRTs to be used with the NVIS. Note that the red acceptable for use with class B is not a 'deep red', as might be expected in a full-colour display, but more an orange red.

3. The class C filter was introduced to make HUD symbology visible through the NVIS by incorporating a 'notch' or 'leak' in the green part of the spectrum.

For further description of the NVIS and NVIS compatibility, see Jukes (2004 – Chapter 12).

References

Bartlett, C.T. Head up displays and helmet mounted displays. a comparison of accuracy and integrity.

Birch, S. (2001) Eurofighter ergonomics. *Aerospace Engineering*, June.

Casey, C.J. (2002) *Emergence of Solid State Helmet Mounted Displays in Military Applications*, Kaiser Electronics, SPIE, Vol. 4711.

Collins, L. (2003) Roll-up displays: fact or fiction. *IEE Review*, February.

Fisher, J.F. *The Operation of Wide Angle Raster HUDs*, Marconi Avionics, Rochester, Kent, UK.

Foote, B. Design impacts to the HMD when used as the primary display.

Garland, D.J., Wise, J.A., Hopkin, V.D. (1994) *Handbook of Aviation Human Factors*, Lawrence Erlbaum Associates.

Greeley, K. and Schwartz, R. *F-22 Cockpit Avionics – A Systems Integration Success Story*, Lockheed Martin, Marietta, GA, USA.

Hornbeck, L.J. *Digital Light Processing and MEMS: Timely Convergence for a Bright Future*, Texas Instruments.

Hussey, D.W. (1981) Wide angle raster HUD design and application to future single seat fighters. AGARD 42nd Symposium, Andreas, Greece, October.

JSSG-2010-3, Joint service specification guide – crew systems – cockpit/crew-station/cabin handbook, US Department of Defense.

JSSG-2010-5, Joint service specification guide – crew systems–aircraft lighting handbook, US Department of Defense.

Jukes, M. (1997) Designing displays for severe environment military applications. SPIE (The International Society for Optical Engineering) Conference Proceedings, Vol. 3057, *Cockpit Displays IV*, Orlando, FL, USA.

Jukes, M. (2004) *Aircraft Display Systems*, Professional Engineering Publishing, ISBN 1 86058 406 3.

MIL-HDBK-87213, Electronically/optically generated airborne displays, US Department of Defense.

MIL-STD-1784, Head-up display symbology, US Department of Defense.

MIL-STD-3009, Lighting, aircraft, night-vision imaging system (NVIS) Compatible, US Department of Defense.

Quaranta, P. (2002) The evolution of visionics systems, military technology, MILTECH 7/2002.

Tisdale, G. and Billings, M. (2001) Tomorrow's HUD. *Flight Deck International*, February.

Howard, W.E. eMagin Corporation.

Bibliography

Adamy, D.A. (2003) *EW 101 A first Course in Electronic Warfare*, Artech House.

Airey, T.E. and Berlin, G.L. (1985) *Fundamentals of Remote Sensing and Airphoto Interpretation*, Prentice-Hall.

Bamford, J. (2001) *Body of Secrets*, Century.

Bartlett, C.T. Head Up Displays and Helmet Mounted Displays. A Comparison of Accuracy and Integrity.

Beaver, P. (ed.) (1989) *The Encyclopaedia of Aviation*, Octopus Books Ltd.

Birch, S. (2001) Eurofighter ergonomics. *Aerospace Engineering*, June.

Bryson Jr, R.E. (1994) *Control of Spacecraft and Aircraft*, Princeton University Press.

Budiansky, S. (2004) *Air Power*, Viking.

Burberry, R.A. (1992) *VHF and UHF Antennas*, Peter Pergrinus.

Casey, C.J. (2002) *Emergence of Solid State Helmet Mounted Displays in Military Applications*, Kaiser Electronics.

Cebrowski, A.K. and Garstka, J.J. (1998) *Network-centric warfare: its origin and future*, *Proceedings of Naval Institute*, January.

Collins, L. (2003) *Roll-up displays: fact or fiction. IEE Review*, February.

Conway, H.G. (1957) *Landing Gear Design*, Chapman and Hall.

Currey, N.S. (1984) *Landing Gear Design Handbook*, Lockheed Martin.

Fisher, J.F. *The Operation of Wide Angle Raster HUDs*, Marconi Avionics, Rochester, Kent, UK.

Foote, Bob. *Design impacts to the HMD when used as the primary display*,

Gardner, W.J.R. (1996) *Anti-submarine Warfare*, Brassey's.

Garland, D.J., Wise, J.A., and Hopkin, V.D. (1999) *Handbook of Aviation Human Factors*, Lawrence Erlbaum Associates.

Greeley, K. and Schwartz, R. *F-22 Cockpit Avionics – A Systems Integration Success Story*. Lockheed Martin, Marietta, GA, USA.

Hornbeck, L.J. *Digital Light Processing and MEMS: Timely Convergence for a Bright Future*, Texas Instruments.

Hughes-Wilson, J. (1999) *Military Intelligence Blunders*, Robinson Publishing Ltd.

Hunt, T. and Vaughan, N. (1996). *Hydraulic Handbook* 9th edn, Elsevier.

Hussey, D.W. (1981) *Wide angle raster HUD design and application to future single seat fighters*. AGARD 42nd Symposium, Andreas, Greece, October.

Jukes, M. (1997) Designing displays for severe environment military applications, SPIE (The International Society for Optical Engineering) Conference Proceedings Vol. 3057, *Cockpit Displays IV*, Orlando.

Jukes, M. (2004) *Aircraft Display Systems*, Professional Engineering Publishing.

Kayton, M. and Fried, W.R. (1997) *Avionics Navigation Systems*, John Wiley.

Lloyd, E. and Tye, W. (1982) *Systematic Safety*, Taylor Young Ltd.

Lynch Jr, D. (2004) *Introduction to RF STEALTH*. SciTech Publishing Inc.

McDaid, H. and Oliver, D. (1997) *Robot Warriors – The Top Secret History of the Pilotless Plane*, Orion Media

Moir, I. and Seabridge, A.G. (1986) Utility systems management. *Royal Aeronautical Society Journal – Aerospace*, **13**(7), September.

Moir, I. and Seabridge, A.G. (2001) *Aircraft Systems*. Professional Engineering Publishing.

Moir, I. and Seabridge, A.G. (2003) *Civil Avionics Systems*, Professional Engineering Publishing Ltd/ American Institute of Aeronautics and Astronautics.

Oxlee, G.J. (1997) *Aerospace Reconnaissance*, Brassey's.

Pallett, E.H.J. (1987) *Aircraft Electrical Systems*, Longmans Group Limited.

Pallett, E.H.J. (1992) *Aircraft Instruments and Integrated Systems*, Longmans Group Limited.

Poisel, R.A. (2003) *Introduction to Communication Electronic Warfare Systems*, Artech House.

Pratt, R. 2000, *Flight Control Systems: Practical Issues in Design and implementation*. IEE Publishing.

Quaranta, P. (2002) *The Evolution of Visionics Systems*, Military Technology MILTECH 7/2002.

Raymond, E.T. and Chenoweth, C.C. (1993) *Aircraft Flight Control Actuation System Design*, Society of Automotive Engineers.

Schleher, C.D. (1978) *MTI Radar*, Artech House.

Schleher, C.D. (1999) *Electronic Warfare in the Information Age*, Artech House.

Skolnik, M.I. (1980) *Introduction to Radar Systems*, McGraw-Hill.

Stimson, G.W. (1998). *Introduction to Airborne Radar*, 2nd edn, SciTech Publishing Inc.

Thornborough, A.M. and Mormillo, F.B. (2002) *Iron Hand – Smashing the Enemy's Air Defences*, Patrick Stephens.

Tisdale, G. and Billings, M. (2001). Tomorrow's HUD. *Flight Deck International*, February.

Urick, R.J. (1983) *Sound Propagation in the Sea*, Peninsula Publishers.

Urick, R.J. (1982) *Principles of Underwater Sound*, Peninsula Publishers.

Van Brunt, L.B. (1995) *Applied ECM*, EW Engineering Inc.

Walton, J.D. (1970) *Radome Engineering Handbook*, Marcel Dekker.

Wilcock, G., Totten, T. and Wilson, R. (2001) The application of COTS technology in future modular avionics systems. *Electronics and Communications Engineering*, August.

AIR 4271 Handbook of System Data Communications.

AIR 4288 Linear Token Passing Multiple Data Bus User's Handbook.

AIR 4290 Validation Test Plan for AS4074.

Guide to digital interface standards for military avionics applications, Avionics Systems Standardisation Committee, ASSC/110/6/2 issue 2, September 2003.

Joint Advanced Strike Technology Program, Avionics architecture definition – issues/decisions/rationale document, version 1, 8 August 1994.

Joint Advanced Strike Technology Program, Avionics architecture definition – Appendices, version 1, 8 August 1994.

Principles of avionics data buses, Avionics Communications Inc., 1995.

Howard, W.E., eMagin website.

EFJ-R-EFA-000-107, Cockpit lighting standard, Eurofighter.

Commercial Handbooks, Standards and Specifications

ARINC 429, Mk 33 digital information transfer system, 1977.

ARINC 629, Multi-transmitter data bus.

ARINC 653, Specification standard for time and system partition.

IEEE 1149.

IEEE 1394.

RTCA DO-178B, Software considerations on airborne systems and equipment certification.

Military Handbooks, Standards and Specifications

JSSG-2010-3, Joint service specification guide – crew systems – cockpit/crew-station/cabin handbook,
US Department of Defense.

JSSG-2010-5, Joint service specification guide – crew systems – aircraft lighting handbook, US
Department of Defense.

MIL-HDBK-87213, Electronically/optically generated airborne displays, US Department of Defense.

MIL-STD-1553B, Digital time division command response multiplex data bus, Notice 2, 8 September
1986.

MIL-STD-1760, Department of Defense interface standard for aircraft/store electrical interconnection
system.

MIL-STD-1784, Head-up display symbology, US Department of Defense.

MIL-STD-3009, Lighting, aircraft, night vision imaging system (NVIS) compatible, US Department of
Defense.

Stanag 3910, A fibre-optic derivation of MIL-STD-1553 used in NATO/UK.

Advisory Circulars

Advisory circular AC 00-31A, National aviation standard for the very high frequency omni-directional
radio range (VOR)/distance measuring equipment (DME)/tactical air navigation (TACAN) systems,
20 September 1982.

Advisory circular AC 20-67B, Airborne VHF communications installations, 16 January 1986.

Advisory circular AC 20-121A, Airworthiness approval of LORAN-C navigation systems for use in the
US national airspace Systems (NAS) and Alaska, 24 August 1988.

Advisory circular AC 20-130A, Airworthiness approval of navigation or flight management systems
integrating multiple sensors, 14 June 1995.

Advisory circular AC 20-131A, Airworthiness approval of traffic alert and collision avoidance systems
(TCAS II) and mode S transponders, 29 March 1993.

Advisory circular AC 25-4, Inertial navigation systems (INS), 18 February 1966.

Advisory circular AC 25-23, Airworthiness criteria for the installation approval of a terrain awareness
and warning system (TAWS) for part (25) airplanes, 22 May 2000.

Advisory circular AC 90-45A, Approval of area navigation systems for use in the US national airspace
system, 21 February 1975.

Advisory circular AC 90-94, Guidelines for using global positioning system equipment for IFR en-route
and terminal operations and for non-precision approaches in the US national airspace system, 14
December 1994.

Advisory circular AC 90-96, Approval of US operators and aircraft to operate under instrument flight
rules (IFR) in European airspace, March 1998.

Advisory circular AC 90-97, Use of barometric vertical navigation (VNAV) for instrument approach
operations using decision altitude, 19 October 2000.

Advisory circular AC 120-33, Operational approval for airborne long-range navigation systems for flight
within the North Atlantic minimum navigation performance specifications airspace, 24 June 1977.

Advisory circular AC 121-13, Self-contained navigation systems (long range), 14 October 1969.

Advisory circular 129-55A, Air carrier operational approval and use of TCAS II, 27 August 1993.

Standards

RTCA DO-181, Minimum operational performance standards for air traffic control radar beacon system/ mode select (ATCRBS/mode S) airborne equipment.

RTCA DO-185, Minimum operational performance standards for traffic alert and collision avoidance systems (TCAS) airborne equipment.

RTCA DO-186, Minimum operational performance standards (MOPS) for radio communications equipment operating with the radio frequency range 117.975 to 137.000 MHz, dated 20 January 1984.

Technical Standing Orders

Technical standing order (TSO) C-129a, Airborne supplementary navigation equipment using global positioning system (GPS), 20 February 1996.

Useful Websites

Federal Aviation Authority (FAA), RVSM website: www.faa.gov.ats/ato/rvsm1.htm

Glossary

AAA	anti-aircraft artillery or triple A
ABL	airborne laser
AC, ac	alternating current
AC	advisory circular
ACARS	aircraft communications and reporting system
ACE	actuator control electronics
ACES	advanced concept ejection seat
ACM	air cycle machine
ACT	active control technology
Ada	a high-order software language
AD	attack display
A to D, A/D	analogue-to-digital
ADC	air data computer
ADF	automatic direction finding
ADI	attitude direction indicator
ADIRU	air data and inertial reference unit
ADM	air data modules
ADS	avionics display set
ADS	automatic dependent surveillance
ADS-A	automatic dependent surveillance – address mode
ADS-B	automatic dependent surveillance – broadcast mode
AE	avionics environment
AESA	active electronically scanned array
AEU	antenna electronic unit
AEW	airborne early warning
AFDC	autopilot flight director computer
AGC	automatic gain control
AHARS	attitude and heading reference system

Military Avionics Systems   I. Moir and A. Seabridge
© 2006 John Wiley & Sons, Ltd

AI	airborne interception
AIFF	advanced IFF
AlGaAs	aluminium gallium arsenide
ALARM	air-launched anti-radar missile
AM	amplitude modulation
AMAD	airframe-mounted accessory gearbox
AMLCD	active matrix liquid crystal display
AMP	avionics modification programme
AMRAAM	advanced medium-range air-to-air missile
AMS	avionics management system
AMSU	aircraft motion sensing unit
ANG	Air National Guard
ANP	actual navigation performance
AoA	angle of attack
AoA	angle of arrival
AOC	airline operational centre
APGS	auxiliary power generation system
API	application programming interface
APU	auxiliary power unit
ARI	air radio installation
ARINC	Aeronautical Radio Inc.
ARM	anti-radar missile, anti-radiation missile
ASE	aircraft survivability equipment
ASI	aircraft station interface
ASIC	application-specific integrated circuits
ASR	air sea rescue
ASR	anonymous subscriber messaging
ASUW	anti-surface unit warfare
ASW	anti-submarine warfare
ATA	advanced tactical aircraft
ATC	Air Traffic Control
ATF	advanced tactical fighter
ATM	asynchronous transfer mode
ATM	air targeting mode
ATM	air transport management
ATR	air transport racking
AWACS	airborne early warning and command system
Az	azimuth
BAe	British Aerospace – now BAE SYSTEMS
BC	bus controller
BCD	binary coded decimal
BFoV	binocular field of view
BILL	beacon illuminating laser
BIT	built-in test
BIU	bus interface unit
BRNAV	basic area navigation in RNP

BROACH	bomb Royal Ordnance augmented charge
BSC	beam-steering computer
BVR	beyond visual range
C	Celsius scale of temperature
C++	a programming language
C3	command, control and communication
C&C C^2	command and control
CA	course acquisition
CAP	combat air patrol
CAS	close air support
CBIT	continuous built-in test
CCD	charge coupled device
CCDTV	charge coupled device television
CCIP	continuously computed impact point
CDROM	CD read-only memory
CCRP	continuously computed release point
CDU	control and display unit
CEC	cooperative engagement capability
CEP	circular error probability
CFAR	constant false alarm rate
CFIT	controlled flight into terrain
CG/cg	centre of gravity
CHBDL	common high-band data link
CIE	Commission Internationale de l'Eclairage
CIP	core integrated processor
CLA	creeping line ahead – a maritime patrol search pattern
CMD	countermeasure dispenser
CMT	cadmium mercury telluride (CdHgTe)
CNI	communications, navigation, identification
CNS	communications, navigation and surveillance
CNS/ATM	communications, navigation, surveillance/air traffic management
COIL	chemical oxygen iodine laser
COMINT	communications intelligence
COTS	commercial off-the-shelf
CP	centre of pressure
CPDLC	controller to pilot data link communications
CPG	copilot gunner
CRT	cathode ray tube
CSI	carriage store interface
CSSI	carriage store station interface
CW	continuous wave
CWMS	common missile warning system
DA	decision altitude
DAPS	data access protocol system
DAS	distributed aperture system

DASS	defensive aids subsystem
DBS	Doppler beam sharpening
DC, dc	direct current
DD	defence display
DefAids	defensive aids subsystem
Def Stan	defence standard
DF	direction finding
DGPS	differential GPS
DHA	display head assembly
DHUD, DOHUD	diffractive optics HUD
DIRCM	direct infrared countermeasures
DLP	digital light projector
DMD	digital micromirror device
DME	distance-measuring equipment
DoD	department of defense (US)
DPS	data processing set
DSP	digital signal processor
DVI	direct voice input
DVO	direct vision optics
em, EM	electromagnetic
EAL	evaluation assurance level
EAP	experimental aircraft programme
EBR	enhanced bit rate
EC	European Community
ECAM	electronic checkout and maintenance system
ECCM	electronic counter countermeasures
ECL	electronic check list
ECM	electronic countermeasures
ECS	environmental control system
EDP	engine-driven pump
EEZ	economic exclusion zone
EFAB	extended forward avionics bay
EFIS	electronic flight instrument system
EGI	embedded GPS inertial
EGNOS	european geostationary navigation overlay system
EGPWS	enhanced ground proximity warning system
EICAS	engine indication and crew alerting system
EIS	entry into service
El	elevation
EL	electroluminescent
ELINT	electronic intelligence
ELM	extended length messages
EMCON	emission control
EMD	engineering manufacturing and development
EMP	electromagnetic pulse
EO	electrooptical

EOB, EoB	electronic order of battle
EOTS	electrooptic targeting system
EPU	emergency power unit
ESA	electronically steered array
ESA	European Space Agency
ESM	electronic support measures
ETA	estimated time of arrival
Eurocontrol	European Organisation for the Safety of Air Navigation
EU	European Union
EUROCAE	European Organisation for Civil Aviation Equipment
EW	electronic warfare
EWMS	electronic warfare management system
EXD	expand display
FAA	Federal Aviation Authority
FADEC	full-authority digital engine control
FANS	future air navigation system
FBW	fly-by-wire
FC	fibre channel
FCC	flight control computer
FCR	fire control radar
FCS	flight control system
FD	flight director
FDDS	flight deck display system
FDX	fast switched ethernet
FFAR	folding fin aerial rocket
FFT	fast Fourier transform
FIS	flight information services
FL	flight level
FLIR	forward looking infrared
FM	frequency modulation
FMS	flight management system
FO	fibre-optic
FOV/FoV	field of view
FPA	focal plane array
FSK	frequency shift key
FTD	flight test display
FTE	flight technical error
GaAs	gallium arsenide
GATM	global air traffic management
GEO	geostationery earth orbit
GLONASS	global navigation satellite system – Russian version of GPS
GM	ground mapping
GMR	ground-mapping radar
GMTI	ground moving target indicator
GNSS	global navigation satellite system

GP	general-purpose
GPIO	general-purpose input/output
GPS	global positioning system
GPWS	ground proximity warning system
GTM	ground targeting mode

HDD	head-down display
HE	high-explosive
hf, HF	high-frequency
HFDL	high-frequency data link
HIDAS	helicopter integrated defensive aids system
HMCS	helmet-mounted cueing system
HMD	helmet-mounted display
HMS	helmet-mounted sight
HMSS	helmet-mounted symbology system
HOJ	home on jam (mode of missile)
HOL	high-order language
HOTAS	hands-on throttle and stick
HPA	high-power amplifier
HQ	headquarters
HSDB	high-speed data bus
HTS	HARM targeting system
HUD	head-up display
HUD/WAC	head-up display/weapon-aiming computer

IAS	indicated airspeed
IBIT	initiated built-in test
IC	integrated circuit
ICAO	International Civil Aviation Organisation
ICNIA	integrated communication, navigation and identification architecture
IDECM	integrated defensive electronic countermeasures
IDG	integrated drive generator
IDM	improved data modem
IDU	integrated display unit
IEEE	institute of electrical and electronic engineers
IFDL	in-flight data link
IFE	in-flight entertainment
IFF	identification friend or foe
IFoV	instantaneous field of view
IFPC	integrated flight and propulsion control
IFR	instrument flight rules
IFTS	integrated FLIR and targeting System
IFU	interface unit
IHADS	integrated helmet and display sight system
ILS	instrument landing system
IMINT	imaging intelligence (photographic)
InSb	indium antinomide

IN	inertial navigation
INEWS	integrated electronic warfare suite
INMARSAT	International Maritime Satellite Organisation
INS	inertial navigation system
INU	inertial navigation unit
I/O	input/output
IOC	initial operational capability
IR	infrared
IRS	inertial reference set
IRST	infrared search and track
IRSTS	infrared search, track and scan
ISAR	inverse synthetic aperture radar
ISS	integrated sensor system
ITU	International Telecommunications Union
IVSC	integrated vehicle subsystem controller
JASM	joint air-to-surface stand-off missile
JAST	joint advanced strike technology
JDAM	joint direct attack munition
JHMCS	joint helmet-mounted cueing system
JIAWG	Joint Industrial Avionics Working Group
JOVIAL	a software high-order language
JSF	joint strike fighter
JSOW	joint stand-off weapon
JTIDS	joint tactical information distribution system
JTRS	joint tactical radio system
K	Kelvin scale of temperature
LAAS	local area augmentation system
LANTIRN	low-altitude navigation and targeting infrared for night
Laser	light amplification by stimulated emission of radiation
LC	liquid crystal
LC	load centre
LCoS	liquid crystal on silicon
LED	light-emitting diode
LGB	laser-guided bomb
Link 11	naval tactical data link
Link 16	tactical data link (basis for JTIDS)
LLTV	low-light television
LNA	low-noise amplifier
LNAV	lateral navigation
LoRaN	long-range navigation
LO	low observability
LOC	localiser
LORAN	a hyperbolic navigation beacon system
LOS	line of sight

LOX	liquid oxygen
LP	lightweight protocol
LP	log periodic
LP cock	low-pressure (fuel) cock
LPI	low probability of intercept
LRG	laser rate gyro
LRM	line replaceable module
LRU	line replaceable unit
LTPB	linear token passing bus
LTT	lead tin telluride
LWF	lightweight fighter
LWIR	long-wave infrared
M	Mach (number)
MAC	media access control
MAD	magnetic anomaly detector
MASA	multiarm spiral array
MAW	missile approach warning
MCDU	multifunction control and display unit
MCOS	multicomputing operating system
MCU	modular concept unit
MDA	minimum decision attitude
MDC	miniature detonating cord
MDE	mission data entry
MDHC	McDonnell Douglas Helicopter Company
MFD	multifunction display
MFOV	medium field of view
MHDD	multifunction head-down display
MILS	multiple independent levels of security
MIL-STD	military standard
MLC	main lobe clutter
MLS	microwave landing system
MMA	multirole maritime aircraft
MMC	modular mission computer
MMIC	monolithic microwave integrated circuits
MMR	multimode receiver
MM/SI	miniature munitions/store interface
MMU	memory management unit
MMW	multimode millimetric wave
MNPS	minimum navigation performance specifications
Mode A	ATC mode A (range and bearing)
Mode C	ATC mode C (range, bearing and altitude)
Mode S	ATC mode S (range, bearing, altitude and unique identification)
MoD	Ministry of Defence (UK)
MOPS	million operations per second
MOPS	minimum operational performance
MOTS	military off-the-shelf

MPA	maritime patrol aircraft
MPCD	multipurpose CRT display
MPD	multipurpose display
MRA	maritime reconnaissance and attack
MSL	mean sea level
MTBF	mean time between failures
MTL	multiturn loop
MTR	marked target receiver
MTT	multiple target tracking
MUX	Multiplex
MWIR	mid-wave infrared
MWS	missile warning system
nm	nautical mile
Navaids	navigation aids
NAS	national airspace system
NAT	North Atlantic
NATO	North Atlantic Treaty Organisation
NBC	nuclear, biological, chemical
Nd:YAG	neodymium ytrrium gallium arsenide
ND	navigation display
NDB	non-directional beacon
NEMP	nuclear electromagnetic pulse
NFOV	narrow field of view
NIIRS	national imagery interpretability rating scale
NTSC	National TV Standards Committee (TV standard)
NVD	night-vision devices
NVE	night-vision equipment
NVG	night-vision goggles
NVIS	night-vision imaging systems
OBOGS	on-board oxygen generation system
OCU	operational conversion unit
OLED	organic light-emitting diode
OS	operating system
OSB	option select button
OTH	over the horizon
P3I	preprogrammed product improvement
Pave Pace	US Air Force avionics technology programme (1990s)
Pave Pillar	US Air Force avionics technology programme (ca 1987)
PAL	phase alternating line (TV standard)
PbS	lead sulphide
PbSi	lead silicide
PBIT	power-up built-in test
PC	personal computer
PD	pulse Doppler

PDC	power distribution centre
PDG	pilot's display group
P-DME	precision DME
PDU	pilot's display unit
PFD	primary flight display
PFR	primary flight reference
PIRATE	passive infrared airborne tracking equipment
PIU	pylon interface unit
PJND	perceived just noticeable differences
PLB	personal locator beacon
PMFD	primary multifunction display
PNVS	pilot's night-vision sight
POA	poly-α-olefin
PPI	plan position indicator
PPS	precise positioning service (a GPS service)
PRF	pulse repetition frequency
PRNAV	precision navigation
PRSOV	pressure-reducing shut-off valve
PS	power set
PSP	presignal processor
PSR	primary surveillance radar
QWIP	quantum well infrared detector
rms	root mean square
RadAlt	radar altimeter
RA	resolution advisory
RADINT	radar intelligence
RAeS	royal aeronautical society
RAF	Royal Air Force
RAIM	receiver autonomous integrity monitor
RAM	radar-absorbent material
RAT	ram air turbine
RCS	radar cross-section
RDMA	remote direct memory access
RDP	radar data processor
RF	radio frequency
RFI	radio-frequency interferometer
RFU	radio-frequency unit
RIO	remote input output
RLG	ring laser gyro
RMI	radio magnetic indicator
RNAV	area navigation
RNP	required navigation performance
RoE	rules of engagement
RP	radar processor
RPV	remotely piloted vehicle

RSS	root sum squares
RT	remote terminal
RTA	required time of arrival
RTCA	Radio Technical Committee Association
RTOS	real-time operating system
RTZ	return to zero
RVSM	reduced vertical separation minima
RWR	radar warning receiver
RWY	runway
RX	receive
SatCom	satellite communications
SA	situational awareness
SA	synthetic aperture
SAARU	secondary attitude air data reference unit
SAC	scene of action commander
SAE	Society of Automobile Engineers
SAHRS	secondary attitude and heading reference system
SAM	surface-to-air missile
SAR	search and rescue
SAR	synthetic aperture radar
SAW	surface acoustic wave
SCI	scalable coherent interface
SCSI	small computer system interface
SD	situation display
SD-S	situation display – secondary
SDU	satellite data unit
SEAD	suppression of enemy air defences
SEAM	sidewinder expanded acquisition mode
SFG	standby flight group
SHF	superhigh-frequency
SIAP	standard instrument approach procedure
SID	standard instrument departure
SIGINT	signals intelligence
SINCGARS	single-channel ground and airborne radar system
SLAR	sideways looking airborne radar
SLC	sidelobe clutter
SMD	stores management display
SMFD	secondary multifunction display
SMS	stores management system
SOV	shut-off valve
SPD	surface position display
SPS	secondary power system
SPS	standard positioning service (a GPS service)
SSB	single sideband
SSR	secondary surveillance radar
STA/Sta	station

S-TADILJ	satellite tactical data link J
STALO	stable local oscillator
STAR	standard terminal arrival routes
STC	sensitivity time control
STT	single-target tracking
STTI	single-target tracking and identification
SWIR	shortwave infrared
TA	traffic advisory
TAB	tape automated bonding
TACAN	tactical air navigation
TADS	target acquisition and designation sight
TAS	true airspeed
TAT	total air temperature
TAWS	terrain awareness and warning system
TBD	to be determined
TCAS	traffic collision avoidance system
TCP/IP	transport control protocol/Internet protocol
TDMA	time division multiple access
TERPROM	terrain profile matching
TFoV	total field of view
TFT	thin-film transistor
TFR	terrain-following radar
TIALD	thermal imaging and laser designation
TILL	target-illuminating laser
T-LCD	transmissive LCD
TPM	terrain-profiling mode
T/R	transmit receive
TSO	technical standing order
TV	television
TWS	track while scan
TWT	travelling wave tube
TX	transmit
UAE	United Arab Emirates
UAV	unmanned air vehicle, uninhabited air vehicle
UCAV	unmanned combat air vehicle
UFD	up-front display
UHF	ultrahigh-frequency
UK	United Kingdom
ULP	upper-level protocol
UMS	utilities management system
US	United States (as in American)
USA	United States of America
USMC	United States Marine Corps
USN	United States Navy
USNR	United States Navy Reserve

USSR	Union of Soviet Socialist Republics
UV	ultraviolet
VDL	very high-frequency data link
VGA	video graphics adaptor
VHF	very high-frequency
VHFDL	very high-frequency data link
VLF	very low-frequency
VMC	visual meteorological conditions
VME	versa module Europe – a flexible open ended data bus system
VMS	vehicle management system
VNAV	vertical navigation
VORTAC	VOR/TACAN
VRD	virtual retinal display
VSCF	variable-speed constant-frequency
VSI	vertical situation indicator
WAAS	wide-area augmentation system
WATRS	West Atlantic route system
WBSA	wide-band synthetic array
WCMD	wind-corrected munitions dispenser
WGS-84	world geodetic system of 1984
WMS	wide-area master system
WRS	wide-area reference system
WWII	World War II (1939 to 1945)
XVGA	X video graphics adaptor
YAG	yttrium aluminium garnet

Units

Å	angstrom units
cd/m^2	candela per metre squared
dB	decibel
ddBsm	decibel square metre
ft^3	cubic foot
ftc	foot candle
ft-L	foot Lambert
h	hour
Hz	Hertz
kbit/s, kbps	kilobits per second
kHz	kilohertz
kV	kilovolt
Gbit/s	Gigabits per second
GHz	Gigahertz

GOPS	Giga operations per second
lb	pound
micron	1 millionth of a metre
mJ	millijoule
mrad	milliradian
Mbit/s	megabits per second
MHz	megahertz
μm	micron
nm	nautical mile
ns	nanosecond
THz	terrahertz
W	watt
$	US dollars

Aircraft Types

A-10	Fairchild A-10 Thunderbolt
A-12	Lockheed A-12 high-speed reconnaissance aircraft
A300	Airbus A300 civil aircraft
A310	Airbus A310 civil aircraft
A330	Airbus A330 civil aircraft
AH-64	Boeing AH-64 Apache attack helicopter
Alphajet	Dassault/Dornier Alphajet trainer
ATA	advanced tactical aircraft
ATF	advanced tactical fighter
B-1	Rockwell B-1 Lancer – long-range multirole bomber
B-2	Northrop B-2 Spirit – multirole bomber
B-52	B-52 Stratofortress – long-range strategic bomber
B707	Boeing 707 civil aircraft
B737	Boeing 737 civil aircraft
B7657	Boeing 757 civil aircraft
B767	Boeing 767 civil aircraft
B777	Boeing 777 civil aircraft
BAC 1-11	BAE SYSTEMS 1-11 civil aircraft
Buccaneer	Blackburn Buccaneer bomber
C130	Lockheed Martin C-130 Hercules transport – basis for many types
Challenger	Bombardier Challenger 300 business jet
DC-9	McDonnel Douglas DC-9 civil aircraft
E-2C	Northrop-Grumman E-2C Hawkeye carrier-based AEW
E-3	Boeing E-3 Sentry AEW aircraft
E-4	Boeing 747 National Airborne Operational Centre
E-6	E-6 Mercury airborne communications system
EAP	BAe Experimental Aircraft Project
EMB-312	Embraer EMB-312 Tucano Trainer
EMB-326	Embraer EMB-326 Trainer
EMB-339	Embraer EMB-339 Trainer

F-4	McDonnell Douglas F4 Phantom fighter/bomber
F-15	Boeing (McDonnell) F-15 Eagle – all-weather tactical fighter
F-16	Lockheed Martin F-16 Fighting Falcon – multirole fighter
F/A-18	F/A-18 Hornet
F-22	Raptor
F-35	Lockheed Martin F-35
F-86	LTV F-86 Sabre fighter
F-89	Lockheed F-89 Scorpion
F-94	Lockheed F-94 Starfire
F117	Lockheed Martin F-117 Nighthawk stealth attack aircraft
G-V	Gulfstream G-V
Global Express	Bombardier Global Express
Gripen	SAAB Gripen
Jaguar	Sepecat Jaguar
KC-135	Boeing KC-135 Stratotanker Tanker conversion from B707
Mig-21	Mikoyan-Gurevich Mig-21 Fishbed fighter/bomber
Mig-23	Mikoyan-Gurevich Mig-23 Flogger fighter/bomber
Mig-29	Mikoyan-Gurevich Mig-29 fighter
Mirage 2000	Dassault Mirage 2000 fighter
Nimrod MR2	BAE SYSTEMS Nimrod maritime patrol aircraft
Nimrod MRA4	BAE SYSTEMS Nimrod maritime patrol aircraft
PC-9	Pilatus/BAE SYSTEMS PC-9 primary trainer
Rafale	Dassault Rafale – two-engine multirole combat aircraft
RAH-66	Commanche helicopter
S-3	Lockheed S-3 Viking ASW aircraft
S-92	Sikorsky S-92 utility helicopter
Sea Harrier	BAE SYSTEMS Sea Harrier
SR-71	Lockheed Martin Blackbird – high-altitude reconnaissance aircraft
Su-24	Sukhoi Su-24 Fencer
Su-27	Sukhoi Su-27
T-38	Northrop T-38 Talon trainer
T-46	Fairchild Republic T-36 trainer
Tornado	Panavia Tornado
TU-20	Tupolev TU-20 Bear bomber/maritime patrol aircraft
TU-22	Tupolev TU-22 Backfire bomber
TU-126	Tupolev TU-126 AEW
U-2	Lockheed Martin U-2 high-altitude reconnaissance aircraft
VC10	BAE SYSTEMS (Vickers) VC10
YF-22	Lockheed Martin
YF-23	Northrop

Index

Military Avionics Systems   I. Moir and A. Seabridge
© 2006 John Wiley & Sons, Ltd